S0-ANP-398

POWER PLANT
COST ESCALATION

POWER PLANT COST ESCALATION

Nuclear and Coal Capital Costs, Regulation, and Economics

CHARLES KOMANOFF

Foreword by
I. C. BUPP

VNR VAN NOSTRAND REINHOLD COMPANY
NEW YORK CINCINNATI TORONTO LONDON MELBOURNE

Copyright © 1981 by Charles Komanoff

Library of Congress Catalog Card Number: 82-8625
ISBN: 0-442-24903-9

All rights reserved. No part of this work covered by the copyright hereon may
be reproduced or used in any form or by any means–graphic, electronic, or
mechanical, including photocopying, recording, taping, or information storage
and retrieval systems–without permission of the publisher.

Manufactured in the United States of America

Published by Van Nostrand Reinhold Company Inc.
135 West 50th Street, New York, N.Y. 10020

Van Nostrand Reinhold Publishing
1410 Birchmount Road
Scarborough, Ontario M1P 2E7, Canada

Van Nostrand Reinhold Australia Pty. Ltd.
17 Queen Street
Mitcham, Victoria 3132, Australia

Van Nostrand Reinhold Company Limited
Molly Millars Lane
Wokingham, Berkshire, England

15 14 13 12 11 10 9 8 7 6 5 4 3 2 1

Library of Congress Cataloging in Publication Data

Komanoff, Charles.
    Power plant cost escalation.

    Reprint. Originally published: 1st ed. New York:
Komanoff Energy Associates, c1981. (Publication /
Komanoff Energy Associates; KEA-12)
    1. Atomic power-plants–Costs.   2. Coal fire power
plants–Costs.   I. Title.   II. Series: Publication
(Komanoff Energy Associates); KEA-12.
[TK1078.K648 1982]   338.4'3621312132   82-8628
ISBN 0-442-24903-9   AACR2

Acknowledgments

This book originated during a series of discussions in February 1978 with Vince Taylor, then a senior economist with Pan-Heuristics in Los Angeles and now with the Union of Concerned Scientists. Vince and I were examining nuclear capital cost data compiled by Bill Mooz of the Rand Corporation, and we were struck by the strong correlation Bill had found between capital costs and the date of construction start, even when costs were adjusted for inflation. We were dissatisfied, however, with Bill's use of the passage of *time* to explain increases in capital costs, and we began to ruminate over the underlying factors that, acting over time, were leading to cost increases. This led us to consider the relationships among costs, regulatory criteria, and growth of the nuclear power sector, bringing us to the *sector-size* hypothesis which is developed in Chapter 3 and which permeates much of the book.

Because Vince was occupied with writing *Energy: The Easy Path* and moving back to Vermont, it fell to me to develop a data base and cost analysis covering coal plants as well as nuclear, a task that consumed the rest of 1978 and much of 1979. Research on regulatory and design changes in nuclear and coal plants, both past and prospective, began in late 1979 and continued through final manuscript drafting in early 1981. Vince's guidance and ideas were invaluable during most of this work, particularly in refining the sector-size hypothesis, developing the statistical analysis, writing Chapter 2, and conceptualizing the book as a whole.

Many other persons devoted considerable time to reviewing draft material, answering technical questions, and improving style and presentation. Foremost among them was Bill Steigelmann of Synergic Resources Corp., whose detailed comments significantly improved Chapter 4, especially the section on environmental qualification. Ernst Habicht, Jr., formerly energy projects director of the Environmental Defense Fund and now a private consultant on Long Island, enhanced the manuscript's rigor and clarity through his persistent constructive criticisms. Bob Pollard of the Union of Concerned Scientists reviewed several chapters and answered numerous technical questions. Others who improved the manuscript significantly or otherwise helped nurture the book include Stephen Baruch, Dale Bridenbaugh, Steve Cohn, Nancy Folbre, Richard Grossman, Dick Hubbard, Amory Lovins, Greg Minor, Ken Semmel, Louis Slesin, and Ellyn Weiss.

The book owes much to the research of others. Bill Mooz's work helped me develop the concept of *steam-plant dollars* for measuring increases in power plant costs apart from inflation in construction factors. Bill also supplied the nuclear capital cost data that sparked Vince's and my early speculations, and his exhaustive and original regression modelling provided a firm base for mine here. Bill also generously furnished construction cash-flow equations and other key data in the early stages of the project. Chip Bupp and his colleagues at Harvard and MIT published exceedingly valuable and prescient work on nuclear costs during 1974-75 that anticipated many of the conclusions of this book. David Okrent, a long-time member of the Advisory Committee on Reactor Safeguards and professor of nuclear engineering at UCLA, provided a draft copy of his new book, *Nuclear Reactor Safety: On the History of the Regulatory Process* (University of Wisconsin Press, Madison, WI, 1981) which added greatly to my understanding of the evolution of nuclear power regulation.

My information-gathering also required the help of many people. Scores of Nuclear Regulatory Commission staff members graciously sat through innumerable interviews and telephone calls to answer questions concerning past and future regulatory changes. NRC regulatory reports and research studies were an invaluable source of information, as were articles from *Nuclear Safety*, some of which editor Bill Cottrell kindly supplied when copies could not be obtained in New York. The staff of the Advisory Committee on Reactor Safeguards, particularly secretary Carol Ann Rowe, also provided helpful information and materials.

My secretary, Janice Kay Young, patiently and expertly typed the manuscript through many drafts, organized my files, and procured research materials, all with an irreverent but supportive spirit that helped sustain my sanity at difficult times. Computer whiz Walter Bourne wrote the programs and file specifications that enabled me to develop the regression analyses which underlie much of the book. My production editor and erstwhile colleague, Wendy Schwartz, shepherded manuscript into book with her customary skill and aplomb, and the staff of Journal Graphics ably handled typesetting and graphics.

Finally, I gratefully acknowledge the support I have received over the years from many individuals and groups within the safe energy movement. And above all, I wish to thank my companion, Gail Daneker, for her constant strength, love and encouragement throughout the writing of this book.

Charles Komanoff
March 1981

Contents

Foreword

Less than five years ago, few business leaders, politicians, or scholars doubted that cheap electricity from nuclear power would be an important bulwark in defending the oil-importing countries against further oil price increases or supply interruptions. Today, it is obvious that nuclear power is in deep trouble nearly everywhere. Only in France are nuclear power plants being built more or less on the scale projected in the mid-1970s. And even the French nuclear construction program is, for the time being at least, proceeding in spite of skyrocketing costs.

Since the mid-1970s, it has been increasingly apparent that the "post-OPEC" expectations for rapid and cheap expansion of nuclear power were at odds with some harsh realities. From the beginning of commercial-scale nuclear power two decades ago, costs have always risen, never fallen.

One might have expected documentation of these cost increases to come from the nuclear power industry, or at least from the electric utilities that purchase and operate nuclear power plants. The utilities have the cost data and anecdotal experience necessary for such analysis, and presumably they would be ardently interested in learning industry cost trends. But, reluctant to buck what my colleague, Jean-Claude Derian, and I have called the "extravagance of prophesy" that has long prevailed among nuclear power supporters, and wary about offering ammunition to its critics, the nuclear industry has produced remarkably little analysis of its economic misfortunes. Nor have the industry's official government and academic sponsors produced any objective analysis of nuclear costs.

That task has had to be assumed by outsiders. Derian and I were the first to show, in late 1974, that the cost of building reactors was increasing much faster than both the overall rate of inflation and the costs of other heavy construction projects. Two years later, Charles Komanoff demonstrated that nuclear plants were failing by a wide margin to meet their generating performance targets, with the largest reactors suffering the largest shortfalls. In 1978, Bill Mooz at Rand showed that nuclear construction cost increases were continuing into the late seventies at an accelerating pace. Although each of these studies bore important messages for the nuclear industry, supporters of nuclear power largely ignored them.

This book, although more pessimistic about nuclear power than its predecessors, is less likely to be ignored. For one thing, it appears at a time when some government and business leaders are beginning to question the need for a major expansion of nuclear power. It is now evident that energy consumers in the industrial countries are responding to the steeply rising prices of the 1970s with substantial efforts to conserve. Growth in electricity use has slowed significantly throughout the West, even though government policy in many countries still encourages electrification. In addition, the technology to reduce the most blatant environmental effects of coal burning is now apparently at hand. This development will be of little interest in countries that lack dependable supplies of coal, but elsewhere it is beginning to blunt what has been an important argument in favor of nuclear power.

This book will also be read, one hopes, because it brings us several strides closer to the actual reality of nuclear and coal economics. It rests upon a thorough and highly competent quantitative analysis of actual nuclear and coal capital cost experience. Indeed, one of the most impressive things about this book is how far a single analyst has been able to move beyond the lavishly funded efforts of large institutions like the National Academy of Sciences in basic data collection and analysis. Komanoff's robust estimates of the actual rates of nuclear and coal capital cost increases in the 1970s are a major contribution.

Komanoff's account of past and prospective design changes for nuclear and coal plants is also both thorough and lucid. His review of cost increases and emission reductions for coal plants is especially valuable, establishing for the first time the approximate costs to build clean coal plants—or at least cleaner than oil-burning plants. Although these coal plants may not satisfy the most ardent environmentalists, they are likely to be clean enough to be licensed and built in most parts of the U.S. and almost anywhere else.

The most difficult—and most important—problem in nuclear power economics is to project the cost of building new reactors. The electric utilities have failed badly at this and have lost precious credibility as a result—first among the public, and increasingly on Wall Street. Most utilities have been unable to anticipate the great many changes in design criteria for their plants, let alone the costs of the changes.

Derian and I have argued that there is a direct causal link between nuclear power's rising costs and the controversy over reactor safety.* In our opinion, concern over reactor safety has become a profound political and economic force that has exerted a powerful, perhaps decisive, influence over the development of nuclear power technology. We have predicted since the mid-1970s that the costs of nuclear power were unlikely to stabilize as long as nuclear safety concerns—whether "rational" or "irrational"—were not appeased.

Komanoff has extended this argument here, with a major new twist: he proposes that the capital cost increases in the nuclear sector are primarily the result of efforts to contain total accident and environmental risks that would otherwise have expanded in proportion with the growth of the sector. This is an

---

*I.C. Bupp and J. C. Derian, *The Failed Promise of Nuclear Power*. Basic Books, Inc./Harper Colophon Books (New York, NY, 1981).

important and challenging hypothesis, supported by both a quantitative analysis of costs and an historical review of nuclear regulation. As a method of projecting nuclear capital costs, it may well come much closer to the mark than engineering estimation. But—and this is my only important analytic difference with the author—I question whether any model—statistical, engineering, or otherwise—can capture enough of the political, regulatory, and technological factors that together will determine the costs of new reactors.

Pending further research and discussion—which I hope this book will stimulate—I continue to prefer my own hypothesis: the basic cause of the cost increases that are documented here is a breakdown of the democratic political process. The collapse of nuclear power we are now witnessing is at least in large part and perhaps nearly completely due to a basic change in the values of the governing middle classes of wealthy western society. Technological progress, long taken for granted as the ultimate source of growing wealth, has come to be identified as the cause of some negative byproducts of that growth.

Although Komanoff might not agree, I think the nuclear advocates are essentially correct in contending that the shifts in political power that followed this change in values profoundly affected the realization of their own dreams. Whether Komanoff's hypothesis about the future costs of nuclear power is confirmed depends, in my opinion, on whether the 1980s will bring new changes in perceptions, values, and behavior.

As we wait for the future to bring the answer, we should be grateful to the author of this book for painstaking work in drawing such a rich picture of the past and for providing us with an excellent vantage point for observing changing events.

I.C. Bupp
Associate Professor of Business Administration
Harvard Business School

1
Introduction: Nuclear And Coal Capital Costs And Economics

This book examines the upheaval in the economics of nuclear and coal electrical generation that occurred in the 1970s, and explores further changes likely in the next decade. It is the product of six years spent grappling with the economics of nuclear and coal power, the last three focusing on the costs of efforts to reduce the health, environmental, and societal hazards of electricity generation.

The book investigates increases in nuclear and coal capital (construction) costs on three levels:

- an *empirical* level, through the first published statistical analysis of the actual construction costs of U.S. nuclear and coal-fired power plants completed throughout the 1970s;
- an *engineering* level, through an analysis of design and construction changes that contributed to past cost increases and those that can be anticipated to cause future increases;
- an *etiological* (underlying causal) level, through development of an hypothesis to explain the growth in regulatory standards that spurred the design changes which increased plant costs.

Section 1.1: Major Findings

The conclusions of this book differ radically from the views generally held within the federal government, the electric power industry (utilities, reactor vendors, and architect-engineers), and the technical and policy communities. To clarify the differences, the views are summarized next to the corresponding and opposing major findings of this book below. (The bases for these findings are presented in the next section.)

1. Past Capital Cost Increases

The Conventional View: Inflation and regulation affected nuclear and coal capital costs approximately equally during the 1970s. Any difference between nuclear and coal cost increase rates is solely attributable to the lack of sulfur dioxide "scrubbers" on some recent coal plants.

Findings Of This Book: Nuclear capital costs increased over twice as fast as coal capital costs, in "real" (inflation-adjusted) terms, between 1971 and 1978. The real increases averaged 142% for nuclear plants (13.5% per year); 68% for coal plants including the cost of scrubbers (7.7%/year); and 33% for coal plants if scrubbers are excluded (4.2%/year); *all figures are in addition to construction-sector inflation.*

2. Relative Capital Costs Of Recently Completed Plants

The Conventional View: Recently completed reactors did not cost substantially more to construct than equivalent coal-fired plants with scrubbers.

Findings Of This Book: Because nuclear capital costs increased much more rapidly than coal plant costs in the 1970s, the average excess of nuclear capital costs over coal plant costs increased from 6% for 1971 completions to 52% for 1978 completions. The latter figure assumes that 1978 coal plants have scrubbers; without coal plant scrubbers, the average excess of nuclear over coal capital costs for 1978 completions is 91%.

3. Causes Of Capital Cost Increases

The Conventional View: a) Capital costs increased because Congress and regulatory agencies, under pressure from environmentalists and nuclear power opponents, imposed excessive, burdensome safety and environmental requirements.
b) In addition, construction de-

Findings Of This Book: a) Capital costs increased in real terms in both the nuclear and coal sectors primarily because of efforts to prevent total accident and environmental risks from expanding in proportion to the growth of either sector. These efforts involved major design changes to contain nuclear accident hazards and substantial equipment imlays caused by licensing interventions and protests drove up costs, especially for nuclear plants.

provements to reduce emissions from coal-burning. Nuclear power, technically less mature and more prone to catastrophic accidents than coal combustion, was especially susceptible to design changes to correct safety problems revealed through operating experience.

b) Licensing interventions and protests appear to have contributed to past cost increases *only when costs are expressed in "current" (inflated) dollars*. When costs are measured in *real* terms (adjusted for inflation), plants which took longer to license did *not* have higher costs than other plants built at the same time.

4. Future Nuclear Capital Cost Increases

The Conventional View: a) Nuclear power is now technically mature in most respects, as evidenced by its excellent safety record to date. Current designs and equipment do not require significant modification, and therefore future nuclear plants will not cost significantly more to build, in real terms, than recently completed reactors.

b) The "bottom line" of the Three Mile Island (TMI) accident—no identifiable deaths or injuries despite multiple design defects, equipment failures, and operator errors — demonstrated that reactors are fundamentally safe. The future of nuclear power is imperiled, however, by "licensing instability"(*i.e.* continually changing regulatory standards) and the lack of a commitment to

Findings Of This Book: a) The capital costs of nuclear plants will probably increase in the 1980s by an amount equal to the increase in the 1970s on a dollar-per-kilowatt (kW), inflation-adjusted basis (although the increases will be smaller in percentage terms). Increases will be required to implement regulatory design requirements promulgated too late to affect recently completed plants, to correct the many outstanding "unresolved safety issues," and to address new safety problems that are still being detected at record rates through operating experience.

b) The TMI accident has added greatly to upward pressures on costs by discrediting fundamental regulatory premises and demonstrating the need for major design changes and new safety features, thus effectively eliminating the possibility of regulatory stability.

nuclear generation by the Nuclear Regulatory Commission (NRC).

5. Future Coal Capital Cost Increases

The Conventional View: Coal-fired plants are only now beginning to climb the "regulatory hump" that drove up nuclear construction costs in the 1970s. Scrubbers and other advanced emission control equipment will affect the costs of coal plants more than new safety requirements will affect nuclear costs in the 1980s. Accordingly, the capital costs of new reactors will exceed those of new coal plants by no more than 30%, and probably by less.

Findings Of This Book: Emission rates of "criteria pollutants"—sulfur dioxide, nitrogen oxides, and particulates, from coal plants completed in 1978 average two-thirds less than those from 1971 plants and are no greater, on average, than those from typical oil-fired power plants. Nevertheless, new coal plants will incur some further cost increases to meet new, more stringent standards. For approximately the same real increase in per-kilowatt costs as that experienced in the 1970s, it will be possible to reduce emissions from new coal plants to one-third of the new Environmental Protection Agency (EPA) standards for new plants, nine-tenths less than emissions from coal plants completed in 1971. Coal plants meeting these targets would be cleaner than oil-fired plants burning very low-sulfur oil. Assuming that these costs are incurred, and even discounting most of the impact of TMI, the average capital cost of nuclear plants completed in the late 1980s is still likely to be 75% higher than that of equivalent coal plants.

6. Total Generating Costs

The Conventional View: Because new nuclear plants will cost, at worst, only 30% more to build than new coal plants, will operate at 65-75% "capacity factors" (performance relia-

Findings Of This Book: Because the capital costs of new nuclear plants will exceed those of new coal plants by a wide margin (an estimated 75%), the lifetime generating costs of new nuclear plants will exceed those of new coal

bility), and will have relatively low fuel costs, their total generating costs will equal or slightly undercut those of new coal plants on a national-average basis.

plants by 20-25%, on average, even assuming:

- an improvement in the capacity factors of large reactors to 60% from the historical 54% average;
- coal plants cleaner than plants burning low-sulfur oil (as described in 5, above;
- real escalation of 2-2½% per year in coal fuel prices over the next 40 years to pay for reducing the health, safety, and environmental costs of coal-mining;
- decommissioning and disposal of radioactive wastes accounting for only 8% of total nuclear costs;
- little or no incorporation of the effect of the TMI accident in nuclear cost calculations.

Plausible estimates of the impact of TMI on nuclear capital costs, capacity factors, and financing problems raise the likely average excess of nuclear generating costs *vis-a-vis* coal for future plants to the 35-50% range. This suggests that many reactors under construction could be scrapped in favor of new coal-fired plants with advanced emission controls with little or no economic penalty.

The conventional view portrayed above is no caricature. Dozens of capital cost projections are currently being circulated by federal energy officials, electric utilities, architect-engineers, power plant manufacturers, and "policy establishment" research groups. None forecasts that nuclear capital costs will exceed those of coal by more than 35%, and most put the difference between 10% and 25%.[a] (Several particularly prominent projections, by the

a. Most of these projections employ "net capability" ratings to measure the generating capacity of coal-fired plants, whereas this study uses higher "generator nameplate" ratings for both internal consistency and compatibility with the author's past measurements of coal capacity factors.[1] The average 4-5% difference between the two measures accounts for only a small fraction of the 40-60% gap in respective projections of the difference between nuclear and coal capital costs. It is offset in any case by the resultant lower capacity factors used to calculate total coal-electric generating costs here.

Department of Energy, the CONAES panel of the National Academy of Sciences, and Commonwealth Edison of Chicago, among others, are cited in Chapter 2.)

Capital costs will account for two-thirds of the lifetime generating costs of future reactors and one-third for new coal plants.[b] The assumption that future nuclear and coal capital costs will differ only modestly is therefore crucial to arguments for nuclear power's economic superiority over coal-generated electricity. And indeed, every organization within the institutional groupings referred to above projects that new nuclear plants will be more economical, on average, than new coal facilities.

The Three Mile Island nuclear accident has not significantly changed these forecasts. Bechtel, the world's largest reactor builder, predicted before the accident in early 1979 that nuclear capital costs would be 21% greater than coal costs; it now projects only a 25% difference.[2] Similarly, Westinghouse stated in late 1979, six months *after* the accident, that nuclear power had attained ''regulatory maturity'' and was no longer vulnerable to significant cost escalation, whereas coal was now commencing its passage through regulatory turbulence.[3] These judgments are voiced even by most utilities that are cancelling reactors. They customarily cite nuclear power's ''financing uncertainty'' or its ''regulatory instability.'' Rarely, if ever, do they note its probable high cost relative to coal.[4]

Section 1.2:   Basis Of This Study's Findings

What, then, is the basis for this study's sharp divergence from the mainstream?

Statistical Analysis: Most importantly, the study's conclusions regarding capital cost increase rates in the 1970s and relative costs of recently completed plants arise from a comprehensive statistical analysis of *actual nuclear and coal capital cost experience.* Only two statistical investigations of capital costs have been published previously, and although much less comprehensive than the present study, both were supportive of the conclusions reached here. Strikingly, no nuclear power proponent has published a comprehensive statistical analysis to support the claim that nuclear and coal capital costs differ by only a small margin.

The data base used here is sufficiently up-to-date to reflect the impact of most current regulations. The nuclear plant sample comprises all reactors completed on a commercial basis following the vendor-subsidized ''turnkey'' era and before the accident at Three Mile Island: 46 units completed from

---

b. The importance of capital costs to nuclear generating costs is magnified by the lower capacity factor and higher fixed charge rate assumed for nuclear plants (see Chapter 12).

Chapter 1

December 31, 1971 through December 31, 1978. The coal plant sample encompasses all 116 coal-fired plants above 100 megawatts (MW) capacity completed from January 1, 1972 through December 31, 1977, including 15 plants with costly sulfur dioxide scrubbers. (These data were extrapolated through 1978 for comparability with the nuclear data, as explained in Section 2.3).

The statistical analysis corrects for inflation in construction labor and materials and employs synthesized "standard" plants to ensure that nuclear and coal costs are compared on an equal footing, in constant dollars and without geographical bias. For the first time, then, we have a definitive measurement of nuclear and coal capital cost increases in the 1970s. It reveals that nuclear plant capital costs grew more than twice as rapidly as those of coal plants during the past decade and are today one-and-a-half to two times as great as coal plant costs (with and without scrubbers, respectively). This confirms the findings of Bupp's 1974 nuclear and coal analysis and Mooz's 1978-79 nuclear studies,[5] neither of which measured recent rates of cost increase for coal plants.

In contrast, the one attempt by supporters of nuclear power purporting to compare actual nuclear and coal generating costs—the annual cost surveys by the Atomic Industrial Forum (AIF)—not only lumped new plants together with less costly older plants and made no adjustment for inflation, but also used extremely unrepresentative samples. The two most recent AIF surveys excluded 12 of the 14 reactors with the highest capital costs as well as a majority of reactors suffering the longest shutdowns. The surveys also covered only 15% of total U.S. coal-fired generation, excluding, for example, two giant coal-burning utilities which together generated more power from coal, at 19% lower cost, than all the coal plants included in the surveys.[6]

Analysis Of Design Changes: This study has developed the most complete picture to date of the design and construction changes that have brought. about the increases in capital costs. It demonstrates that nine-tenths of the real (inflation-adjusted) capital cost increases in coal plants in the 1970s was attributable to new equipment to reduce environmental pollution: sulfur dioxide scrubbers, particulate control upgradings, and miscellaneous improvements ranging from ash disposal to noise abatement.

No such itemization of cost increases was possible for nuclear plants due to their considerable engineering variability and the pervasive interconnections among internal plant systems affected by design changes. The comprehensive account of nuclear design and construction changes developed in Chapter 4 strongly suggests, however, that design and equipment changes to reduce the hazards of nuclear power generation underlay most of the real past increases in nuclear costs.

Analysis Of Regulatory Changes: This study examines currently outstanding environmental and safety issues whose resolution will further increase the costs of nuclear and coal plants. It establishes that future coal plants should be able to reduce emissions of sulfur dioxide, particulates, and oxides of nitrogen to less than one-fourth of the average emissions from coal plants completed in 1978, which in turn are almost two-thirds less than 1971-plant emissions. This will require approximately the same increase in real per-kilowatt costs as that experienced during 1971-78 (this conclusion assigns the cost of "first-generation" scrubbers to 1978 plants).

The costs of future nuclear safety improvements could not be itemized, because of the reasons given above and also because some improvements will probably be required to remedy defects that will be revealed through future operating experience and licensing reviews. Nevertheless, Chapters 5 and 6 demonstrate that implementing existing requirements, disposing of "unresolved safety issues," and rehabilitating nuclear power regulation in the wake of the Three Mile Island accident will prevent design stabilization and cause large new cost increases.

Development Of A Statistical Model Of Capital Costs: The statistical analysis of past cost increases and consideration of the impetus behind new safety and environmental requirements together suggest a new approach for estimating future cost increases for nuclear plants. The power industry's cost projection method, *engineering estimation*, is applicable only to coal plants, for which regulatory changes are reasonably predictable and are confined to specific plant systems. It has failed for over a decade to predict nuclear costs reliably, and for good reason: the accurate itemization of construction labor, materials, and equipment which engineering estimation requires is impossible when continual engineering and regulatory changes cause "as-built" plants to differ radically from the original conceptual designs upon which the itemizations are based.

Instead, in this study future nuclear costs are projected from the hypothesis that real increases in nuclear capital costs occur more or less in concert with expansion of the nuclear generating sector. This "sector-size" hypothesis is supported by a detailed review of the development of nuclear regulatory requirements (Chapter 3). This review shows that expansion of the nuclear sector created the conditions for the past increases in regulatory standards—increases which, in turn, were responsible for most of the past rise in real reactor costs. The sector-size hypothesis is also supported by the statistical analysis of past nuclear costs: using nuclear sector size as a proxy for regulatory change, the analysis obtained an extraordinarily high "goodness of fit"—a 92% correlation between projected and actual costs. This exceeds not only that in Bupp's or Mooz's analyses but also the fit obtained when time rather than sector size is used to represent the increase in safety standards.[c] Fi-

c. The calculations of 1971-78 nuclear and coal plant cost increases do not depend upon

nally, use of the sector size hypothesis to project nuclear costs is supported by the coal cost analysis: future coal capital costs projected by extrapolating the past, observed relationship between coal plant costs and coal sector size coincide with the projections derived through engineering estimates of the costs of advanced new control equipment required to meet anticipated more stringent environmental standards.

To be sure, the sector-size model does not replicate the complete array of forces leading to new standards and higher costs for nuclear plants. It does not explicitly reflect: regulatory lag between detection and correction of safety problems; oil price rises that appear to make alternative energy sources more vulnerable to regulatory action; or the singular impact of Three Mile Island, which will almost certainly swell the already considerable tendency of reactor operating experience to add to nuclear costs. Nevertheless, the model captures, more successfully than prior explanations, the processes that brought forth costly regulatory requirements during nuclear power's expansion in the 1970s. Thus it provides the best existing tool for estimating future nuclear capital costs so long as the nuclear sector continues to grow. It is certainly superior to engineering estimation, which is inherently unsuited to nuclear power's dynamic regulatory situation.

Moreover, the sector-size model's projection that nuclear plants undertaken today will cost approximately 75% more to construct than comparable coal-fired plants represents a modest extrapolation of cost experience in the 1970s. The implied 1978-88 increase in the ratio of nuclear to coal capital costs is 16%; between 1971 and 1978, the actual, measured increase in the cost ratio was 43%. In order to fulfill the conventional forecasts by nuclear power advocates of a 10-25% differential between future nuclear and coal capital costs, nuclear construction costs would have to *fall* by 20-30% relative to those of coal. No credible evidence has been offered to support this highly improbable reversal of past experience.

Analysis Of Other Cost Factors: Finally, cost factors other than capital costs that will affect total plant generating costs—capacity factors, fuel costs, operating and maintenance costs, and financing charges—were projected here on the basis of careful study of empirical data. Moreover, considerable conservatism (assumptions favorable to nuclear power) was built into the projection of most cost factors, as can be seen from examining the assumed values of two especially significant variables: nuclear capacity factor and coal fuel cost.

Because nuclear plants are very expensive to build but relatively inexpensive to operate, their generating costs are sensitive to their on-line performance rate, or capacity factor. Future nuclear plants, which will all exceed 800 megawatts capacity and will average 1150 MW, are assumed here

---

the sector size formulation. Statistical models employing the date of construction start yield comparable results.

to generate at 60% capacity factors over their 30-year life. A lower projection could easily be justified on the basis of actual experience. Nuclear plants over 800 MW capacity now account for almost two-thirds of commercial reactors and well over half of commercial reactor operating experience, but they have averaged only 54% capacity factor to date, and only 55.3% in the five years since the author published his projection of 55% performance.[7] Projections of 65-75% capacity factors by nuclear power advocates—the same institutions that formerly anticipated 75-80% performance and derided the author's 55% projection[8]—assume that reactors will "mature" at rates far exceeding the very modest improvements evident in actual experience.

Coal-electric generating costs are particularly sensitive to fuel costs. A mere 1% deviation from the assumed annual rate of real escalation in the delivered cost of coal translates into an approximate 10% difference in the total lifetime generating cost of a new coal-fired power plant. For purposes of projecting total generating costs in this study, it was assumed that the annual rate of increase from the actual 1979 average cost of mining and transporting coal would be 2.3% greater than the average inflation rate for other industrial commodities over the next 40 years. This would duplicate coal's real rate of cost increase during 1974-79, a period marked by large increases in coal mining, chronic labor-management strife, and significant investment and operating expenditures to adapt to new health, safety, and environmental requirements. These factors will undoubtedly cause costs to increase in the future, but probably at diminishing rates, as evidenced by the real *declines* in the average cost of coal delivered to utilities in both 1979 and 1980. (The 1980 decline was ignored in projecting costs here; including it in the post-1974 increase rate used for extrapolation would have reduced projected average coal generating costs by 6%, as discussed in Section 11.3.) Indeed, the future cost of coal projected here is greater than that assumed by most nuclear proponents (see Table 11.3).

Section 1.3:   Scope Of This Book

This study has deliberately been limited to the capital costs, and, to a lesser extent, the total generating costs, of new nuclear and coal-fired power plants. Other important issues in nuclear and coal economics are *not* addressed here, including:

- costs and extent of alternatives to central-station power generation, including improved end-use efficiency, renewable energy sources, and cogeneration;
- costs of retrofitting emission control systems and accident-prevention or mitigation equipment onto existing coal and nuclear plants;

• past and present subsidization of nuclear and coal electrical generation through government research and development, tax and depreciation allowances, and federal assumption of indirect fuel-cycle costs such as black lung compensation, "orphan" strip mines and mill tailings, uranium enrichment, etc.

The estimates developed here of the direct costs to electric utilities of investments in new nuclear and coal-fired plants provide benchmarks for assessing these issues, however. The author particularly hopes that these estimates will improve the rigor of cost comparisons of electricity supply with alternative, non-central-station technologies for providing energy services, such as end-use efficiency improvements.

Similarly, this study has not attempted to recommend whether public policy should favor coal-fired power plants over nuclear facilities under construction or being planned. Such a determination requires consideration of many critical factors that elude expression in a simple per-kWh formulation of generating costs and are outside the scope of this book. They include: the health, safety, and environmental effects of nuclear and coal generation; the contribution of fossil-fuel burning to the build-up of carbon dioxide in the atmosphere; the linkage between nuclear power growth in the United States and the development of nuclear weapons technology here and abroad; and the comparative employment, social, and political implications of energy growth in the form of central-station power supplies *vis-a-vis* growth through improved efficiency of energy use and development of renewable resources.

The study has found that future coal-fired plants can be expected to generate electricity at considerably less cost than future nuclear plants, although both types of plants will be much more expensive, in real terms, than their counterparts today. This finding applies throughout the United States, as Chapter 12 shows. Nevertheless, readers should adapt the generic approach of the study to local conditions, as appropriate. This is true particularly for the projected cost of coal fuel, which varies considerably among regions. Conversely, where utilities project that new nuclear plants will cost much less than estimated here, readers are urged to examine the extent to which those projections require, contrary to the evidence presented in Chapters 5 and 6, that nuclear regulatory requirements more or less stabilize in the future.

Although the analysis presented here implies that many reactors in early construction could probably be abandoned and replaced by new coal facilities with little or no economic penalty, the book does not attempt to assess the impact of such substitutions on electricity supply or coal markets. These effects of nuclear cancellations are largely outside the scope of this study and merit further assessment, although data developed here, such as construction durations of nuclear and coal plants, are relevant to such analyses.[d]

d. The effect on air quality of replacing planned nuclear capacity would also be of

Finally, the issues addressed here could be explored profitably in greater detail, and the data and analysis extended and revised as events change and new perceptions emerge. The statistical correlations can be updated and refined as more plants are completed and new causal hypotheses are advanced. The costs of recent design and regulatory changes will become more apparent as time passes, and prospective further changes will come into sharper focus. The underlying processes that give rise to increased regulatory requirements particularly deserve greater study.[9] Projections of other relevant cost factors, such as capacity factors, fuel costs, and reactor decommissioning, need continual checking against actual experience.

The author has endeavored throughout to state all his methods and assumptions, to express costs in constant dollars for comparability with other estimates, and, most importantly, to project future trends on the basis of hard, empirical evidence. Future efforts by others should do the same.

References

1. C. Komanoff, *Power Plant Performance* (Council on Economic Priorities, New York, NY, 1976).

2. W.K. Davis, Vice President, Bechtel Power Corp., "Statement on U.S. Nuclear Power Program" before the Select Committee on Energy of the U.K. House of Commons, 4 June 1980. The 25% capital cost difference is inferred from his Table 7, unescalated levelized costs, fixed charges, 1980 dollars, 1.41¢/kWh for nuclear and 1.13¢/kWh for coal (high- and low-sulfur average with scrubber). The prior 21% difference is in W.K. Davis and R.O. Sandberg, "Light Water Reactors: Economics and Prospects" (February 1979), Table 6, 1.24¢/kWh nuclear and 1.025¢/kWh coal, same basis as above except 1979 dollars. Davis also estimates (1980, p. 9) that the TMI accident will add 5% to nuclear capital costs.

3. L.G. Hauser, Manager, Energy-Systems Applications, Westinghouse Power Systems Co., "Generation Planning: Consider the Cycles—II," in *Electrical World, 192* (No. 11), 20-22 (1 December 1979).

4. For a recent example, see *MacNeil/Lehrer Report*, "The Nuclear Marketplace," aired 25 November 1980 (transcript available through Public Broadcasting System). Four authorities were interviewed concerning cancellation of a nuclear plant order by Virginia Electric & Power (VEPCO): executives of VEPCO, Commonwealth Edison, and the Atomic Industrial Forum, and a leading investment broker. All insisted that new nuclear plants would be cheaper than coal plants in the long run, provided they could be licensed.

5. Bupp's and Mooz's studies are discussed and referenced in Section 2.2.

6. C. Komanoff, *Power Propaganda* (Environmental Action Foundation, Washington, DC, 1980). The two coal-burning utilities noted in text are the Tennessee Valley Authority and American Electric Power. This report analyzed AIF's survey of 1978 costs. The author subse-

concern. A calculation in Section 7.3, however, indicates that replacement of all planned reactors less than 40% complete—65,000 MW of capacity—by new coal plants meeting the stringent emission rates assumed here for 1988 plants, would add surprisingly little to total nationwide emissions: approximately 1% to sulfur dioxide, 3-4% to nitrogen oxides, and .2% to particulates.

quently discovered that AIF's 1977 survey too omitted 12 of the 14 costliest reactors and the same two coal-burning utilities. The AIF 1977 and 1978 survey results were reported in many prestigious publications, including *The New York Times* on 8 April 1979 (Section 3) and 30 December 1979 (Section 4); *The Wall Street Journal* on 24 April 1979; and *Science* at *204*, 596 (11 May 1979) and *207*, 724 (15 February 1980). In some instances the results were attributed to the Edison Electric Institute, which also circulates the AIF figures. As of this writing, AIF has not issued a survey for 1979 or 1980.

7. The 54% figure extends through June 1980 and is in C. Komanoff, *Bulletin of the Atomic Scientists*, *36* (No. 9), 51-54 (November 1980). The 55.3% figure is author's calculation covering 1976 through October 1980. The 55% performance projection is in Reference 1, p. 8.

8. See C. Komanoff and N.A. Boxer, *Nuclear Plant Performance Update* (Council on Economic Priorities, New York, NY 1977), Appendix D, for citations of critiques of Reference 1 by the Atomic Industrial Forum, the U.S. Energy Research and Development Administration (now Department of Energy), and Commonwealth Edison, and rebuttals thereto. Other critiques were subsequently published by the Edison Electric Institute and Oak Ridge National Laboratory.

9. D. Okrent, *Nuclear Reactor Safety: On The History Of The Regulatory Process* (University of Wisconsin Press, Madison, WI, 1981), a major study of nuclear regulation by a long-time member of the Nuclear Regulatory Commission's Advisory Committee on Reactor Safeguards, arrived as this book was going to press.

2
Summary:
Cost Escalation
At Nuclear And
Coal Power Plants

Efforts to contain the societal costs of electricity generation from coal and nuclear power have grown as the coal and nuclear generating sectors have expanded. Measures to limit coal-generated pollution and to reduce nuclear accident hazards added significantly to the costs of building power plants in the United States during the 1970s. These measures have had a far greater effect on nuclear plants, causing the average ratio of nuclear to coal capital (construction) costs to increase from roughly 1.05 in 1971 to over 1.5 in 1978. If past relationships between sector expansion and regulatory stringency continue, by the late 1980s nuclear plants will cost at least 75% more to build than will coal plants incorporating advanced emission controls; nuclear generating costs would then significantly exceed those of coal.

Section 2.1:   Introduction

Over the lifetime of new electric-generating plants, capital costs will account for approximately two-thirds of total generating costs for nuclear and one-third for coal plants. Accordingly, capital costs are central to the relative economics of new nuclear and coal plants. They also strongly affect the competitiveness of both nuclear and coal plants with cogeneration, renewable energy sources, and improved end-use efficiency — a critical issue, but beyond the scope of this book.

The capital costs of the first nuclear plants completed on a commercial basis in the early 1970s were only slightly greater than those of contemporaneous coal plants, but costs for both plant types have increased rapidly since then. Although inflation in construction wages and material prices has been a major contributory factor, increased environmental and safety standards have played a particularly important role.

Most representatives of the power industry, the federal government, and the technical and policy communities have argued that future environmental and safety standards will affect nuclear and coal plants equally and will not cause major differences in capital costs. Of 12 predicted nuclear-coal capital cost ratios cited in the 1977 Ford-Mitre report on nuclear power, the highest was only 1.23.[1] Commonwealth Edison of Chicago, the nation's largest nuclear utility, believes that its new nuclear plants will cost only 8% more to build than coal plants.[2] The federal Department of Energy (DOE) projects an 18% difference,[3] and the Committee on Nuclear and Alternative Energy Systems of the National Academy of Sciences projects the difference to be in the 0-25% range.[4]

If these forecasts were accurate, new nuclear plants, which will cost less to operate than new coal units, would probably have slightly lower life-cycle generating costs. Such forecasts underlie the present plans of U.S. utilities to almost triple nuclear generating capacity by 1990 from the 54 gigawatts (GW) operating in early 1981, and they suggest that new orders will be forthcoming when the current surplus of generating capacity is used up.

The projections of nearly equal future nuclear and coal capital costs are, however, belied by experience: the average ratio of nuclear to coal capital costs of plants *completed in the late 1970s*, after adjustment to eliminate the effects of inflation and add the costs of sulfur dioxide scrubbers to coal plants, *was more than 1.5 to 1*. Moreover, the ratio increased steadily throughout the 1970s, and an examination of the impetus behind new regulatory standards suggests that the trend toward higher nuclear capital costs relative to coal will continue for some time.

Section 2.2: Prior Capital Cost Analyses

Virtually all nuclear plants completed in the 1970s cost much more to construct than originally estimated by reactor manufacturers and electric utilities, even after adjusting for the effects of inflation.[5] The failure to predict accurately nuclear capital costs reflects a fundamental limitation in the power industry's technique of *engineering estimation*, which employs conceptual plant designs to calculate the labor, materials, equipment, and engineering effort needed to build a plant. The technique requires that the scope of work be known at the start of construction, yet nuclear plants, as the largest reactor builder has noted, are subject to "new requirements ... imposed after the design and construction are well advanced, requiring substantial rework that increases both schedule and cost."[6] The 1979 nuclear accident at Three Mile Island (TMI), moreover, has provoked a thorough and still-continuing reappraisal of nuclear regulation that promises to produce many new, more stringent standards.

Only two major empirical studies of nuclear capital costs were published in the 1970s, neither by the power industry. Both demonstrated that nuclear costs to date were increasing rapidly over time (after adjusting for inflation) and, thus, cast doubt upon engineering estimates which anticipate little or no future real increases in capital costs. But neither study can be used to project future relative coal and nuclear costs. Bupp *et al.*[7] found far higher rates of cost increase for nuclear than for coal plants, but their data base is now six years old and includes no scrubbers for coal plants, which some claim have offset increases in nuclear costs. Mooz's more recent analyses[8] of nuclear costs employed more explanatory variables and a larger data base, but the absence of a coal plant analysis precluded comparison of escalation rates.

Moreover, both Bupp and Mooz attributed all nuclear cost increases not accounted for by construction inflation to the passage of time, assuming that costs increase linearly (by a constant amount) with calendar year. Their models contained no explanatory variables that would allow one to predict whether the cost increases would continue at past rates, accelerate, or diminish. The authors implicitly recognized this limitation by stating that they did not believe their analyses had predictive value.

This book extends Bupp's and Mooz's work in three major respects:

- It uses a larger, more current data base for both coal and nuclear plants, one which reflects the impacts of scrubbers and other recent changes in regulatory standards. Accordingly, it provides a comprehensive, up-to-date view of power plant cost experience during the 1970s and permits calculation of nuclear and coal capital costs based on 1978 regulatory standards and construction costs.
- It analyzes the changes in plant design requirements and construction conditions that accounted for most of the ''real'' (inflation-adjusted) cost increases in the 1970s and those that will probably cause further increases in the 1980s. The analysis is particularly useful in estimating the costs of future coal-fired plants with advanced pollution controls. It also demonstrates that many unresolved nuclear safety issues will make it difficult to stabilize plant designs.
- It formulates an hypothesis to explain the underlying factors that caused coal and nuclear regulatory standards, and hence costs, to increase in the 1970s. The hypothesis is developed in considerable detail for nuclear plants and is used to generate a first-order projection of future nuclear capital costs.

Section 2.3:   Cost Increases, 1971-1978

To measure past increases in capital costs, a cost data base was compiled for all U.S. nuclear and coal units greater than 100 megawatts (MW) capacity that achieved commercial operation from December 31, 1971 to December 31, 1977 (for coal) and to December 31, 1978 (for nuclear).[a] It includes all reactors constructed on a commercial basis through the end of 1978: 46 units totaling 39,265 MW capacity, ranging from 514 to 1130 MW each. The 116 coal units total 70,509 MW (114-1300 MW).

Capital cost data were obtained from utility records published by the federal Energy Information Administration, as described in Chapters 8 and 9. Costs are expressed there in "mixed current dollars," *i.e.*, as the sum of all dollars expended in the different years of construction, including interest during construction (IDC) paid by the utility on capital borrowed during construction. To control for widely varying rates of inflation and interest during the sample period, costs were converted to 1979 dollars without IDC, using the Handy-Whitman Index of Public Utility Construction Costs, a semi-annual compilation of wages, material prices, and other factor costs in steam-electric plant construction (see Appendix 3).[9] This converts all costs to comparable "steam-plant construction dollars," permitting isolation of cost increases caused by changes in the characteristics of plants — a major focus of this analysis.

Even with costs expressed in constant 1979 steam-plant dollars, the nuclear and coal samples are not fully comparable because of differences in a variety of variables. Therefore, *multiple regression analyses* were employed to control for the effects of unit size, construction of multiple units, geographical location, use of cooling towers and scrubbers, and the number of reactors built by the architect-engineer (A-E). Only variables with a statistical significance level of 95% or better were included, with one minor exception (see Section 9.1). The dependent variable — that is, the variable correlated with these factors — was capital cost-per-kilowatt of design capacity,[b] adjusted to constant steam-plant dollars as described above.

After all adjustments, including that for inflation, capital costs were found to have increased persistently throughout the period of the study (1971-78). To allow for this phenomenon in the regression analysis, two alternative formulations were examined. In one model, capital costs were assumed to be related to *time* in addition to all the other variables; in the other model, they were assumed to be related to the *size of the respective power-generating*

---

a. The data base excludes four reactors completed pursuant to fixed-fee contracts during 1972-73, because some costs were absorbed by the vendors and did not appear in the utilities' costs. Inclusion of these units would increase the apparent nuclear cost increase rate computed below. 1978 coal data were not available for this study.

b. Design capacity is maximum nameplate generator rating (coal) and original design electrical ratings prior to their late-1976 revisions by the NRC (nuclear). These ratings are used in Chapter 11 to estimate capacity factors. See Sections 8.1 and 9.1 for further discussion of capacity ratings.

sector. Time was measured by the date each nuclear plant received its construction permit and each coal-fired boiler was ordered from a vendor. Sector size was defined as the megawatts of nuclear or coal capacity operating or being built on these same dates.

Both formulations reflect the effect on costs of the upgrading of nuclear and coal regulatory standards over time. The formulation with sector size was devised because it appears to capture more of the societal processes that give rise to new standards, as discussed in Section 2.5 and Chapter 3. *The calculated costs of 1971 and 1978 plants are relatively independent of which formulation is employed*, however. The choice of sector size or time is important primarily for estimating *future* capital costs.

The results of the regression analysis are described in detail in Chapters 8 and 9. The models with sector size, shown in Tables 8.1 and 9.1, explain most of the variance in costs—92% for nuclear, and 68% for coal—considerably more than in Bupp's or Mooz's analyses.[c] Key findings are as follows:

Time or Sector Size: Costs measured in the model employing calendar year grew by 24% per year for nuclear plants and by 6% annually for coal. Alternatively, costs were proportional to the .58 power of sector size for nuclear and the .61 power for coal, so that an approximately 50% real cost increase was associated with each nuclear or coal sectoral doubling. This is shown for nuclear plants in Figure 2.1. However, the coal sector was more than 14 times as large as the nuclear sector in 1971, so its 75 GW expansion to 1978 represents a far smaller percentage increase (53%) than the 46 GW increase for nuclear (467%); thus coal expansion was associated with a much smaller increase in capital costs.

The rates of increase exclude mitigating effects such as greater architect-engineer experience (for nuclear) and exacerbating effects such as addition of scrubbers (for coal). Composite nuclear and coal cost increase rates that incorporate these effects are presented below.

Unit Size: Nuclear per-kW costs were proportional to the $-.20$ power of unit size, *i.e.*, declining 13% for each doubling in size[d]—a cost reduction less than half that projected by the power industry and the federal government.[10] Coal plant costs were not affected by unit size, notwithstanding prevailing estimates of 10-15% cost reductions per size doubling.[11]

Multiple Units: Identical nuclear or coal units built consecutively at one

c. the highest r^2 (goodness of fit) values in Bupp *et al.* were 64% for nuclear and 43% for coal. Mooz reached 76% in his nuclear analysis.

d. Incorporating the greater cost of interest during construction (adjusted for inflation) for larger reactors reduces the nuclear capital cost reduction per size doubling to 10%.

Chapter 2

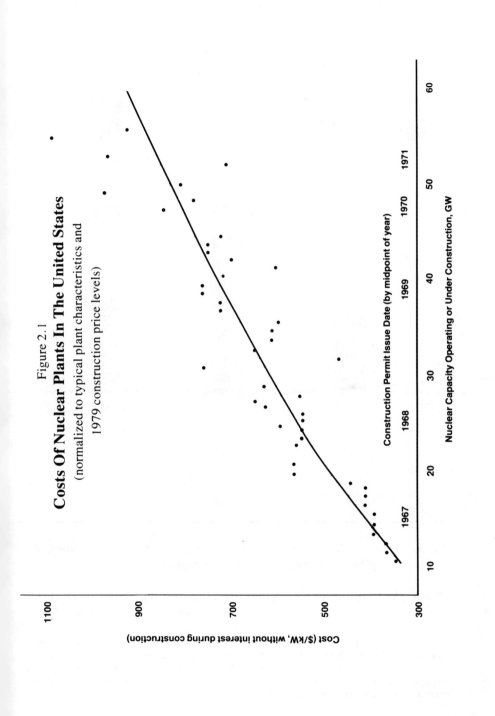

Figure 2.1

Costs Of Nuclear Plants In The United States

(normalized to typical plant characteristics and
1979 construction price levels)

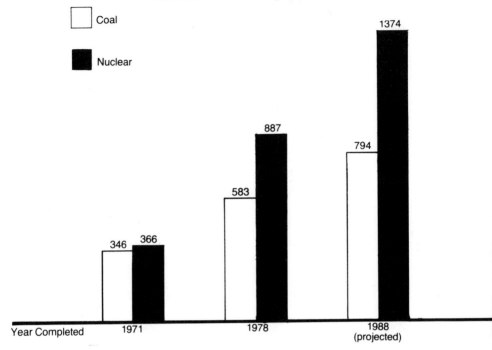

Figure 2.2
Power Plant Capital Costs
(in 1979 steam-plant $/kW,
with real interest during construction)

Coal

Nuclear

1374

887

794

583

346   366

Year Completed   1971   1978   1988
                                                                (projected)

Costs were synthesized by applying nuclear and coal cost regressions to "standard plants" as described in text. Costs of 1971 and 1978 plants vary little with the regression model employed. 1978 and 1988 coal plants include scrubbers. See Table 2.1 for breakdown of costs of coal plant pollution controls. Real interest during construction (IDC) accounts for between 6% and 15% of total plant costs. See Table 10.10 for IDC's share of each plant's cost.

site cost 10% less than other units, evidently because of shared planning and facilities as well as "learning" in design and construction.[e]

Architect-Engineer Experience: Nuclear costs declined by 7% for each

e. The cost savings from building multiple units and the limited savings from building large units imply that, during the 1970s, large single plants could have been replaced by twin half-size units at no increase in capital costs (nuclear) and at a 10% savings (coal). These results do not reflect a further advantage of small units—their higher average generating performance compared to large plants (see Sections 11.1 and 11.2).

Figure 2.3
Power Plant Construction Durations
(in years)

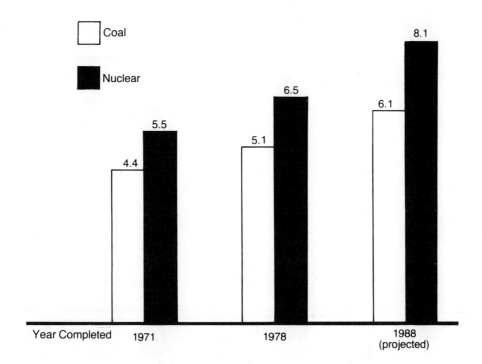

Durations are measured from construction permit award (nuclear) or boiler order (coal) to commercial service date. They were synthesized by applying nuclear and coal duration regressions to "standard plants" as described in text. Durations of 1971 and 1978 plants vary little with the regression model employed.

doubling in the number of plants constructed by an A-E, a relatively modest "learning" effect.

Scrubbers: Systems for controlling sulfur dioxide added 26% to coal plant costs, or $120/kilowatt (kW) (in 1979 steam-plant dollars).[12]

Costs of Standard Plants: To measure the combined effects of the above cost factors during the 1970s, two sets of "standard plants" were defined, one representing those completed in late 1971, just prior to the start of the study period, and the second corresponding to the last plants in the sample, com-

Table 2.1

Pollution Control Costs For New Coal Plants

(in mid-1979 steam-plant $/kW)

| Pollutant | Actual | | Projected |
	1971	1978	1988
Particulates	20	60	65-80
SO₂		120	140-180
NOₓ		10	60-90
Solid Waste	0-5	5	30-45
Other	5	45	65-75
TOTAL	25-30	240	360-470
INCREASE		210-215	120-230

See Chapter 7 for derivation of costs. All costs are calculated using 1978 standard plant characteristics (Table 10.3) for consistency. Real interest during construction accounts for 8% of total costs.

pleted in late 1978.[13] Their characteristics, shown in Table 10.3, mirror the data base, except that geographical characteristics were merged to avoid regional bias. A scrubber was also specified for the 1978 standard coal plant (whereas about half of recent coal plants lack scrubbers, using low-sulfur coal instead to meet emission standards).

Costs calculated for plants with these characteristics are shown in Figure 2.2. Approximately 6-10% of their costs are contributed by real interest during construction (Section 10.4), which is proportional to plant construction time and the real cost of money for utilities.[14] Typical construction periods for 1971 and 1978 plants were calculated from separate regressions for construction duration; they increased from 4½ to 5 years for coal and from 5½ to 6½ years for nuclear, as shown in Figure 2.3.

Costs shown in Figure 2.2 are expressed in 1979 steam-plant dollars and thus appear greater than costs of most completed plants since the latter are reported in mixed-current dollars which reflect earlier, cheaper material prices and wage rates. From 1971 to 1978, the capital cost of the standard nuclear plant increased by 142% (13.5% annually) or $520/kW in 1979 steam-plant dollars. For coal plants, the typical 1971-78 real increase was 68% (7.7% annually), or approximately $240/kW, of which scrubbers accounted for half.

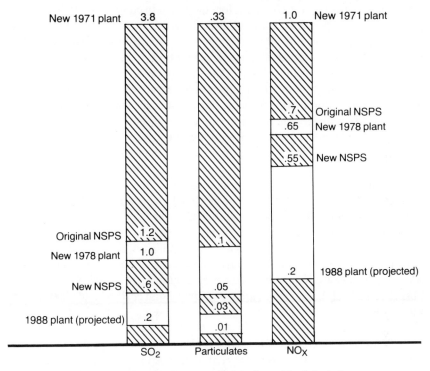

Figure 2.4

**Emissions Of Criteria Air Pollutants
By Typical New Coal Plants**

(pounds of pollutant per million Btu of coal burned)

See Chapter 7 for basis of estimates and comparison with oil-fired plants.

Thus, the ratio of typical nuclear to coal capital costs increased from 1.06 in 1971 to 1.52 in 1978, assuming the addition of scrubbers to 1978 coal plants; without scrubbers, the 1978 ratio was 1.91. The 1978 cost ratio far exceeds the highest nuclear-coal capital cost ratio projected by industry and government for future plants. It indicates that recently completed nuclear plants may not show life-cycle cost advantages over comparable coal facilities.[15]

Section 2.4: Cost Increases And Regulatory Changes, 1971-1978

Because the cost increase figures above are in addition to wage and material inflation, they had to result from increases in the *quantity* of labor, materials, equipment, and engineering required to build power plants In turn, these increases were occasioned almost entirely by the application of more stringent environmental and safety standards to coal and nuclear plants.

Coal: Most of the real (inflation-adjusted) increase in the cost to construct coal plants in the 1970s went for equipment to control the three major air pollutants produced in coal combustion: particulates, sulfur dioxide, and nitrogen oxides. The collection efficiencies of electrostatic precipitators were improved from the former 97% average to 99.5% to enable them to capture smaller particulates, adding an average of $40/kW to capital costs, as Table 2.1 shows. Installation of "first-generation" scrubbers that remove an average of 75% of sulfur dioxide from the stack added an average of $120/kW, and boiler modifications reducing nitrogen oxide emissions by 35% added about $10/kW. In return for these expenditures, the typical 1978-completed coal plant produces 64% less "criteria pollutants" than a typical 1971 plant — surpassing the average 55% reduction required by the original New Source Performance Standards (NSPS) promulgated by the federal Environmental Protection Agency (Figure 2.4). The new coal plants are slightly cleaner than typical oil-burning plants, as discussed in Chapter 7.

Other environmental improvements — better emission monitoring, safer disposal of waste ash, increased use of cooling towers, and abatement of construction pollution, liquid effluent, and noise — cost an average of $45/kW. All told, new equipment to reduce the environmental impact of coal plants absorbed 90% of the total increase in the real capital costs of coal plants during the 1970s.[f]

Nuclear: The costs of increased regulatory standards cannot be itemized for nuclear plants as they were for coal. Changes in nuclear design or equipment requirements tend to cause indirect, site-specific changes in diverse supporting equipment that are hard to identify and even more difficult to quantify. But although cost estimates are not available for each specific change, nuclear design and construction changes in the 1970s were nevertheless extensive and costly.

One key indicator of regulatory standards, the number of Atomic Energy Commission (AEC) and Nuclear Regulatory Commission (NRC) "regulatory guides" stipulating acceptable design and construction practices for reactor systems and equipment, grew almost seven-fold, from 21 at the end of 1971 to 143 at the end of 1978. Professional engineering societies developed new nuclear standards at an even faster rate (often in anticipation of AEC/NRC

---

f. The remaining 10% is reasonably attributable to the modest real increase in IDC caused by lengthened construction and higher real interest rates and to design and construction improvements to raise operating reliability.

regulations).[16] These led to more stringent (and costly) manufacturing, testing, and performance criteria for structural materials such as concrete and steel, and for basic components such as valves, pumps, and cables.

Requirements such as these had a profound effect on nuclear plants during the 1970s. Major structures were strengthened and pipe restraints added to absorb seismic shocks and other postulated "loads" identified in accident analyses. Barriers were installed and distances increased to prevent fires, flooding, and other "common-mode" accidents from incapacitating both primary and back-up groups of vital equipment. Similar measures were taken to shield equipment from high-speed missile fragments that might be loosed from rotating machinery or from the pressure and fluid effects of possible pipe ruptures. Instrumentation, control, and power systems were expanded to monitor more plant factors under a broadened range of operating situations and to improve the reliability of safety systems. Components deemed important to safety were "qualified" to perform under more demanding conditions, requiring more rigorous fabrication, testing, and documentation of their manufacturing history.

Over the course of the 1970s, these changes approximately doubled the amounts of materials, equipment, and labor and tripled the design engineering effort required per unit of nuclear capacity, according to the Atomic Industrial Forum.[17] They also increased the real costs of many construction commodities. Moreover, because many changes were mandated *during* construction — as new information relevant to safety emerged — much construction lacked a fixed scope and had to be let under cost-plus contracts that undercut efforts to economize. Completed work was sometimes modified or removed, often with a "ripple effect" on related systems. Construction sequences were frequently altered and schedules for equipment delivery were upset, contributing to poor labor productivity and hampering management efforts to improve construction efficiency. In general, reactors in the 1970s were built increasingly in an "environment of constant change"[18] that precluded control or even estimation of costs and which magnified the direct cost impacts of new regulations and design changes.

Section 2.5:   The Impetus Behind New Regulatory Requirements

The major force behind the real increase in power plant costs in the 1970s was the imposition of increasingly stringent regulatory requirements. The same will be true for the future. Thus if we can anticipate future regulatory trends, we can gain substantial insight into the probable future course of power plant costs.

This is a fairly straightforward matter for coal: most new environmental regulations are reasonably predictable, and, since they affect specific plant systems, their costs are roughly quantifiable. Nuclear plants, however, present a major difficulty: many safety requirements that will affect future reactors have not yet been promulgated — indeed, some will be occasioned by problems identified during future operating experience. Moreover, even the costs of those future requirements known today are not calculable on a generic basis since they frequently affect so many internal plant systems that their impact will vary both with plant design and the stage of construction.

But, even if *specific* future nuclear regulatory standards and their costs cannot be determined in advance, prior experience may nevertheless provide a means of approximating their *overall rate* of application. Increased regulatory stringency appears to be closely related to expansion of the *nuclear sector* — the amount of nuclear capacity operating and under construction — and to increased reactor operating experience. This conclusion is suggested by the remarkably high (92%) "goodness of fit" (agreement between predicted and actual costs) attained when increases in nuclear costs in the 1970s are correlated with increases in nuclear sector size (Section 2.3). It is also supported by an historical analysis, reported in Chapter 3 and summarized here, of growth in nuclear regulations.

One factor motivating this growth has been regulators' desire to reduce the risks per reactor as the number of plants increases. As nuclear pioneer and proponent Alvin Weinberg wrote following the Three Mile Island accident in 1979, "[f]or nuclear energy to grow in usefulness, the accident probability *per reactor* will simply have to diminish."[19] Otherwise, nuclear expansion could lead to such a high rate of accidents *per year* that the public's confidence in nuclear power would collapse and plants would be forced to close.

Weinberg's prescription has many antecedents. The Advisory Committee on Reactor Safeguards, an influential body of senior nuclear scientists attached to the AEC, wrote in recommending improved inspection of reactor vessels in 1965,

> [T]he orderly growth of the industry, with concomitant increase in the number, size, power level and proximity of nuclear power reactors to large population centers will, in the future, make desirable, even prudent, incorporating [stricter design standards] in many reactors.[20]

Similarly, AEC staff, in recommending back-up shutdown systems to prevent events in which the control rods fail to "scram" (shut down) the reactor during an accident (events known as Anticipated Transients Without Scram, or ATWS), wrote in 1973,

> The present likelihood of a severe ATWS event is . . . acceptably small, *in view of the limited number of plants now in operation . . . As more plants are built, however, the overall chance of ATWS will increase*, and the staff believes that design improvements are appropriate to maintain and to

improve further the safety margins provided for the protection of the public.[21]

A second factor contributing to new regulations is the public's increasing awareness of and concern about nuclear hazards. This has put pressure on elected officials and regulators to reduce nuclear risks in order to make nuclear expansion more acceptable to the public. Citizen interventions and protests have been blamed for increasing costs through delay, but most such increases were negligible in real (inflation-adjusted) terms for reactors completed in the 1970s; delays caused by citizen challenges generally affected reactor *licensing*, not construction, and reactors which took longer to license did not have inordinately high capital costs (see p. 205). Far more importantly, intervenors and expert critics have identified new safety concerns, such as emergency cooling criteria and cable separation, and have generally made the regulatory staff "considerably more cautious and conservative"[22] in licensing and rule-making hearings and in internal formulation of regulatory policy.

New regulations also arise from detection of previously unrecognized reactor defects. Reviews of new plants by the reactor manufacturers and AEC/NRC have provided one such means of detection. For example, General Electric and Westinghouse discovered potentially large dynamic forces that could affect reactor containment structures and reactor vessel supports, respectively, in accident analyses performed for licensing reviews of new plants in the mid-1970s. Other safety issues leading to new regulatory standards, including seismic and tornado protection, quality assurance problems, main steamline breaks, and intermingling of systems for reactor operation and shutdown, have been identified in reviewing individual reactor applications and have been applied subsequently to other plants.

Even more importantly, many unanticipated safety problems have been revealed by operating experience. Contrary to early expectations, increased reactor operation has generally warranted widening rather than reducing design margins. The "lack of perfection in design, construction and operation" of early reactors prompted the Advisory Committee on Reactor Safeguards to advocate use of more back-up safety systems.[23] Fuel leaks, pipe cracks, and malfunctioning components later formed what NRC called a "considerable body of operating reactor experience [indicating] the need for expanded technical review in areas previously thought to be not sufficiently important to warrant much attention."[24]

Adverse operating experience has also given rise to numerous regulatory guides and "unresolved safety issues." Major examples are the 1975 Browns Ferry fire, which led to costly new rules for fireproof construction and ventilation; reactor control breakdowns in 1978-80 due to power failures to instruments that have prompted consideration of increased separation of "safety" from "non-safety" instruments (Section 5.4); and the 1979 TMI accident, which has sparked an across-the-board review of fundamental regulatory premises.

The origins of nuclear regulatory standards are varied and complex, but the above discussion suggests that nuclear capacity growth and increased reactor operating experience have been the major causes of new requirements. Growth in the number of plants has engendered greater public concern over nuclear power hazards, has generated new licensing reviews in which safety problems were discovered and has forced regulators to endeavor to reduce per-plant risks to contain the industry-wide probability of a serious accident. Increased operating experience, similarly, has unearthed safety defects requiring remedial regulatory action. In addition, the enlarged regulatory effort required to oversee an expanding nuclear sector has necessitated greater documentation and standardization of regulatory requirements, generally at a more stringent level.

Growth in size of the sector also underlay the rise in real costs of coal plants. Expansion of coal-fired generating capacity in the 1960s and 1970s aggravated that sector's environmental impact, both nationwide and in new regions previously without large-scale coal use, such as the Four Corners region in the Southwest. Public outcry over these potential impacts led to more stringent and costly pollution control requirements. Similarly, concern over acid rain, which is also a product of expanded electricity generation from coal (although disproportionately from earlier, poorly controlled plants), may cause coal emission control requirements to be tightened further.

Section 2.6:   Regulatory Standards And Capital Costs In The 1980s

The preceding discussion indicates that the real capital costs of new nuclear and coal plants will increase further (in addition to increases caused by general inflation) as regulatory standards become still more stringent in the future. New standards and their costs for typical plants assumed to be completed in 1988 are investigated in Chapters 5-7 and are outlined below.

Coal Standards and Costs: EPA has promulgated stricter New Source Performance Standards for coal plants ordered after September 1978 (Figure 2.4). But for purposes of estimating 1988 coal plant costs here, future emission rates have been geared to emerging control technology and assumed to be approximately *two-thirds less* than those specified by the new NSPS. This implies a 76% emission reduction compared to plants completed in 1978, and 91% compared to 1971 plants (see Table 2.2). This striking improvement — to a level cleaner than existing plants burning low-sulfur oil (Figs. 7.2 and 7.3) — would enable utilities to significantly expand coal-generated electricity without exacerbating acid rain and most other emission-related effects of burning coal.

Table 2.2
Percentage Reductions In Emissions
By Typical New Coal Plants

Pollutant	Actual From 1971 to 1978	Projected From 1978 to 1988	Projected From 1971 to 1988
Sulfur Dioxide	74%	80%	95%
Particulates	83%	80%	97%
Nitrogen Oxides	35%	69%	80%
Average Reduction	64%	76%	91%

Obtaining these and other pollutant reductions is likely to add between $120/kW and $230/kW (in 1979 steam-plant dollars) to the costs of 1978 coal plants, as estimated in Chapter 7 and shown in Table 2.1. Most of the increase will go for nitrogen oxide control devices and improved solid waste disposal methods, since upgrading particulate and sulfur dioxide removal will probably cost less than past (1971-78) improvements. Baghouse filters appear capable of capturing 99.9% of particulate emissions for only 10-30% higher cost than typical 99.5%-efficient electrostatic precipitators today, and 95%-efficient sulfur dioxide scrubbers producing saleable wastes (sulfur, gypsum, or sulfuric acid) should cost only 15-50% more than current 75%-efficient waste-generating scrubbers.

The average estimated incremental cost of these measures for 1988 coal plants is $175/kW above the cost of a 1978 plant with scrubbers. A $210/kW increase — about the same as the 1971-78 cost rise — is obtained by extrapolating the 1971-78 relationship between coal capital costs and sector size, based on assumed coal sector growth to 1988.[g] The resulting projected cost of about $800/kW, in 1979 steam-plant dollars, for coal plants beginning operation in 1988, would be sufficient to allow: (i) $15/kW for possible problems with new control devices or for currently unanticipated minor standards; (ii) $10/kW for the increased real cost of interest during construction anticipated as coal construction durations grow from five years (1978 average) to six years for 1988 plants; and (iii) $10/kW for investments to

g. This calculation is performed in Chapter 10 with the coal capital cost regression equation (Table 9.1). The six-year construction duration for 1988 plants mentioned in the text is also derived in Chapter 10, using the coal construction duration regression (Table 9.7).

improve plant operating reliability.

The close agreement between the two methods of projecting coal capital costs—engineering estimation and econometric, or regression, analysis—lends considerable credence to the $800/kW estimate. Moreover, both methods include substantial conservatisms, the former through very strict emission targets and generous cost estimates, the latter through a high estimate of 1978-88 coal sector growth (13 GW per year[25]) that allows for an upturn in coal orders due to reactor cancellations. These conservatisms and the long lead times for most coal plant regulatory requirements make it unlikely that actual costs for 1988 plants will significantly exceed these estimates.

Nuclear Standards: Future reactors will have to institute a myriad of new safety measures not reflected in the approximately $900/kW average cost of plants completed in 1978—a figure that already exceeds the $800/kW projected cost of *1988* coal plants (both figures in 1979 steam-plant dollars). They will first have to make extensive design changes and equipment upgradings already promulgated but not incorporated in most 1978 units. These include: greater physical separation of redundant safety-related equipment such as electrical cables; more durable electrical components and wide-range instrumentation designed to function under accident conditions of high temperature, humidity, pressure, and radiation; and improved quality assurance programs to reduce design and construction defects such as seismic deficiencies found belatedly in many operating reactors during 1979-80 (see Section 5.1).

Further requirements will result from NRC's efforts to correct the 17 *unresolved safety issues*—"matter[s] affecting a number of nuclear power plants that pose important questions concerning the adequacy of existing safety requirements for which a final resolution has not yet been developed . . ."[26] Many of these issues involve fundamental design considerations. For example, *Asymmetric Blowdown Loads* might require modifying pressurized water reactors to prevent postulated ruptures in reactor coolant pipes from over-stressing reactor vessel supports, a condition that could damage both normal and emergency cooling systems while impeding insertion of control rods to stop nuclear fissioning. Another unresolved issue, *Systems Interactions*, might require segregating all components and circuits of certain safety systems in costly "bunkered" housings to avert inter-system interferences that could degrade performance of vital equipment.

Moreover, NRC's list of unresolved safety issues is growing, not shrinking (see Section 5.3), fed by new licensing reviews and the growing inventory of operating experience—experience that will more than triple by 1988 if utilities realize their current expansion plans. Past practice indicates that these issues will be addressed incrementally, thwarting "licensing stabilization" and frustrating efforts to optimize design and construction. High costs and declining growth in electricity sales may force some utilities to extend construction schedules to conserve cash and preserve credit ratings. Although the effects of stretch-outs are usually expressed in current dollars, and, thus, appear inflated, they do increase real interest costs, disrupt construction logistics, and expose plants to additional regulatory requirements.

Estimating Future Nuclear Costs: The additional costs of nuclear regulatory standards for future plants cannot be estimated directly, for reasons discussed at the start of Section 2.5. Not only do nuclear design changes tend to "ripple" differently through reactor systems at different plants, but also many prospective standards have not yet been promulgated (most stemming from the Three Mile Island accident will not be determined for several years, as discussed below). The alternative used here is to estimate future nuclear costs *by assuming that the expansion of the nuclear sector now in progress will have the same proportional effect on reactor capital costs as in the past.* That is, other things being equal, a doubling of nuclear capacity operating or under construction will lead to an approximately 50% rise in real reactor capital costs.

This "sector-size hypothesis" is based on the historical relationships observed among growth in the nuclear sector, increases in regulatory stringency, and increases in nuclear costs. It is advanced here not as a means of projecting future nuclear costs precisely, but as an alternative to engineering-based estimates that are invariably overrun by ever-changing regulatory and design criteria. Linking costs to nuclear sector size is also superior to relating costs only to the passage of time, as has been done in the two econometric models cited earlier, since it more closely reflects the forces which, acting over time, add to reactor regulations and costs.

The sector-size approach to estimating future costs is likely to evoke criticism from supporters of nuclear power, since it implies that nuclear economics will worsen steadily as the nuclear sector expands. Hence, it is important to clarify what the approach implies and what it does not.

- The sector-size approach implies that costs will rise as more reactors are licensed, due to the many strong connections between growth in sector size and increased regulatory stringency; it does *not* suggest that there is a simple, explicit causal link between the number of reactors and costs.
- It implies that licensing interventions are among the factors engendering increased regulatory stringency; *not* that construction delays caused by nuclear opponents have been a major factor in rising real capital costs.
- Use of the sector-size hypothesis to project costs also implies that, for the foreseeable future, increases in the number of operating plants will reveal new safety and operating problems, leading to new regulatory requirements and design changes which, in turn, will cause costs to rise. It does not deny that

reactor construction has evidenced a ''learning curve.'' Rather, it implies that, as in the past, increasing architect-engineer familiarity with nuclear construction will be outweighed by advancing regulatory criteria requiring A-Es to employ more labor, equipment, and materials to build nuclear plants.

In summary, the sector-size model provides an indirect way to anticipate the future rate of increase in nuclear regulations and costs, provided that the conditions under which sector growth stimulated new regulations in the past remain in effect. This proviso is likely to hold so long as utilities seek to complete the approximately 90 GW of nuclear capacity with construction permits. These plants long ago gave impetus to new regulations that will raise reactor costs, when they were awarded construction permits in the 1970s, signalling to regulators the need for greater safety measures to prevent increases in the sector-wide probability of a serious accident. In theory, the impetus could be defused if public attitudes toward nuclear risks change substantially or if it proves possible to dismiss outstanding regulatory issues without heightening accident risks. Neither event seems plausible, however. The growing number of genuine safety issues and the continued widespread mistrust of nuclear power strengthen the presumption that more stringent and expensive requirements are in the offing.

The one eventuality that might be expected to slow the anticipated rate of increase in nuclear regulations and costs is cancellation of a large number of reactors with construction permits. (Most cancellations in recent years have befallen plants that merely were in the planning stage; because they had not received construction permits, they were not counted in tallying 1988 sector size.) Large-scale cancellations of plants being built would ease public concern and also enable regulators to relax growth in safety requirements somewhat without driving up the sector-wide accident probability. Readjustment to reduced future capacity would be constrained, however, by continued detection of safety problems through the operation of existing reactors. Indeed, as measured by the rate of issuance of NRC generic Bulletins and Circulars, the detection rate per *reactor-year* reached an all-time high during 1979-80 (see Figure 2.5). Even if the detection rate per reactor were to decline, the total number of safety problems detected per *year* could well continue to increase for some time to come as the number of plants in operation rises.

Ideally, one would prefer to use both reactor operating experience and sector size as explanatory variables for projecting future costs. Unfortunately, the two are too closely correlated to permit valid statistical estimation of their joint relationship to capital costs. Thus, statistical necessity forced the choice of one of the two. Sector size was chosen because it appears to have had the more telling effect on costs, judging from both its greater statistical power, as discussed in Section 8.1, and from the historical analysis of the regulatory process presented in Section 2.5.

Figure 2.5
NRC Bulletins And Circulars Per Operating Reactor

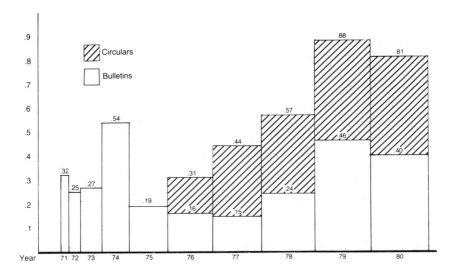

Bar widths represent numbers of licensed operating reactors (including partial years from commercial start, excluding units under 400 MW). See Figure 5.1 and p. 144 for explanation of Bulletins and Circulars.

The omission of operating experience as an explanatory variable does not appear serious so long as the nuclear sector continues to expand rapidly, since growth in sector size will coincide closely with, and thus can subsume statistically, growth in operating experience. If, however, nuclear growth slows significantly due to cancellations of many plants being built, reactor operating experience might become the dominant source of new regulations, and it would be necessary to add it to statistical models used to project future costs.

For purposes of estimating the cost of new nuclear plants here, it is assumed that all reactors now holding construction permits are completed—a near-tripling of nuclear capacity to 150 GW, implying a continuation of nuclear power's rapid growth, and, thus, warranting use of the sector-size model to project costs. When Table 8.1, embodying the 1971-78 correlation between actual reactor costs and cumulative nuclear capacity,[h] is applied to a

---

h. Cumulative nuclear capacity, rather than plant number, is employed as the proxy for the impetus behind more stringent standards, since the perceived hazard and the corresponding effort to reduce risks depend on plant size as well as number.

1988 "standard reactor" (Table 10.1), the result is a calculated cost of approximately $1400/kW (in 1979 steam-plant dollars), 55% above the actual cost of a typical 1978 reactor. This cost includes real interest during construction, whose cost share will increase due to the rise in projected construction time to an average of 8 years from 6½ years for 1978 plants (Figure 2.3).

Given the $800/kW cost projected for 1988 coal plants, the ratio between typical 1988 nuclear and coal capital costs would be 1.73. The 73% cost differential, although triple that in the establishment forecasts cited earlier, represents an increase of only 21 percentage points from the 52% differential for 1978 plants. It is also less than the estimates yielded by alternative regression models. When nuclear cost is represented as a function of both sector size and reactor operating experience (a statistically questionable procedure—see above), the calculated 1988 nuclear cost is $1450/kW, 6% above the nuclear cost projected here and 83% greater than the coal projection. Alternatively, if the 1971-78 annual rates of increase in real costs are extrapolated to 1988, the resulting costs are $3140/kW for nuclear and $1230/kW for coal (in 1979 steam-plant dollars)—a 2.6 nuclear-coal ratio with over three times the cost differential projected here.

Section 2.7:   The Impact Of Three Mile Island

The foregoing 1988 nuclear cost projection and outline of regulatory issues affecting reactor design and equipment are both based upon nuclear power as regulated through 1978. Thus, they do not explicitly reflect the effects of the March 1979 accident at Three Mile Island. The accident's impact has been profound, "permanently alter[ing] the regulatory process for nuclear power," according to a Department of Energy analysis.[27] Because of extensive core damage, overheating of fuel rods, and shut-off of safety systems, the accident "exceeded many of the present design bases by a wide margin [and was] evidently a significant precursor of a core-melt accident," according to NRC's "Lessons Learned Task Force."[28] Moreover, nuclear regulation and management as a whole were implicated in TMI, most tellingly in NRC's and industry's failure to learn from prior "dress rehearsal" accidents at two other reactors.[29]

As a result, post-TMI nuclear regulation will almost certainly reflect greater willingness than previously to pay more to obtain greater safety. Past regulation placed "the burden of proof . . . on the regulators to justify negative findings on safety matters,"[30] admits NRC Commissioner Gilinsky, and mandated only "the most conservative requirements consistent with the commercial viability of nuclear power."[31] But the trauma of TMI has now "shattered [NRC's] complacency"[32] about reactor hazards, says Gilinsky, and has led NRC Chairman Hendrie to put the nuclear industry on notice that "safety [not costs] must be the dominant element in our considerations."[33] Such attitudinal changes would help fulfill the calls from the President's Commission on the TMI Accident (the Kemeny Commission) and the NRC's own investigation for "fundamental changes" in the "practices and . . . attitudes of the NRC"[34] and "in the way commercial nuclear reactors are built, operated and regulated."[35]

Accordingly, NRC is expanding its licensing process to make "explicit consideration of accidents involving severely damaged or molten cores."[36] Among the many changes already ordered or being considered are: upgrading the "single-failure criterion" so that some safety equipment must function despite multiple equipment failures; providing capability to cool "degraded" reactor cores; strengthening reactor containments to mitigate previously "incredible" core-melt accidents; expanding the "safety-related" classification of equipment subject to exacting quality assurance requirements; and requiring each reactor to install training simulators and no less than three heavily instrumented facilities for accident management (for technical support, operational support, and emergency coordination). Each proposal has the potential to change plant designs significantly and to impose large costs and delays, as discussed in Chapter 6. Indeed, the pending rulemaking on degraded cores alone may take three years to complete and add 40 regulatory guides to the approximately 150 now in effect.[37]

Section 2.8: Total Generating Costs

Capital costs will strongly affect total generating costs of future plants, especially nuclear plants, but other cost factors will also play a part. Projected capital costs are combined with estimates of fuel costs and other factors in Table 2.3 to yield estimated life-cycle generating costs per-kilowatt-hour (kWh) for typical 1988 plants. The plants are assumed to be 1150-MW capacity in the case of nuclear and 300-MW for coal. The calculated capital costs reflect the mix of multiple units, cooling towers, geographical location, etc. shown in Table 10.3. Operating costs are calculated for assumed 30-year plant lives spanning 1988-2017 (calculational methods are explained in Section 12.1). All costs are national-average estimates drawn from the most recent experience available and modified for likely future changes; they are *expressed in 1979 constant dollars with assumed escalation rates stated relative to the producer price index*.[38] The estimate bases are described fully in Chapter 11 and are merely summarized here without citations.

Capital Costs: The costs of power plant construction inputs are as-

Table 2.3
Projected Costs, 1988 Plants, U.S. Average
. (in 1979 constant dollars)

	Nuclear	Coal
Unit Size	1150 MW	300 MW
Capital Cost	$1460/kW	$838/kW
Decommissioning	$138/kW	
Real Fixed Charge Rate	10.3%	9.8%
Capacity Factor	60%	70%
Capital Cost Fixed Charges	2.86¢/kWh	1.34¢/kWh
Decommissioning Fixed Charges	.21¢/kWh	
Fuel	1.09¢/kWh	1.96¢/kWh
Operating and Maintenance	.62¢/kWh	.62¢/kWh
TOTAL	4.78¢/kWh	3.92¢/kWh
Nuclear/Coal Cost Ratio	1.22	

sumed to inflate by 1% annually relative to the producer price index (the approximate historical rate[39]). The costs of new nuclear and coal plants expressed in 1979 *constant* dollars deflated by producer price inflation are $1460/kW and $838/kW, respectively (see Section 12.1 for calculations). (These are slightly higher than the costs calculated earlier in 1979 *steam-plant* dollars deflated by construction factor inflation.) For comparison with estimates expressed in unadjusted (as-spent) dollars, respective nuclear and coal costs in mixed current dollars would be approximately $2700/kW and $1600/ kW, assuming annual inflation and interest rates of 9%.[40]

Capacity Factors: A 60% capacity factor is assumed for 1150-MW nuclear units, equal to the average for all commercial reactors through mid-1980. This assumes substantial improvement from the historical experience for large reactors: units over 800 MW, which account for over half of all operating experience, have averaged only 54%. Increased size is significantly correlated with reduced nuclear performance; and increased age, later vintages, and duplicate designs have brought only modest improvements. The projection does not reflect the full brunt of prospective closer regulation of reactor operations after TMI.

A 70% capacity factor is assumed for base-loaded 300-MW coal units, equalling the capability shown in performance data over the past decade for 200-400 MW coal units, absent "load-following." Mediocre capacity factors (55-65%) commonly cited for coal plants are for *large* units (over 400 MW) and also reflect intentional cutbacks dictated by excess capacity. Fully loaded· 600-MW units would be expected to have 65% capacity factors, but because coal plants have relatively low capital costs, even 60% performance would add only 7% to coal generating costs.

Fuel Costs: Refined uranium ore (U_3O_8) is assumed to cost $35 per pound (1979 dollars), with ore depletion and environmental and occupational health regulations causing 2% annual escalation relative to the producer price index. Enrichment cost of $94 per "separative work unit" is assumed to escalate at 1.5% per year due to increasing power costs. Spent fuel disposal cost of $650/kilogram is two to three times DOE estimates, in consideration of chronic cost underestimation and continuing technical and institutional uncertainties, but it accounts for only 3% of total nuclear costs.

The cost of coal fuel is extremely important to coal power costs. The cost assumed here is based upon the 1979 average cost of utility-burned coal of 1.2¢/kWh (assuming a 10,000 Btu/kWh heat rate), escalated in real terms at 2.3% per year—the 1974-79 average real rate—for 40 years. The resulting average fuel cost during the 1988-2017 period is 1.96¢/kWh (in 1979 constant dollars), 63% higher in real terms than the 1979 cost. Actual costs may be less than this projection in view of the very large reserves in all major U.S. coal-mining regions and the increasingly successful adaptation of mining design, technology, and management to the environmental and safety regulations that caused much of the real increases in prices in the 1970s. Although temporary shortages in coal supply could cause temporarily higher prices, reductions in the growth rate for electricity—coal's dominant present use— will probably prevent growth in total coal demand from outrunning expansion in mining capacity even if coal exports multiply and if coal makes substantial inroads into oil and gas markets (see discussion in Section 11.3).

Operating and Maintenance Costs in 1979 averaged .4¢/kWh for nuclear plants and slightly over .2¢/kWh for coal. Anticipated safety and environmental requirements, especially coal scrubber reagents and product disposal, are assumed to lift costs to .5¢/kWh in 1979 dollars for both plant types. Assumed 1%/year real escalation gives 1988-2017 "levelized" average O&M costs of about .6¢/kWh.

Financing: Fixed charge rates of 10.3% for nuclear and 9.8% for coal are used to convert capital costs to annual costs. These rates are based on assumed inflation-adjusted costs of capital of 3.8% and 3.6%, respectively, reflecting a slight "risk premium" for nuclear investments. The nuclear rate also includes greater allowance for interim capital replacement—.5% vs. .25% for coal—for anticipated decontamination and regulatory backfits.

Decommissioning: Nuclear plants are assumed to be dismantled at a cost of $125/kW (1979 steam-plant dollars), based on a utility cost estimate which DOE considers "representative [but] subject to potential [real] escalation."[41] This contributes slightly over 4% to calculated nuclear generating costs.

Total Generating Costs: Life-cycle generating costs calculated with the above "base case" assumptions are 3.9¢/kWh for coal and 4.8¢/kWh for nuclear (1979 dollars). The 22% difference implies added costs of $45 million per year per 1000 MW of new nuclear capacity. The difference would stand at 17% if "back-end" nuclear costs (decommissioning and spent fuel disposal) were halved. The difference could also shrink if reactor cancellations ease pressure to strengthen safety requirements, but it is more likely to increase as TMI intensifies regulatory efforts to reduce nuclear hazards, as discussed above.

For projected costs to be equal, 1988 nuclear capital costs would need to be approximately $960/kW (1979 steam-plant dollars)—only 8% above the actual $887/kW cost of a typical 1978 reactor. Alternatively, real escalation of about 4% per year in delivered coal costs, implying that real coal prices would average 133% more during 1988-2017 than in 1979, would also equalize projected costs, as would a 78% nuclear capacity factor. Conversely, the average nuclear generating cost would be 36% above that of coal if, because of TMI or other serious accidents, the 1978-88 real nuclear capital cost increase is 50% greater than projected. Even without this added increase, the nuclear-coal cost spread would be 39% if coal fuel costs escalate by only 1%/year in real terms.

Cost components and the conditions necessary to equalize nuclear and coal costs under widely varying but plausible assumptions are discussed in Chapter 12 and are shown in Table 2.4 and Figure 2.6. The table indicates that the projected nuclear/coal cost ratio is higher when "best" and "worst" cost cases are compared against each other. Moreover, because the best nuclear case has a higher projected cost than the base coal case, a nuclear advantage is possible (on an average basis) only through unanticipated escalation in coal.

Although these are national average results, based on geographical composite plants, new nuclear plants appear equally disadvantaged in every region, as shown in Section 12.2. In the Northeast, for example, the greater nuclear construction cost premium would offset high coal delivery costs. Similarly, western coal mining costs are sufficiently low and the pollution-control specifications reflected in the costs of 1988 coal plants are already sufficiently strict that new California coal plants appear considerably less expensive than new reactors, notwithstanding their long distances from coal

Table 2.4
Breakeven Cost Calculations

Nuclear Cases	Coal Cases	Projected Nuclear-Coal Cost Ratios	Paid-Off Share Of Nuclear Plant For Breakeven	Annual Real Coal Escalation For Breakeven
	BEST	1.49	55%	4.3%
BASE	BASE	1.22	30%	3.9%
	WORST	0.91	−16%	3.2%
	BEST	1.27	32%	3.1%
BEST	BASE	1.03	5%	2.6%
	WORST	0.77	−45%	1.5%
	BEST	1.67	62%	5.0%
CASE 1	BASE	1.36	41%	4.7%
	WORST	1.02	2%	4.1%
	BEST	2.25	84%	7.1%
WORST	BASE	1.84	69%	6.6%
	WORST	1.37	41%	5.3%

Nuclear Cases

BASE: See Table 12.1 (includes little or no TMI impact and 60% capacity factor)

BEST: Little or no TMI impact, 65% capacity factor, and reduced uranium and enrichment costs in Case 6

CASE 1: TMI adds 50% to extrapolated 1978-88 increases in construction cost and duration

WORST: TMI doubles extrapolated 1978-88 increases in construction cost and duration, adds 0.5% to real nuclear rate of return, and reduces capacity factor to 55%; decommissioning charges triple

Coal Cases

BASE: See Table 12.1 (includes 2.3%/year real fuel cost escalation and 70% capacity factor)

BEST: 10% reduction in capital cost, 1%/year real fuel cost escalation, and 75% capacity factor

WORST: 10% addition to capital cost, 4%/year real fuel cost escalation, and 60% capacity factor

---

Paid-off nuclear plant share represents percentage of ultimate nuclear capital cost that utility would need to defray for nuclear generating cost to equal that of coal. Alternatively, annual coal fuel escalation gives coal price increase rate necessary for breakeven.

Nuclear Case 1 adds 50% to extrapolated 1978-88 increases in construction cost and duration as estimate of TMI impact. Best and Worst cases are author's judgments of *plausible* extreme cases and are described in Table 12.4. Base cases are shown in Table 2.3.

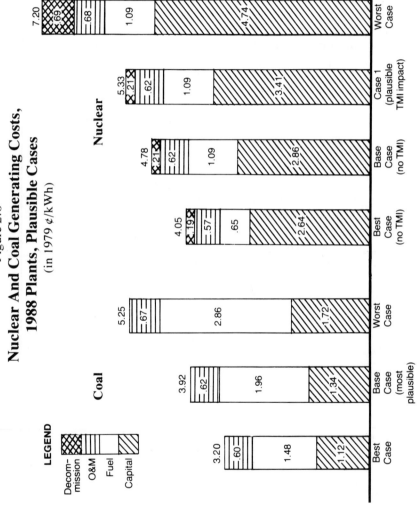

Figure 2.6
Nuclear And Coal Generating Costs,
1988 Plants, Plausible Cases
(in 1979 ¢/kWh)

LEGEND

Decom-mission
O&M
Fuel
Capital

Coal

Nuclear

Best Case
3.20
.60
1.48
1.12

Base Case (most plausible)
3.92
.62
1.96
1.34

Worst Case
5.25
.67
2.86
1.72

Best Case (no TMI)
4.05
.19
.57
.65
2.64

Base Case (no TMI)
4.78
.21
.62
1.09
2.86

Case 1 (plausible TMI impact)
5.33
.21
.62
1.09
3.41

Worst Case
7.20
.69
.68
1.09
4.74

See Table 12.4 for assumptions employed in each case.

fields. Finally, although costs of individual plants will differ from trends, most error bands in the cost regressions are narrow, and the scope is limited for reducing nuclear costs by optimizing plant parameters such as size or architect-engineer experience.

Section 2.9:   Conclusion

Projections that electricity will be cheaper from new nuclear plants than from new coal facilities implicitly require that respective capital costs not differ by more than approximately 25%. However, nuclear capital costs increased more than twice as much as those of coal plants during 1971-78 and already exceed the latter by 50%. In return for increased costs, emissions of criteria air pollutants from new coal plants declined by almost two-thirds, compared to 1971 plants, and could fall by nine-tenths by the late 1980s to levels lower than those from plants burning low-sulfur oil. The far more costly attempts to contain nuclear hazards have produced largely putative, intangible reductions in accident probabilities, reductions whose significance seems questionable in light of the 1979 accident at Three Mile Island—then America's newest reactor.

Efforts to reduce the risks of nuclear power plants grew apace with expansion of the nuclear sector and were the key ingredient in causing nuclear capital costs to rise in real terms in the 1970s. If the past relationship between nuclear sector size and costs continues in the 1980s, the capital costs of reactors now starting construction would exceed those of new coal plants with advanced pollution controls by about 75%. The generating costs of new reactors would then average 20-25% more than those of new coal plants. Although the assumed rate of imposition of new nuclear regulatory standards could fall if many reactors under construction are cancelled, the salutary effect on nuclear cost increases would probably be offset by the added design and construction requirements stemming from TMI.

These figures imply that a typical new nuclear plant with 30% of its capital cost paid off would not be a cheaper source of electricity than a new full-cost coal-fired plant. Accordingly, many reactors in early stages of construction appear to be potential candidates for economical replacement by coal facilities. Cancellation charges would add to the cost of abandoning nuclear projects, but they would be partially offset if some preparatory work at reactor sites, such as excavation and foundations, could be employed by replacement coal plants.

The conclusion that clean coal plants can provide electricity more cheaply than nuclear plants is at variance with current dogma, but it is rooted firmly in empirical cost data, and requires only a modest extension of established regulatory and cost trends. It is incumbent upon nuclear proponents to either demonstrate why their own cost projections should be heeded in light of the evidence presented here, or to acknowledge that construction of new reactors cannot be justified on the basis of an economic comparison with coal.

References

1. S.M. Keeny, ed., *Nuclear Power Issues and Choices* (Ballinger, Cambridge, MA, 1977), p. 123. Although the present author's *Power Plant Performance* (Council on Economic Priorities, New York, NY, 1976, p. 133) was one of the 12 sources cited, the cost ratio therein was based solely on AEC projections. Unlike the estimates in this book, it did not utilize empirical data and research.

2. A.D. Rossin and T.A. Rieck, *Science, 201*, 586 (1978).

3. U.S. Energy Information Administration (EIA), *Annual Report to Congress, 1978*. Mean costs of 1985 plants in 1978 dollars are nuclear, $765/kW (p. 217); coal, $650/kW (p. 386).

4. National Academy of Sciences, *Energy in Transition, 1985-2010* (Washington, DC, 1979), p. 318.

5. C. Blake *et al.*, *Analysis of Projected vs. Actual Costs for Nuclear and Coal Fired Power Plants* (Mitre Corp., McLean, VA, 1976).

6. W.K. Davis and R.O. Sandberg, *Light Water Reactors: Economics and Prospects* (Bechtel Power Corp., San Francisco, CA, February 1979).

7. I.C. Bupp *et al.*, *Trends in Light Water Reactor Capital Costs in the United States: Causes and Consequences*, CAP 74-8 (Center for Policy Alternatives, MIT, Cambridge, MA, 1974). Abridged in *Technology Review, 77* (No. 2), 15-25 (1975).

8. W. Mooz, *Cost Analysis of Light Water Reactor Power Plants*, R-2304-DOE (1978), and *A Second Cost Analysis of Light Water Reactor Power Plants*, R-2504-RC (1979) (RAND Corporation, Santa Monica, CA).

9. Whitman, Requardt and Associates, *Handy-Whitman Index of Public Utility Construction Costs* (Baltimore, MD). The compound average 1965-78 inflation rates, 7.5% coal and 7.1% nuclear, are reflected in the data base steam-plant-dollar capital costs.

10. See Electric Power Research Institute (EPRI), *Technical Assessment Guide*, EPRI PS-1201-SR (Palo Alto, CA, 1979), p. 8-5; U.S. Energy Research and Development Administration (now DOE), *Concept, Phase IV, User's Manual*, ERDA-108 (1975), Table 3.2.

11. Reference 10, EPRI, p. 8-3.

12. Costs of scrubbers were measured separately because few sample plants have them (15 of 116), and thus their effect would not have been adequately reflected in the sector size coefficient. Omitting the scrubber variable raises that coefficient to .86 but yields lower 1978 and 1988 coal plant costs than obtained here. See Section 10.2.

13. Although the coal sample extends only through 1977, it includes several plants ordered later than all 1978-completed plants, so its sector size value (measured according to date ordered) reflects 1978 installations. See Appendix 2.

14. Assumed annual real costs of capital are 2.8% for 1971 plant and 3.3% for 1978, from NRC, *Treatment of Inflation in the Development of Discount Rates and Levelized Costs in NEPA Analyses For the Electric Utility Industry*, NUREG-0607 (1980), Table 1. Weighted cost of capital, with common stock yield based on price earnings ratio, without inflation (implicit price

deflator), was averaged for 1967-71 for 1971 plant and 1972-77 for 1978 plant.

15. The nuclear industry's widely cited 1978 U.S. per-kWh generating cost averages of 1.5¢ for nuclear plants and 2.3¢ for coal plants rest on omission of nearly all the costliest nuclear plants and cheapest coal plants. See Section 1.2 and C. Komanoff, *Power Propaganda* (Environmental Action Foundation, Washington, DC, 1980).

16. J. Crowley in *Nuclear Engineering International, 23* (No. 7), 39 (1978).

17. Atomic Industrial Forum (AIF), *Licensing, Design and Construction Problems: Priorities for Solution* (Washington, DC, 1978), Exhibits 1 and 9.

18. Reference 17, p. 1.

19. A.M. Weinberg, *The Wilson Quarterly*, III (No. 3), 96 (1979). Emphasis in original.

20. ACRS Chairman W.D. Manly, letter to AEC Chairman Glenn T. Seaborg, 24 November 1965. Reprinted in Joint Committee on Atomic Energy (JCAE), *Hearings on Licensing and Regulation of Nuclear Reactors*, 1967, p. 119.

21. AEC, *Technical Report on Anticipated Transients Without Scram for Water-Cooled Reactors*, WASH-1270 (1973), p. 6. Emphasis added.

22. E.S. Rolph, *Nuclear Power and the Public Safety* (Lexington Books, Lexington, MA, 1979), p. 124.

23. Statement by ACRS Chairmen N.J. Palladino (1967) and D. Okrent (1966) before the JCAE, 5 April 1967. Reprinted in Reference 20, p. 92.

24. NRC, *Nuclear Power Plant Licensing: Opportunities for Improvement*, NUREG-0292 (1977), p. 3-3.

25. The assumed growth in coal sector size, from 217 GW in 1978 to 347 GW in 1988, approximates the growth projections in National Electric Reliability Council (NERC), *9th Annual Review of Overall Reliability and Adequacy of the North American Bulk Power System* (Princeton, NJ, 1979).

26. NRC, *Identification of Unresolved Safety Issues Pertaining to Nuclear Power Plants*, NUREG-0510 (1979), p. 10.

27. DOE, *Nuclear Power Regulation*, DOE/EIA-0201/10 (1980), p. xiv.

28. NRC, *TMI-2 Lessons Learned Task Force Final Report*, NUREG-0585 (1979), p. 3-5.

29. See M. Rogovin *et al., Three Mile Island: A Report to the Commissioners and to the Public* (1980), Vol. 1, p. 94ff.

30. V. Gilinsky, Speech at Brown University, Providence, RI, 15 November 1979.

31. Reference 22, p. 77

32. Reference 30.

33. J.M. Hendrie, Speech at the AIF International Conference on Financing Nuclear Power, Copenhagen, 24 September 1979.

34. J.G. Kemeny *et al., Report of the President's Commission on the Accident at Three Mile Island* (1979), p. 7.

35. Reference 29, Vol. 1, p. 91.

36. NRC, *NRC Action Plan Developed As A Result of the TMI-Accident*, NUREG-0660, Vol. 1 (1980), p. II-1.

37. Reference 36, p. II.B-12; and *Inside N.R.C.*, 2 (No. 15), 7 (1980).

38. The producer price index for industrial commodities has exceeded the GNP implicit price deflator by about ½% per year in recent years.

39. Average annual compound 1965-79 Handy-Whitman escalation, deflated with the producer price index, was .64% for nuclear plants, .98% for fossil.

40. Calculated from costs in 1979 steam-plant dollars without IDC, assuming 1980-88 construction for nuclear and 1982-88 for coal, 13% construction inflation from 1979 to 1980 and 9% annually thereafter, and a 9% annual IDC rate.

41. Reference 27, p. 172.

3

The Sources Of Nuclear Regulatory Requirements[a]

From where do new nuclear safety requirements arise? This question is critical to the future of nuclear power in the United States. Reactor construction costs have risen rapidly, even after adjusting for the effects of inflation, as Chapters 8 and 10 show. Much of this real increase in costs resulted from promulgation of more stringent regulatory standards. Yet future standards and their costs have proven difficult to predict. An analysis of the processes involved in developing safety requirements may provide insight into the future rate of their imposition, and, thus, into future nuclear costs.

This chapter reviews the evolution of commercial nuclear power regulation and licensing in the United States. It indicates that increased regulatory stringency arises primarily from three phenomena: the attempt to reduce the permissible risk to public health and safety per reactor; new information indicating that current standards are insufficient to reduce risks to the desired levels; and the greater documentation and standardization of regulatory requirements that accompany expansion of the regulatory effort. These phenomena, in turn, are fed primarily by six distinct motivating sources:

1. increases in reactor population requiring that the per-plant risk be reduced in order to limit the overall accident probability;
2. increases in reactor size expanding the potential consequences and probabilities of accidents;
3. government and industry design and licensing reviews discovering new safety issues;
4. reactor operating experience also uncovering previously undetected safety problems and underscoring the severity of known, unresolved issues;

a. A slightly altered version of this chapter has been accepted for publication in *Nuclear Safety, 22* (No. 4) (July/August 1981).

5. public concern contributing to both reducing permissible risks and to the unearthing of new safety problems; and

6. increases in the size and purview of regulatory staff resulting from the increase in reactor population and the foregoing five factors.

The chapter begins with a brief history of nuclear power commercialization in the United States and then explores each of the six sources of increased regulatory stringency listed above. It concludes by considering the relationship between increased stringency and nuclear power expansion and by assessing the prospects for stabilizing regulatory requirements.

Section 3.1:   The Expansion of Nuclear Power

The modern era of nuclear power operation in the United States began 13 years ago, with declarations on New Year's Day 1968 that "commercial operation" status had been achieved at America's first 500-megawatt (MW)-class reactors — the Haddam Neck (CT) and San Onofre (CA) plants. Previously, U.S. power reactors comprised only the dozen 10-75 MW reactors that started entering service in the late 1950s and three 200-MW-class plants that began operating in the early 1960s.[1] All were subsidized under the Power Reactor Demonstration Program managed by the Atomic Energy Commission (AEC) from 1955 to 1963 to promote commercialization of nuclear power.

These reactors were succeeded by a dozen so-called "turnkey" plants provided at a fixed price (and at a loss) by Westinghouse and General Electric (GE). The order for the first of these, 650-MW Oyster Creek in 1963, was heralded as proof that nuclear power could compete economically with fossil generation of electricity, without direct government support of plant construction or the fuel cycle. The nearly identical Nine Mile Point 1 unit was ordered several months later and built on a commercial (non-turnkey) basis. After a two-year pause, the other turnkey plants, also in the 500-800 MW range, were ordered in 1965 and early 1966. Most entered service between late 1969 and 1972.

Following (and occasionally overlapping with) the turnkey contracts came a second and much larger wave of reactor orders. Fifteen additional units were ordered in 1966, and twice that number in the following year. These reactors averaged several hundred megawatts more than their immediate predecessors. Some surpassed 1000 MW, up to five times the capacity of the largest operating reactors. This not only violated the power industry's precept against large jumps in unit sizes, but also raised significant new safety issues, as discussed below.

Nevertheless, the nuclear rush was on. After over a decade of striving, with limited success, to advance nuclear power as a commercial power source, the AEC was suddenly being required in the mid-to-late sixties to license and to ensure the safety of dozens of large reactors, with many more expected to follow shortly.[2]

Section 3.2:   Reducing The Accident Probability Per Plant

Following the Three Mile Island accident in March 1979, nuclear pioneer Alvin Weinberg wrote that "For nuclear energy to grow in usefulness, the accident probability *per reactor* will simply have to diminish."[3] Otherwise, nuclear expansion could lead to such a high rate of accidents *per year* that the public's confidence in nuclear power would collapse and plants would be forced to close.

Weinberg's prescription appears to have been followed historically. As nuclear power has expanded, nuclear regulators have tried ever harder to reduce new reactors' risks to prevent the overall nuclear accident probability from expanding as fast as the reactor population.

Although it is not codified in regulations, this effort informs much of the advocacy of improvements in nuclear safety within the regulatory community over the past decade and a half. It is particularly pronounced in the recommendations of the Advisory Committee on Reactor Safeguards (ACRS), an influential body of senior nuclear safety experts that advises the Nuclear Regulatory Commission (NRC, formerly AEC) on safety matters and individual reactor licensing applications.[4] In November 1965, for example, the ACRS called upon AEC to upgrade standards for reactor pressure vessels on the grounds that

> [T]he orderly growth of the industry, with concomitant increase in number, size, power level and proximity of nuclear power reactors to large population centers will, in the future, make desirable, even prudent, incorporating stricter design standards in many reactors.[5]

The letter stimulated major efforts by AEC and the nuclear industry to improve the design, fabrication, and "in-service inspection" of reactor vessels.[6] It also led to a 1967 AEC report on emergency core cooling systems (ECCS) that recommended improvements in the manufacture and inspection of nuclear piping, valves, and pumps because "the large number of plants now being constructed and planned for the future makes it prudent that even greater assurance be provided henceforth."[7] Also in that year, the ACRS urged that greater attention be paid to reactor safety problems because "large increases in the number of reactors lead to the desire to make still smaller the already small probability per reactor that an accident of any significance will occur."[8]

Similar sentiments were expressed by the ACRS during the 1970s to support more stringent standards. Two ACRS chairmen, one of whom subsequently chaired the NRC, told the Congressional Joint Committee on Atomic Energy in 1971 that "the high degree of conservatism used in both nuclear plant designs and in safety reviews" was justified by "the increased number of reactors soon to be operating and...the trend toward large reactors of higher power densities."[9] Two years later, the ACRS asked the AEC chairwoman to seek improvements in ECCS on the grounds that "...for an expanding nuclear industry, the cumulative effects of the added improvements represent prudent goals."[10]

The AEC regulatory staff also appears to have been guided by considerations of the total sector accident frequency. In 1973, when recommending back-up shut-down systems to prevent events in which the control rods fail to *scram* (shut down) the reactor during sudden interruptions in its normal operation (events known as Anticipated Transients Without Scram, or ATWS), the staff wrote that "since larger safety margins are appropriate as increasing numbers of power reactors are built and operated, design improvements should be made to reduce the probability of ATWS in new plants to a negligible level..."[11] Staff further wrote,

> The present likelihood of a severe ATWS event is considered by the staff to be acceptably small, *in view of the limited number of plants now in operation,* the reliability of current protection system designs, and the expected occurrence rate of anticipated transients of potential safety significance. *As more plants are built,* however, *the overall chance of ATWS will increase,* and the staff believes that design improvements are appropriate to maintain and to improve further the safety margins provided for the protection of the public.[12]

In other words, as more reactors come into operation, the per-reactor probability of an accident must be reduced in order to control the overall accident frequency. Similarly, in 1975 the staff mandated improvements in leakage control systems for main steam isolation valves because "there is a need for design improvements to provide appropriate safety margins for the large number of plants now planned."[13] Conversely but consistently, "the limited number of operating nuclear power reactors" in 1980 following the slowdown in reactor licensing and construction in the late 1970s was cited by the staff as a reason to grant utilities several years to phase in ATWS-mitigating design changes rather than requiring them immediately.[14]

Section 3.3:   Increases In Reactor Size

Many of the statements quoted above cite increases in reactor *power level* as well as in the *number* of reactors as a source of concern. And indeed, the rapid increases in reactor generating capacities — from 200 MW for plants licensed in the mid-1950s to 600 MW in the mid-1960s and 1000 MW shortly thereafter — induced regulators to seek more stringent safety measures. Accidents at larger reactors could have more serious consequences since they carry more fuel with a proportionately greater fission product inventory which is subject to release. In addition, greater preventive measures were required to constrain large units' accident probabilities. As AEC regulatory staff stated in 1967,

> The increase in this potential hazard [from larger reactors] must be matched by corresponding improvement in the safety precautions and requirements if the safety status is to keep pace with advancing technology. The protective systems must have shorter response times, larger capacities and greater reliability to cope with the more rigorous demands presented by the large reactors. [15]

Although the need for shorter emergency response times has receded somewhat since thinner fuel rods with reduced fuel center-line temperatures began to be introduced in the early to mid-1970s, larger plants do generate proportionately more ''decay heat'' following reactor shutdown. Removal of decay heat is a particular concern in many postulated accidents, including small-break loss-of-coolant accidents (LOCAs). As the ACRS noted in 1967, ''the *decay heat* production from a large reactor such as...Browns Ferry [1098 MW] begins to approach a level comparable to the original *full load* power level of the Shippingport Reactor [70 MW].''[16]

Moreover, starting with the commercial-size plants first licensed in the early 1960s, new reactors required concrete shielding for the containment walls to reduce exposure to nearby persons in the event of accidents. This further cut down the rate of heat dissipation through the containment. As a result, whereas reactor vessels were believed capable of containing a 100-MW molten core, a 1000-MW core ''would eat its way right through the pressure vessel'' and perhaps through the containment as well.[17] The result was that as plant sizes grew, increased consideration was given to the ECCS, to systems for removing decay heat from containment, and to systems for removing radioactivity from the containment atmosphere to reduce leakage.[18]

Concerns over increasing reactor sizes and increasing reactor numbers combined to produce more stringent regulatory standards for ECCS and primary reactor piping in the late 1960s. The ACRS was particularly troubled by uncertainties in ECCS performance for the first 800-MW reactors. In 1966 it drafted a letter to the AEC chairman asserting that

> [A]s more and more reactors come into existence, particularly reactors of larger size and higher power density, the consequences of failure of emergency core cooling systems take on increased importance.[19]

Although never formally sent, the letter apparently contributed to General Electric's decision to expand the ECCS at the Dresden reactors under construction by adding a separate core flooding system to the two ECCS core spray systems.[20] Perhaps more importantly for the long-term, the letter led AEC to establish the study group on ECCS referred to earlier. The stress in the group's report on the difficulty of delivering cooling water to a partially-melted and reshaped core highlighted the importance of equipment to prevent or promptly halt any loss-of-coolant-accident.

This and other recommendations in the report prompted the reactor vendors to increase in-service inspection and leak-detection tests for primary system piping and to further expand the diversity and reliability of the ECCS to ensure that it could respond to a wider range of system pressures corresponding to a greater number of potential accidents.[21] GE substantially increased the capacities of the ECCS core flooding and core spray systems for Browns Ferry (1100-MW class) compared to those for Dresden or Quad Cities (800-MW class.)[22] Finally, the report's implicit finding that current containment designs might not suffice to contain a melted core sparked concern over ECCS performance, both within and outside the AEC, and helped lay the basis for the tumultuous ECCS rulemaking hearings in the early 1970s.

Section 3.4:   Growth Of The NRC Licensing Effort

The increase in the number of applications for reactor construction permits and operating licenses has also contributed to growth in regulatory requirements. The increase has necessitated a larger NRC staff that in turn permitted a broader range of safety issues to be examined. It also has led to standardized review procedures that have tended to raise the stringency of standards applied to all plants. Many specific applications have raised new safety concerns, provoking development of new criteria that have been extended to other reactors.

AEC staff reviews of the first commercial-size reactors were generally limited and haphazard. Each construction-permit application was reviewed by several engineers from the Hazards Evaluation Branch within AEC's licensing division. Experts outside the division had to be called upon for technical support, and staff positions on specific design issues were frequently casually codified and documented. Although the first draft reactor *general design criteria* were issued for comment in 1965,[23] few standards had been developed to determine whether proposed designs and equipment satisfied the criteria. According to licensing specialist B. N. Naft of NUS Corp., "In the days of the earliest commercial plants, guidance from AEC was based on direct communications and what 'the last applicant had been through.' ''[24]

By early 1967, with a dozen large reactors under construction and over a dozen more construction-permit applications docketed, AEC began to significantly expand its licensing division to cope with the expanded caseload. The larger staff included reactor specialists who could question applicants more thoroughly on plant designs and construction methods. "Word-of-mouth" approvals of design approaches were superseded by detailed examinations requiring documentation of engineering assumptions, analyses, and tests.

The AEC also moved to formalize licensing reviews in order to equalize the scrutiny applied to different reactors, to expedite applicants' responses, and to establish uniform procedures to be followed by the growing staff. Licensing positions on specific safety issues were detailed in "Branch Technical Positions." In 1970, staff inaugurated a series of Regulatory Guides (RGs, initially called Safety Guides) specifying acceptable approaches to problematic design and construction issues. Early guides often did not contain new approaches but rather codified previously-developed positions to provide documentation needed by both staff and applicants. Subsequent guides were generated so that guidelines that had been formulated and used in individual reactor reviews could be employed for other applications. As the staff and the licensing effort grew, all construction permit reviews began to be pegged to the highest common denominator. This process elicited greater conservatism in areas such as design of engineered safeguards and quality assurance (QA) programs.

The number of Regulatory Guides grew from three in 1970 to 21 in 1971 and 33 in 1972. Some were innocuous from a cost standpoint, but others — those pertaining to construction methods, seismic criteria, and engineered safeguards — engendered design changes and cost increases. The guides' status also evolved from guidelines to requirements. Staff usually insisted upon close adherence to the practices outlined in the guides, and applicants "volunteered" to conform rather than engage in time-consuming negotiations. As a consultant report to NRC noted, "Utilities often conclude that proposing alternatives to solutions and approaches identified in NRC guidance would be too costly. In these cases the NRC guidance serves as *de facto* regulation."[25]

The Standard Review Plan, a compilation of internal review procedures begun by AEC in 1972 and initially issued by NRC in 1975, has also tended to raise regulatory practice to the highest common denominator. The plan contains the criteria new plants must satisfy and staff procedures for assessing whether the criteria are met. It was developed to provide a handbook of requirements for the growing staff, to serve as a benchmark for evaluating changes in regulatory practice, and to standardize licensing criteria. It has had to be updated continuously to keep up with proliferating new requirements. The Standard Review Plan now references approximately 150 Regulatory Guides (many in their second or third edition) and the number is still growing, fed by new staff reviews and expanding reactor operating experience.

Many Regulatory Guides have been prompted by particular license applications. As the AEC deputy director of regulation stated in 1967,

[W]hen any safety problem is first encountered by our regulatory staff...
we first encounter it on a case-by-case basis. In that process...we might
come to understand the issues involved, the effects that might result with
respect to reactor safety and what the protective mechanisms might be.[26]

Seismic issues, for example, were first raised in the early 1960s in applications
to build reactors near earthquake faults in California, at Bodega Bay on the
northern coast and at Malibu near Los Angeles. Prior to these applications,
AEC had not considered seismic phenomena in licensing and had no familiarity with them. (Two small California reactors licensed in the 1950s, GE's
Vallecitos test facility and the Humboldt Bay plant, received no detailed
seismic review at that time and subsequently shut down in the late 1970s rather
than upgrade seismic safeguards at high cost.) Shortly thereafter, AEC commissioned seismological and geological research demonstrating that the eastern United States also has considerable seismic potential. In addition, new
research in soil mechanics and in structural seismic response led to improved
understanding of the transmittal of seismic loadings to reactor equipment. This
information led to eight Regulatory Guides delineating methods of calculating
earthquake forces and specifying the instrumentation, structural reinforcement, and component reliability necessary to reduce susceptibility to damage
and accidents, which apply, in varying degrees, to all U.S. reactors.

Similarly, concern over intermingling instrumentation for reactor control (operation) and safety (shutdown) first emerged as a significant issue in the
ACRS review of the pressurized water reactor (PWR), Diablo Canyon 1 in
1967. This Westinghouse-designed unit "was to be one of the first of the
high-power PWRs built...which made it a logical reactor on which not only to
look for new, previously unanticipated issues, but to resolve some that had
been ongoing," according to one long-time ACRS member.[27] The questions
raised in the Diablo Canyon review were also directed at the next Westinghouse reactor in line for a construction permit, Prairie Island, even though the
much smaller Prairie Island design had already been accepted by AEC staff.[28]
These and all succeeding Westinghouse plants were ultimately required to
increase the separation of control and safety circuits, although not to the extent
desired by some ACRS members.[29]

Specific license applications have brought other safety issues to the fore.
Hurricanes were first considered in the construction permit review for Turkey
Point in Florida in 1966 and subsequently were factored into the reviews of all
East and Gulf Coast sites, and even many flood-prone inland sites.[30] Tornado-
protection requirements began to be applied to new reactors in the late 1960s
after the first review for a reactor in a high-tornado area — GE's Southwest
Experimental Fast Oxide Reactor (SEFOR) in Arkansas — established that
tornadoes occurred sufficiently frequently in most parts of the country to
warrant uniform defenses in design.[31] Similarly, some sites with relatively
high population densities appear to have acted as "magnets" for greater

regulatory emphasis on engineered safeguards which then spread to other reactors.[32]

Operating license reviews have also uncovered generic issues leading to changes in regulations. During late construction at Oyster Creek in 1967, cracks resulting from a combined design and welding deficiency were discovered in most of the control rod housings. When AEC staff examined the utility's quality assurance program, it found widespread deficiencies in field construction, installation of instrumentation and power circuits, and equipment procurement, including installation of secondhand valves of unknown condition. These findings provided greater impetus for increased AEC inspections and audits and for promulgation of quality assurance regulations.[33]

Section 3.5: Industry Reviews

Reviews of new reactors by utilities, reactor vendors, and architect-engineers (A-Es) have also developed information contributing to changing standards. These reviews are usually considerably more detailed than those by NRC, which primarily audits industry's analyses. Accordingly, industry analyses of new designs or of new reactors employing previously approved designs sometimes uncover safety problems that NRC staff failed to unearth independently.

An example is the "pressure-suppression" issue for boiling water reactor (BWR) containment structures. BWR containments have progressed through three stages: Mark I containments are used at most operating BWRs and at several nearing completion; most BWRs now in advanced construction employ Mark II designs, while most in design or early construction will use Mark III. All three containments employ a pool of water as a heat sink located in, below, or around the primary containment wall. They differ with regard to the materials employed, the geometry of the pool and the configuration for venting air or steam during accidents.

As part of its development of engineering data required by NRC to approve the Mark III, GE constructed a test facility in 1975 to measure the pressures that would be exerted on structures within containment during postulated accidents. These tests showed that very large vibratory pressures could result from the rush of air and steam from the reactor into the surrounding pool during a LOCA. This led GE and NRC to question the adequacy of Mark I and II designs along with that of the less-developed Mark III, and to bolster all of the containments to reduce the chances of equipment failure during possible accidents. At plants under construction or in design, extra steel has been added to reinforced concrete containment walls, all-steel containment walls have been thickened or further ribbed, and supports have been strengthened for equipment located within the pool area.[34]

An analogous example affecting PWRs concerns possible "asymmetric loading'' on reactor vessel supports. In 1975, Westinghouse and the Virginia Electric & Power Company (VEPCO) notified NRC of findings from improved analytical models being applied to the North Anna plant then under construction: certain postulated LOCAs could create ''pressure transients'' in the reactor vessel that could overstress the vessel supports. The resulting displacement of the vessel could compress the fuel assemblies and prevent control-rod insertion, disable the ECCS, and damage supports for the reactor coolant pumps and steam generators.[35] This previously unidentified scenario was subsequently established as an *unresolved safety issue* — a pending, generic problem whose resolution may require formulation of new regulatory requirements (see Section 5.3).

These examples illustrate NRC's involvement in industry's contribution to increased standards. Also relevant are the standards and codes developed by technical societies such as the American Society of Mechanical Engineers (ASME) or the Institute of Electrical and Electronics Engineers (IEEE), whose members work in various facets of reactor technology and safety. Over a hundred of these documents issued since the late 1960s have led to more stringent manufacturing, testing, and performance criteria for structural materials such as concrete and steel and for basic components such as valves, pumps, and cables. Most of the codes have been ''endorsed'' by NRC Regulatory Guides; indeed, in some instances the industry-dominated committees drafting new stringent standards have ''justified their stance by stating that unless industry addressed these concerns the NRC would in regulatory guides.''[36]

Section 3.6: Reactor Operating Experience

Reactor technology was initially developed with an expectation that its design, construction, and operation could be rigorously controlled and managed. Nuclear pioneers concentrated their analytical efforts on physics matters such as reactivity accidents, and devoted less attention to the difficult engineering problems of integrating the nuclear steam system with the balance of the power plant, keeping coolant water circulating in the core, and providing safeguards to prevent or mitigate accidents.

These problems began receiving attention in the reviews of the first 500-MW-class reactors in the early 1960s. Criteria considered excessively conservative were specified in some instances because of the paucity of engineering experience. It was anticipated that favorable operating data would ultimately allow some standards to be relaxed. In actuality, reactor components and equipment have frequently failed to achieve intended levels of reliability and performance. Although operating experience has sometimes justified reduced design margins (in fuel performance, for example), it has more often warranted corrective standards and engendered a more conservative overall regulatory approach.

Nuclear operating experience has come in two waves. The first consisted of the three 200-MW reactors and the dozen units under 100 MW licensed in the late 1950s and the 1960s. Although these reactors had accrued only 60 unit-years of operation by early 1967, this experience provided sufficient evidence for the ACRS to conclude that, "based on reactor operating experience...a variety of reactor transients have occurred, a variety of protective features have malfunctioned or been unavailable on occasion, and a variety of defects have been found in operation."[37]

The ACRS cited these specific failures:

1. loss of normal and emergency power in the same incident;
2. simultaneous loss of all (as many as five) incoming power lines;
3. blowdown of a primary coolant system;
4. loss of all protection provided by the capability for automatic scramming of control rods;
5. sticking and breaking of control rods;
6. rupture of a poison sparger ring;
7. failure of structural members within the pressure vessel;
8. faulty design of a steam generator support;
9. cracks in large pipes and studs;
10. poor choice of material for vital components;
11. melting of some fuel elements;
12. consecutive procedural errors; and
13. safety systems not wired up in accordance with design criteria even after extensive test programs.

Although it is not possible here to trace specific upgrading of standards to these failures, the "lack of perfection in design, construction and operation"[38] was a major reason for the ACRS's advocacy in the mid-to-late 1960s of conservative design practices and improved safeguards. For example, the ACRS's portentous 1965 letter to AEC on pressure vessel integrity was prompted by one member's concern over failures such as broken stud bolts at the vessel head closure and cracked main control rod shafts. These were "incipient failures which, had complete failure occurred, would have resulted in more serious accidents than any thus far experienced."[39]

Commercial-size (400+ MW) reactors have registered a far larger body of operating experience, beginning with the 1968 start-up of Connecticut Yankee and San Onofre. Experience with these larger reactors accumulated slowly at first, reaching only 11 unit-years at the end of 1970. But it appreciated rapidly as more reactors were completed, reaching 36 unit-years at the end of 1972, and 94 after 1974. This operating experience included fuel leakage, pipe cracks, faulty installation of control rods, operator disabling of shutdown systems, and malfunctioning valves, pumps, and cables. NRC staff later characterized this as a "considerable body of operating reactor experience [which by 1972] indicated the need for expanded technical review in areas previously thought to be not sufficiently important to warrant much attention."[40]

Utilities report operating problems and deviations in "licensee event reports" (originally called "abnormal occurrence reports") to NRC's Office of Inspection and Enforcement (originally AEC's Division of Operating Reactors). Not all reports have stimulated corrective action, however. For example, the stuck pressurizer relief valve that caused a substantial loss of primary coolant at Davis-Besse in 1977 was not corrected at other Babcock & Wilcox reactors, and contributed to the Three Mile Island (TMI) accident. NRC's Special Inquiry Group on TMI inferred from this and other disregarded precursor events to TMI that "NRC and the industry have done almost nothing to evaluate systematically the operation of existing reactors, pinpoint potential safety problems, and eliminate them by requiring changes in design, operator procedures, or control logic."[41]

Notwithstanding the lack at that time of *systematic* evaluation procedures, many adverse operating events have been "incorporated into the safety reviews of new plants,"[42] as a former NRC chairman noted prior to TMI. This process antedates the present-day commercial reactors. For example, a tornado that knocked out all offsite power lines to the Dresden 1 reactor in the early 1960s led to the use of small diesel generators to provide onsite emergency power — a requirement that evolved into much larger diesels to drive safety systems such as the ECCS.[43] More recently, operating experience has arguably become the single largest source of new regulatory requirements. For example, "Fire had been recognized as a potential safety concern of considerable importance for at least a decade before occurrence of the Browns Ferry fire in 1975,"[44] but requirements for cable loading and fire retardancy were upgraded only after serious fires at San Onofre (1967) and Indian Point (1971), and the Browns Ferry fire was the catalyst for major improvements in cable separation and ventilation systems at new plants.

Operating experience has also been cited as the impetus, at least in part, for many Regulatory Guides (see box). Many other guides do not specifically mention operating experience but have also originated from adverse occurrences. They include guides relating to reactor coolant pump flywheel integrity (RG 1.14), protection against pipe whip inside containment (RG 1.46), loose-part detection systems for the reactor primary system (RG 1.133), and many guides concerning quality assurance in component fabrication and plant construction.

Commercial reactor operating experience continues to increase rapidly. The total more than tripled from 140 unit-years at the end of 1975 to 430 at the end of 1980, providing much new fodder for more stringent standards (see box). Some of these problem areas have been incorporated into NRC's roster

Regulatory Guides Citing Operating Experience
(partial listing)

Number	Date*	Title
1.6	3/71	Independence Between Redundant Standby Power Sources
1.31	8/72	Control of Ferrite Content in Stainless Steel Welds
1.43	5/73	Control of Stainless Steel Weld Cladding of Low-Alloy Steel Components
1.44	5/73	Control of the Use of Sensitized Stainless Steel
1.47	5/73	Bypassed and Inoperable Status Indication for Safety Systems
1.55	6/73	Concrete Placement in Category I Structures
1.67	10/73	Installation of Overpressure Protection Devices
1.68.2	1/77	Initial Startup Test Program to Demonstrate Remote Shutdown Capability
1.96	5/75	Design of Main Steam Isolation Valve Leakage Control Systems for BWRs
1.115	3/76	Protection Against Low-Trajectory Turbine Missiles
1.120	6/76	Fire Protection Guidelines

*Most Regulatory Guides are effectively incorporated into the regulatory review process prior to their official publication.

of unresolved safety issues now receiving increased regulatory attention as areas of generic safety concern (*e.g.*, systems interaction, water hammer, residual heat removal). Others have been the subject of NRC Bulletins and Circulars requiring analysis or remedial action by licensees (*e.g.*, loss of high-pressure coolant injection, feedwater weld cracks). The Three Mile Island accident, of course, has "introduced a large number of new or previously non-emphasized generic safety issues"[45] while provoking a sweeping reappraisal of safety regulation transcending the specific design and equipment inadequacies that contributed to the accident.[b]

b. Interestingly, the most important specific issue stemming from the Three Mile Island

Adverse Events From Reactor Operating Experience, 1976-1979*

Serious Events At Individual Reactors

Faulty test procedures eliminating capability to detect a loss-of-coolant-accident (Zion, 1977)

Deep circumferential crack in primary system piping (Duane Arnold, 1978)

Loss-of-coolant accident (Three Mile Island, 1979)

Classes of Events With Multiple Occurrences

Separation of control rods from drive mechanisms at BWRs

DC electrical failures degrading the capability of residual heat removal systems yet requiring their operation

Water hammer and flow-induced vibration causing equipment damage, in some cases to engineered safety features

Systems interaction events compromising independence between presumed redundant safety systems (*e.g.,* Zion event above)

Loss of high-pressure coolant injection (HPCI) capability, due to valve leakage, improper valve lineup, or electrical failure

Leakage between interconnected fluid systems causing loss of residual heat removal systems

Failure to maintain containment isolation

Cracks in welds connecting feedwater piping to steam generators at PWRs

Continued degradation of steam-generator tubes and cracking in steam-generator supports at PWRs

Overpressurization of pressure vessels at PWRs

---

*Compiled from ACRS, *Review of Licensee Event Reports, 1976-1978*, NUREG-0572 (1979).

Section 3.7: The Role Of The Public In Forcing Stronger Safety Standards

Public concern over nuclear power hazards has grown as the nuclear sector has expanded. Increased operating experience has generated additional evidence of actual hazards, while the construction of more nuclear capacity at more sites has enlarged the numbers of people exposed to reactor risks. In turn, rising public apprehension has affected nuclear regulation.

Although citizen interventions have been blamed for causing delays leading to higher costs, that effect was statistically negligible in terms of *real* (inflation-adjusted) dollars for reactors completed in the 1970s. Most delays caused by citizen challenges have affected reactor licensing rather than construction, and reactors which took longer to license did not have inordinately higher capital costs than plants starting construction at the same time, as Section 8.1 shows. Far more importantly, public concern has spawned expert critics who have identified deficiencies in reactor design, construction, and regulation. Public involvement in nuclear regulation has also reinforced conservative tendencies among regulators.

The foremost technically-skilled critic of reactor regulation has been the Union of Concerned Scientists (UCS). UCS was founded in 1969 to examine science and technology policy but soon turned primarily to reactor safety in response to prompting by its members and funders. The group was in the forefront of intervenors at licensing hearings who attacked AEC's 1971 ECCS interim criteria. These challenges helped to force the lengthy rulemaking that led AEC in late 1973 to reduce permissible fuel temperatures and to prod the industry to improve ECCS reliability. At least as importantly, the hearings revealed the presence of dissent within AEC and thereby conferred both publicity and legitimacy upon nuclear critics.

UCS's continued critiques of nuclear regulation have both affected specific issue areas and colored the overall tone of the reactor safety debate. Criticism by UCS was cited for its "important contribution" to a 1980 NRC order upgrading standards governing fire protection for electrical cables and environmental qualification of electrical components.[46] The group's critiques of the Reactor Safety Study (WASH-1400) also helped induce Congress to direct NRC to convene a committee to review the study. That review led the Commission to retract some of its prior support for the study, an action with important consequences for safety regulation, as discussed below.

Intervenors have sometimes brought about design changes in individual licensing hearings. At North Anna, for example, a local environmental group

---

accident — measures to cope with a molten reactor core — was raised several years earlier, although not to its current prominence, in licensing reviews of another new design, that of a floating nuclear plant.

showed that building supports were settling into the ground, causing NRC to order tests and design changes (primarily flexible expansion coupling for piping) that added to costs and delayed plant completion.[47] More generally, intervenor participation in licensing hearings has tended to make the regulatory staff "considerably more cautious and conservative,"[48] according to one observer, by fostering a climate conducive to detailed design review. The prospect of intervenor cross-examination has encouraged applicant and staff witnesses to conduct thorough safety analyses. Similarly, the presence of intervenors tends to reinforce staff concerns with safety and help counteract pressure from the applicant for a speedy review.

Public concern about reactor safety has also reinforced the tendency of regulators to heed the potential for nuclear hazards and thus to add safety requirements designed to limit the overall accident probability. For example,

> The ACRS believes that it is proper that nuclear power be safer than other comparable technologies. The Committee has sought this goal. It believes that the country wants a higher level of safety for nuclear reactors and is willing to pay for it. The ACRS also believes that the country wants a higher degree of assurance as to the level of safety which is being attained.[49]

Statements such as these are a powerful counterweight to the view that different energy sources should have comparable risks, and that nuclear power, with a lower calculated public health impact than some alternatives, should therefore not be subject to further major regulatory requirements. The statement arguably would not have been delivered absent considerable public apprehension over reactor hazards.

Public concern has also affected nuclear regulation through Congress. Although few members of Congress are strongly anti-nuclear, constituents' concerns have led to closer congressional oversight and thence to stricter regulation. In 1977, for example, the chairman of the House Subcommittee on Energy and Environment, Rep. Morris K. Udall, succeeded in attaching a rider to an NRC appropriations bill creating a panel of reactor safety experts to review the Reactor Safety Study. The experts' critical review[50] led to an NRC policy statement in early 1979 withdrawing support for the study's executive summary and restricting staff's use of the study's accident probabilities.[51] The first move has bolstered arguments for stronger regulatory standards, while the second may lead to more conservative design bases in specific licensing issues.

Congress has also required NRC to publish digests of reactor "abnormal occurrences," lists of unresolved safety issues, and Task Action Plans to address key safety concerns. These have raised NRC's priorities for resolving safety problems and thereby enhanced stricter regulation. Moreover, publication of the information has deepened the sense among a large segment of the public that reactors present many potential safety hazards warranting greater attention. In turn, heightened concern has affected the regulatory process through the conduits previously described.

The entire nuclear enterprise, in fact, has been conducted increasingly in a "fishbowl" environment that admits scrutiny of every aspect of nuclear regulation and operation. Operating anomalies are fed back into design and operating reviews; designs are examined in NRC staff reviews and licensing hearings; construction is scrutinized by activists, the press, NRC and, increasingly, by workers themselves.

The latter development was first evidenced in 1971, when a welding supervisor at the Surry plant reported that primary coolant piping contained numerous defective welds. The following year, an anonymous letter to the ACRS, perhaps sent by a reactor design engineer, disclosed that postulated steam-line breaks at the Prairie Island reactors could cause pressure to rise in the auxiliary building to the point that vital electrical and mechanical equipment might fail and impede plant shutdown. Neither the applicant nor AEC staff had evaluated this issue in their reviews.[52] The AEC responded by requiring many plants under construction and in design to conduct further accident analyses, reroute pipes, and modify their auxiliary buildings to provide pressure relief in the event of steam-line failure.[53]

More recently, "heightened public awareness and interest in nuclear power [have] resulted in an increase in the number of allegations received by NRC"[54] of irregularities in plant construction. Since 1977, construction personnel at the Callaway, Wolf Creek, South Texas, and Marble Hill reactors have charged that quality assurance requirements were being bypassed and that designs were being amended in the field by unqualified personnel. These allegations and a critique of NRC construction inspection procedures by the General Accounting Office (GAO)[55] have led the agency to toughen its supervision of construction. At Marble Hill, for example, workers' affidavits led to the discovery of 170 inadequate patching jobs in concrete walls — including voids up to 180 cubic feet in size[56] — and to an NRC suspension of safety-related construction lasting more than a year and a half. NRC has subsequently announced its intention to consider new rules to enhance the independence of QA auditors from construction personnel and to expand its own inspections of reactor construction activities.[57]

Section 3.8: Outlook

The preceding discussion indicates that considerable impetus for new reactor safety requirements in the United States has come from expansion of the *nuclear sector, i.e.,* from increases in the total capacity of reactors operating or under construction. Growth in the population of reactors has required new licensing reviews in which additional safety problems were first discovered or addressed. It has led to a more rigid administration of licensing standards and procedures that has raised the stringency and specificity of safety requirements applied in staff reviews. It has also induced regulators generally to endeavor to reduce per-plant risks to contain the industry-wide probability of a serious accident.

Similarly, increases in reactor generating capacities — which together with reactor population determine sector size — have necessitated greater safeguards in order to maintain desired safety margins. And increased operating experience, which to date has correlated closely with the number of plants, has unearthed new safety defects requiring remedial regulatory action.

Nuclear sector expansion has also broadened and intensified public concern. The increasing number of plants have generated more frequent mishaps which, with more persons living near more reactors, are accorded wider publicity that in turn adds to pressure on regulators to abate perceived hazards. The "fishbowl" environment in which nuclear power must function was not present when only a dozen reactors were operating or planned. Rather, it materialized as the concomitant of a largescale nuclear program undertaken prior to achieving sufficient technical and managerial maturity.

This is not to say that all of the ingredients that have contributed to increased regulatory stringency are subsumed under nuclear sector expansion. Some information leading to more stringent requirements has come from accident-related research by the national laboratories; for example, work at the Idaho National Reactor Testing Station in the early 1970s indicated that ECCS cooling water might not reach the reactor core in some circumstances. (Note, however, that much accident research has been ordered by AEC/NRC in response to new pressures or information originating from the various sources described earlier.)

Separately, although it is likely that increases in the prices of competing energy forms such as coal help create a context in which cost-engendering new requirements are more palatable to nuclear regulators, this factor is not tied to expansion of the nuclear sector. Nor are the regulators' own opinions as to the importance of nuclear power, or those of the president, who appoints the NRC commissioners and can seek to influence their conduct of regulatory policy. Conversely, those responsible for nuclear regulation have generally understood that a serious reactor accident — an event whose probability of occurrence must be proportional to nuclear sector size, absent safety improvements — could spell the end of nuclear power in the United States. This consideration has been reinforced by the Three Mile Island accident.

Accordingly, the linkage of increased regulatory stringency to nuclear sector expansion seems firmly based in both regulatory history and logic. Most of the drive toward greater regulatory requirements appears accounted for by: (1) the need to improve new plants' safety to keep the sector-wide accident frequency at a low level, (2) information concerning safety problems that arises from licensing reviews and plant operation, and (3) the more rigid administration of regulation necessary to license and monitor a large nuclear sector that draws public concern and scrutiny.

This linkage implies that the approximately 90 gigawatts of nuclear capacity under construction face a significant further increase in regulatory standards. These plants long ago gave impetus to new regulations when they were awarded construction permits in the 1970s, signalling to regulators the need for new remedial measures to prevent increases in the sector-wide probability of a serious accident. In theory, the impetus could be defused if public attitudes toward nuclear risks change substantially or if it proves possible to dismiss outstanding regulatory issues without affecting accident risks. Neither event seems plausible, however. The growing number of genuine safety issues and the continued widespread mistrust of nuclear power strengthen the presumption that nuclear regulation will grow more stringent.

The one eventuality that might be expected to slow the rate of increase in nuclear regulations is cancellation of a large number of reactors with construction permits. Large-scale cancellations of plants being built would ease public concern and also enable regulators to relax growth in safety requirements somewhat without forcing up the sector-wide accident probability. Readjustment to reduced future capacity would be constrained, however, by continued detection of safety problems through operating experience at existing reactors. Indeed, judging from the rate of issuance of NRC generic Bulletins and Circulars, the detection rate *per reactor* apparently reached an all-time high during 1979-80 (see Figure 5.1). Major problems detected in 1979-80 include the many systemic deficiencies in design and operation revealed at Three Mile Island, weaknesses in BWR scram systems, inadequate separation of non-safety from safety-grade instrumentation and control systems at Babcock & Wilcox reactors, substandard seismic design and construction procedures, and faulty PWR containment water level controls and indicators, among many others.

At some point, the per-reactor rate of detection of safety problems will almost certainly decline. But even then, the *per-year* rate would fall less rapidly — and might even continue to increase for some time — because of growth in the number of operating plants. New safety issues will thus continually emerge while old ones will be re-emphasized, inhibiting efforts to stabilize reactor design criteria and to standardize plants. Moreover, apart from prospective new standards, plants under construction are subject to many existing requirements from which recently completed plants were exempt, due to "regulatory lag."

Accordingly, the "environment of constant change"[58] that so pervasively complicates nuclear design and construction should not be expected to improve significantly, short of a marked reduction in currently projected nuclear power growth. Such a slowdown would ease, but by no means completely dispel, the pressures that lead to new regulatory requirements.

References

1. See Department of Energy, *U.S. Central Station Nuclear Electric Generating Units: Significant Milestones*, DOE/NE-0030 (quarterly).

2. See I.C. Bupp and J.C. Derian, *Light Water* (Basic Books, New York, 1978); and R. Perry *et al.*, *Development and Commercialization of the Light Water Reactor, 1946-1976* (RAND, Santa Monica, CA, R-2180-NSF, 1977).

3. A.M. Weinberg, *The Wilson Quarterly*, III, 3, 96 (1979). Emphasis in original.

4. See S. Lawroski and D.W. Moeller, "Advisory Committee on Reactor Safeguards: Its Role in Nuclear Safety," *Nuclear Safety*, 20 (No. 4), 387-399 (1979).

5. ACRS Chairman W.D. Manly, letter to AEC Chairman Glenn T. Seaborg, 24 November 1965. Reprinted in Joint Committee on Atomic Energy (JCAE), *Hearings on Licensing and Regulation of Nuclear Reactors*, 1967, p. 119.

6. D. Okrent, *Nuclear Reactor Safety: On The History of the Regulatory Process* (University of Wisconsin Press, Madison, WI, 1981), p.218. This major work by long-time (1963-present) ACRS member Okrent arrived in galley form just before this chapter went to press.

7. W.K. Ergen *et al.*, "Report of the Advisory Task Force on Power Reactor Emergency Cooling," AEC (unnumbered), October 1967.

8. ACRS Chairmen N.J. Palladino (1967) and D. Okrent (1966), statement before the JCAE, 5 April 1967, in Reference 5, JCAE, p. 93.

9. ACRS Chairmen S.H. Bush (1971) and J.M. Hendrie (1970), statement before JCAE, 22 June 1971, reprinted in JCAE, *Hearings on AEC Licensing Procedure and Related Legislation*, 1971, p. 97.

10. ACRS Chairman H.G. Mangelsdorf, letter to AEC Chairwoman D.L. Ray, 10 September 1973.

11. AEC, *Technical Report on Anticipated Transients Without Scram for Water-Cooled Power Reactors*, WASH-1270, September 1973, p. 6.

12. Reference 11, p. 7 (emphasis added).

13. NRC, Regulatory Guide 1.96, "Design of Main Steam Isolation Valve Leakage Control Systems for Boiling Water Reactor Nuclear Power Plants," May 1975, p. 2.

14. NRC, "Proposed Rulemaking to Amend 10 CFR Part 50 Concerning ATWS Events," memorandum SECY-80-409, 4 September 1980, p. A-9.

15. AEC Deputy Director of Regulation C.K. Beck, statement to JCAE, 5 April 1967, in Reference 5, JCAE, p. 64.

16. Reference 8, p. 95 (emphasis added).

17. ACRS, minutes of 73rd meeting, 5-7 May 1966.

18. Reference 6, p. 346.

19. D. Okrent, draft letter to AEC Chairman G.T. Seaborg, 16 August 1966.

20. Reference 6, p. 117.

21. Reference 4, pp. 391-392.

22. ACRS, Report on Browns Ferry Nuclear Power Station, 14 March 1967, p. 2.

23. Published as 10 CFR 50, Appendix A, in 1971 after a revised draft was issued in 1967.

24. *Electrical World, 187* (No. 6), 76 (1977).

25. NRC, *A Study of the Nuclear Regulatory Commission Quality Assurance Program,* NUREG-0321, Sandia Laboratories (1977), p. 22.

26. Reference 15, p. 68.

27. Reference 6, p. 237.

28. Reference 6, pp. 237-238.

29. Reference 6, pp. 236-243.

30. Reference 6, p. 215.

31. Reference 6, pp. 214-215.

32. See ACRS, "Report on Connecticut Yankee Atomic Power Company," 19 February 1964; and ACRS Chairman H.J. Kouts, letter to G.T. Seaborg, 18 November 1964.

33. Reference 6, pp. 230-231.

34. See NRC, *Approved Task Action Plans for Category A Generic Activities,* NUREG-0371, Vol. 1, No. 1 (1977), Technical Activities A-6, A-7 and A-8, for a description of the pressure-suppression issue. See Commonwealth Edison, Docket Nos. 50-373 and 50-374, letter to Chief Olan Parr, NRC LWR Branch No. 3, Division of Project Management, 19 February 1980, for an account of design changes at a Mark II plant.

35. Reference 34, NUREG-0371, Technical Activity A-2, p. 1.

36. E.W. Hagen, "IEEE Nuclear Power Systems Symposium," *Nuclear Safety, 16* (No. 5) 561 (1975).

37. Reference 8, p. 92.

38. Reference 8, pp. 92-93.

39. N.J. Palladino, memorandum to ACRS, 22 November 1965.

40. NRC, *Nuclear Power Plant Licensing: Opportunities for Improvement,* NUREG-0292 (1977), p. 3-3.

41. M. Rogovin, *Three Mile Island: A Report to the Commissioners and to the Public,* NRC, NUREG/CR-1250 (1980), Vol. I, p. 95.

42. J.M. Hendrie, testimony before the House Subcommittee on Energy and the Environment, 26 February 1979.

43. Reference 6, p. 215.

44. Reference 6, p. 331.

45. Reference 6, p. 294.

46. NRC, memorandum and order CLI-80-21, 27 May 1980, p. 3.

47. ACRS, report of 26 October 1976 on North Anna 1 and 2.

48. E.S. Rolph, *Nuclear Power and the Public Safety* (Lexington Books, Lexington, MA, 1979), p. 124.

49. ACRS Chairman M.W. Carbon, letter to NRC Chairman J. Ahearne, 11 December 1979.

50. H.W. Lewis, *et al., Report of the Risk Assessment Review Group,* NUREG/CR-0400 (1978).

51. NRC, "Statement on Risk Assessment and the Reactor Safety Study Report (WASH-1400) in Light of the Risk Assessment Review Group Report," 18 January 1979.

52. Reference 6, p. 234.

53. I.C. Bupp *et al., Trends in Light Water Reactor Capital Costs in the United States: Causes and Consequences,* Massachusetts Institute of Technology, CPA 74-8 (1974), p. III-10.

54. NRC, *1979 Annual Report,* NUREG-0690, p. 168.

55. General Accounting Office, *The Nuclear Regulatory Commission Needs to Aggressively Monitor And Independently Evaluate Nuclear Powerplant Construction,* EMD-78-80, 7 September 1978.

56. *Wall Street Journal,* 24 October 1979.

57. NRC, *NRC Action Plan Developed as a Result of the TMI-2 Accident,* NUREG-0660, Vol. 1, Section II.J.

58. See Chapter 4, Reference 6.

4
Changes In Reactor Design And Construction In The 1970s

The real cost (in constant "steam-plant" dollars per kilowatt) to complete nuclear power plants in the United States increased by an almost incredible amount — 142% — from the end of 1971 to the end of 1978. This escalation occurred even after allowing for inflation in the costs of standard construction inputs — labor, materials, and equipment. It is documented in the analysis of the costs of the 46 U.S. reactors completed during 1972-78 found in Chapters 8 and 10.

The cost increase cannot be explained by changes in the average size, geographical location or builder experience of the reactors, since these changes had only modest effects on costs. Indeed, some of the changes led to *reductions* in cost, as Table 10.4 shows. Nor is it attributable to lengthening reactor construction times, since that average grew by only one year, from 5½ years for a typical 1971 completion to 6½ years in 1978, adding only several percent to real capital costs. Nor did citizen interventions materially affect costs by causing delays: where citizen challenges were effective, they delayed reactor *licensing,* not construction; reactors which took longer to gain a construction permit did not have higher real costs than plants licensed at the same time.

The elimination of construction inflation, lengthened schedules, and factors such as increased plant size as possible causal factors leaves only one plausible explanation for the 142% average real capital cost increase for nuclear plants from 1971 to 1978: design and construction changes that increased the *amounts* of labor, materials, equipment, and engineering effort required to build reactors. Some fraction of this increase was invested in design or equipment changes to improve plant generating performance (capacity factor), as discussed below. But most of it appears to have resulted from the effort to reduce the safety and environmental risks of newer reactors.

The previous chapter traced the processes by which nuclear design requirements and regulatory standards have grown increasingly stringent since the late 1960s. This chapter describes the major changes in nuclear design, construction, and procurement practices that led to increased costs. It gives particular emphasis to the new regulatory standards that appear to have been responsible for most of the changed practices.[a]

The discussion that follows contains few estimates of the costs of complying with new regulations on an individual basis. This is in marked contrast to the itemization of coal plant regulatory standards in Chapter 7. Those standards were clearly responsible for almost all of the real increases in coal capital costs in the 1970s. Most coal pollution controls are upgraded by adding discrete equipment whose costs can be isolated. Nuclear design changes, in contrast, are characterized by "ripple effects" that carry beyond the immediate component or system being altered. As the Atomic Industrial Forum (AIF) noted in a 1978 analysis of nuclear regulatory impacts,

> [I]t is insufficient to identify the cost of material and labor as, for example, an added pipe, or pump or valve . . . [S]ignificant ripples caused by such changes affected not just the changed system but also, for example, supporting structures, normal or emergency power supplies, ventilation systems, radwaste, etc.[1]

Or, as British economist Gordon MacKerron reported,

> [S]ystem designs change in response to new standards, rather than simply involving the simple addition of discrete safety or environmental features to otherwise unchanged designs.[2]

Thus, according to AIF, "Attempts have been made on numerous occasions to pinpoint the full impact of regulatory changes on a nuclear project, and in each case it was found that the total impact was inevitably larger than the sum of the parts."[3] One such attempt by Ebasco Services, an architect-engineering firm that has developed detailed cost estimates for pollution control improvements at coal plants,[b] was able only to rank AEC/NRC regulatory guides according to relative cost impact. Specific cost estimates could not be made because most regulatory guides " . . .have, in addition to [a] direct effect on singular systems or structures, an effect on other related systems and structures."[4]

---

a. The responsibility of nuclear regulation for costly nuclear design changes is a complex and controversial matter. NRC and others contend that considerations of reactor safety would have required that industry apply stricter standards even in the absence of federal regulation, and, moreover, that the nuclear industry has compounded regulatory impacts by contesting and resisting new standards instead of designing for them. For its part, the industry is critical of alleged NRC regulatory delays and inconsistent rulings. Both arguments have substantial validity and are compatible. Neither detracts from this book's contention that increased regulatory standards have created the conditions for higher real nuclear capital costs.

b. These estimates are, unfortunately, presented without correction for inflation. They are referenced in Chapter 7.

Accordingly, this chapter addresses past changes in regulatory standards and their effects on plant design and construction while presenting few cost estimates. The first three sections treat problems of plant construction under changing designs, changes in quality assurance practices, and new equipment qualification procedures designed to ensure that vital equipment can function under extreme conditions. The next three sections treat changes in protective measures against specific accident initiators: "internally initiated" accidents involving pipe rupture, "missile" generation, or crane failure; seismic and other natural phenomena; and fires. The final four sections concern the effects of these measures and of other design and equipment changes upon the four major nuclear plant systems: "fluid systems" comprising reactor core cooling and emergency cooling equipment; the containment structure and systems; instrumentation, control, and electric power systems; and equipment to minimize radiation releases and exposures.

The changes were extensive. Major plant structures were strengthened and additional piping restraints were installed to absorb seismic shocks and other postulated "loads" identified in accident analyses. Barriers were installed and separation distances increased to protect redundant "trains" of safety-related equipment from fires, flooding, and other "common-mode" failures[c] and to shield vital equipment from high-speed missile fragments that might be loosed from rotating machinery and from the pressure, steam, and fluid effects of possible pipe ruptures (including movement of the pipes themselves). Instrumentation, control, and power systems were expanded to monitor more plant parameters under a broadened range of operating conditions and to improve the reliability of safety systems. And many valves, pumps, electrical connectors, and other components considered important to safety were "qualified" to perform under more demanding conditions, such as seismic shocks and loss-of-coolant accidents; this required more rigorous fabrication, testing, and documentation of their manufacturing history.

Changes such as these approximately doubled the amounts of materials and equipment required per unit of nuclear capacity during the 1970s[5] and subjected many construction commodities to large increases in real cost. Moreover, many changes were mandated during construction, as new information relevant to safety emerged. This complicated the evolution of designs and standards as nuclear technology developed. Reactors increasingly were built in an "environment of constant change"[6] that precluded control or even estimation of costs and spawned endemic inefficiency in design and construction. Compared to early-1970s completions, reactors completed in the late seventies

c. A "train" or "division" comprises all of the equipment in one of the two, three, or four parallel sets that comprise a complete system. A "common-mode" failure is the concurrent failure of identical redundant components or equipment trains. It can occur because of shared defects in the fabrication or installation of equipment, "environmental" conditions such as high humidity, or operator error. Common-mode failures are of concern because of their potential to overcome redundancy in safety systems.

required approximately twice as much craft labor and two-thirds more engineering effort per unit of capacity. These increases together added approximately $150/kW (1979 dollars) to plant costs,[7] accounting for one-third of the overall direct increase in reactor capital costs during 1971-78 (exclusive of real interest during construction).

Section 4.1: Construction Under Changing Standards

Reactor designs and regulatory requirements changed markedly in the 1970s, not only from one generation of plants to the next but also between preliminary and final design for each plant, as later sections of this chapter demonstrate. The result was that the nuclear construction process fell prey to logistical problems that magnified the direct impacts of increased standards. These problems continue today. Construction contracts must be let on a "cost-plus" basis, craft workers are inefficiently deployed and labor productivity suffers, backfits during construction are common, and opportunities for innovation or learning from prior construction are narrowed.

Cost-Plus Contracting: "[Because] the work scope at the beginning of a nuclear project cannot be accurately defined to accommodate the seven or eight year construction span the industry is now facing," said Pacific Gas & Electric (PG&E) in explaining cost overruns at its Diablo Canyon plant, "it is not uncommon to have all [construction] work awarded from the beginning on a cost-plus basis."[8] Standard construction contracts provide for a fixed price, perhaps tied to an agreed inflation index. Such arrangements are not feasible, however, when the scope of work is subject to significant changes after the contract is let.

Instead, "recoverable cost-plus" contracts are the norm in safety-related nuclear work. Or, fixed-price arrangements may be changed to cost-plus when a contractor is overwhelmed by design or specification changes and the architect-engineer cannot change contractors or put the revisions out for bids without incurring even greater costs or causing damaging delays. In the process, the architect-engineer sacrifices control over contractor expenditures. In the same way, the utility loses control over the architect-engineer.

A measure of the impact of cost-plus contracts on nuclear costs is provided by Christopher Bassett's analysis of Toledo Edison's Davis-Besse plant, completed in 1977.[9] Bassett, a cost engineer for the Ohio Public Utilities Commission, found that the costs of major equipment items procured from vendors for a fixed amount were only slightly higher than the initial contract awards. The cost of the nuclear steam supply system went from $34 million to $40 million, and the turbine-generator from $22 million to $24 million. In contrast, massive overruns were experienced for tasks involving field construction and the purchase of the myriad of other components that were particularly sensitive to the effects of regulatory changes and, thus, were negotiated on a cost-plus basis. Piping and mechanical work at Davis-Besse increased from $15 million to $80 million, civil and structural from $11 million to $67 million, and electrical from $5 million to $44 million — increases of 440%, 510%, and 780%, respectively.

A large part of these increases was doubtless caused by the direct equipment and engineering costs of the design changes, and would have been incurred under any type of contract. Nevertheless, the cost-plus arrangement almost certainly added to the direct costs. A fixed-price arrangement creates incentives to work efficiently and expeditiously to maximize the contractor's profit. The objective under cost-plus contracts is to maximize total revenues, which shifts the prevailing orientation toward longer schedules and greater expenditures.

Labor Productivity: Constant changes in designs and construction procedures cause inefficient deployment of construction labor for many reasons. First, scope changes lead to frustrating start-and-stop work conditions. New design requirements frequently hold up fabrication and delivery of key components and require altering construction sequences, and "the continuing redefinition of engineering detail . . .requires work stoppages while changes are made."[10] These changes and delays confuse and demoralize workers, compounding cost-plus incentives and creating an environment which discourages attentiveness.

Second, scope changes breed negative expectations about the outcome of work that invariably reduce productivity. Changed requirements sometimes cause equipment installation and field erection to be repeated several times. leading workers to be less careful since they come to anticipate that jobs will be redone anyway. Careless work then forces inspectors to reject jobs, requiring that they be redone and completing the vicious circle.

Increasing quality assurance requirements have contributed to lower productivity in much the same way: by raising the care and time required for some construction; by increasing job-rejection rates; by adding to delays in delivery of equipment; and by necessitating an expansion of supervisory staff, compounding the increase in personnel required to manage construction under design and engineering changes.

Backfitting: Design changes during construction not only dampen labor productivity but also may affect work already in place. When installed equipment fails new standards and must be modified or removed, adjacently located equipment may need to be moved to accommodate the change. "There are significant inefficiencies in trying to design to fit existing buildings and installed components." says PG&E.[11] "Work has to be done out of sequence in a restricted access and work area."

Existing installations must also be protected from construction hazards and occasionally may be damaged. Moreover, contractors previously released from the site may need to be recalled, creating additional inefficiencies as new workforces are assembled and trained.

Overall Construction Logistics: Constructing a modern power plant is a mammoth task. It requires several thousand workers to assemble thousands of equipment items using many thousands of engineering drawings over a multi-year period. The overall management task would be complex even under ideal circumstances, but it is made more difficult by design and regulatory changes.

The ''expectation of design changes'' described by the Atomic Industrial Forum (AIF)[12] forces the architect-engineer to choose between deferring construction tasks until new design requirements are issued or ''over-designing'' jobs to satisfy anticipated regulations. The former strategy often causes delays, especially when critical engineering is being deferred. But the latter approach increases direct costs, not only of the job in question but also of supporting tasks such as structures, piping, cables, and ventilation.

Morevover, management skills which might otherwise be marshalled to improve efficiency and trim costs are instead absorbed by the effort of coping with changing designs and new regulations. The statistical analysis of past nuclear costs in Chapter 8 indicates that ''learning'' by architect-engineers has proceeded slowly in nuclear power construction; each doubling in the number of plants built by an architect-engineer has effected only a 7% average drop in costs. This pace is considerably slower than in other young industries, probably because of the lack of design stability in nuclear plant construction.

Finally, innovation in design and construction is often not feasible in the heavily regulated nuclear power industry. Designers and construction managers often forego potential improvements in favor of less efficient but previously approved methods. In general, with design changes soaking up managerial attention and available capital while potential innovations are closely scrutinized for possible risks, nuclear construction has not offered fertile ground for new approaches that might cut costs.

Performance Improvements: Changes in plant designs or equipment to improve reactor performance (capacity factor) made a modest contribution to real nuclear capital cost increases in the 1970s. Major measures taken at many recently completed plants include improved corrosion-resistant materials for primary system piping, titanium condensers to reduce contaminant in-leakage to steam generators, and increased design margins and parts stocking for some equipment.

Generic cost figures for such measures are not available. Their impacts have been mixed, as later reactors appear to show only slight performance improvements over earlier units, even controlling for size and age differences. Performance trends might have been negative without these efforts, however, due to the apparently heightened effect of nuclear regulation on reactor performance in recent years. This point is pursued in Section 11.1.

Section 4.2: Quality Assurance

The most vital systems and equipment in nuclear power plants are those designed to shut down the fission reaction, maintain reactor cooling, and prevent the off-site release of radioactivity. The design, manufacture, and installation of such *safety-related equipment* are subject to *quality assurance* (QA) procedures to ensure that the equipment can perform effectively as needed. (A particularly critical aspect of these procedures is *equipment qualification* — testing and documentation to ensure that safety-related equipment can function under extreme conditions that may arise during accidents. It is discussed in Section 4.3.)

QA requirements for nuclear power plant construction have expanded greatly since the mid-1960s. Early methods of ensuring quality were largely informal, guided by a modest set of codes and standards. Starting in 1970, however, a new approach has developed requiring adherence to a great number of specific, detailed NRC rules and industry standards. The result was a significant increase in direct equipment costs, in procurement and installation times, and in logistical complexity in nuclear plant construction.

Evolution of Quality Assurance Requirements: The AEC/NRC has habitually defined quality assurance as "All those planned or systematic actions necessary to provide adequate confidence that an item or facility will perform satisfactorily in service."[13] QA is a management system that licensees and their contractors apply to "the design, fabrication, erection, and testing of structures, systems, and components important to safety,"[14] "to assure requisite quality and effective performance."[15]

The AEC first published general requirements for quality assurance programs as Appendix B to Title 10 of the Code of Federal Regulations, Part 50 (10 CFR 50) in 1970, three years after spelling out the general intent of QA in the initial draft of its General Design Criteria for nuclear plants. Also in 1970, the American National Standards Institute (ANSI) established committees to develop formal, explicit procedures for licensees and their contractors to comply with the AEC regulations. ANSI is an umbrella organization for technical societies such as the Institute of Electrical and Electronics Engineers (IEEE) and the American Society of Mechanical Engineers (ASME) whose memberships have a professional interest in and familiarity with nuclear power plants. The ANSI working groups are composed primarily of representatives of electric utilities, power plant designers, and equipment manufacturers but include nominal AEC/NRC representation.

Beginning in 1971 and continuing through the 1970s, the ANSI com-

American National Standards Institute
Quality Assurance Standards

Standard Number	Year Adopted	Subject
N45.2	1971	General QA requirements
N45.2.1	1973	Cleaning of fluid systems
N45.2.2	1972	Packaging, shipping, receiving, storage and handling of equipment
N45.2.3	1973	Housekeeping during construction
N45.2.4	1972	Installation, inspection and testing of instrumentation and electric equipment
N45.2.5	1974	Installation, inspection and testing of structural concrete and steel
N45.2.6	1973	Qualifications of inspection, examination and testing personnel
N45.2.7	1976	Administrative controls for QA
N45.2.8	1975	Installation, inspection and testing of mechanical equipment and systems
N45.2.9	1974	Collection, storage and maintenance of QA records
N45.2.10	1973	QA terms and definitions
N45.2.11	1974	QA requirements in plant design

mittees issued almost two dozen final and draft QA standards pertaining to virtually every phase of plant construction involving safety-related equipment. These include designing, purchasing, fabricating, handling, shipping, storing, cleaning, erecting, installing, inspecting, and testing (see box). All the final and most of the draft standards have been endorsed in AEC/NRC regulatory guides.

Quality Assurance Standards: Some ANSI standards are essentially lengthy lists of procedures for different construction phases. Others delineate precise methods to be employed in specific construction procedures. It is beyond our scope to review each standard, but several examples can illustrate their pervasive impact on construction practices.

ANSI Standard (Std) N45.2.8, issued in 1975, gives requirements for installing and assembling safety-related mechanical items such as valves,

N45.2.12   1977   Requirements for auditing QA programs

N45.2.13   1976   Procurement of items and services

N45.2.14   (pending) Manufacture of safety-related instrumentation and electric equipment

N45.2.15   (pending) Hoisting, rigging and transportation of equipment

N45.2.16   1975   Calibration and control of measuring and test equipment

N45.2.17   (see note) Welding

N45.2.18   (see note) Installation, inspection and testing of concrete

N45.2.19   (see note) Requirements for soil and foundations

N45.2.20   1979   Requirements for subsurface investigations

N45.2.21   (see note) Design and manufacture of nuclear fuel

N45.2.22   (pending) Inspection of dimensional characteristics

N45.2.23   1978   Qualification of audit personnel

Notes: Standards 17, 19, and 21 were withdrawn prior to adoption. Standard 18 was absorbed into Standard 5. Standards 14, 15, and 22 are currently pending approval. Standards 5 and 9 through 13 were re-issued in 1979 as NQA-1, "Quality Assurance Program Requirements for Nuclear Power Plants."

pumps, and piping. Licensees must verify the location and orientation of components; their levelling, alignment, clearances, and tolerances; the tightness of connections and fastenings; all fluid levels and pressures; the absence of leakage; the integrity of any welding operations; the adequacy of measures to prevent damage from installation or from adjacent construction; and the cleanliness of the installation. Following installation, the licensee must check greasing and lubrication, equipment cooling water systems, settings for electrical circuits and relays, instrumentation calibration, piping alignment, valve glands and packing, pump seals and packing, and installation of seismic anchors and restraints.

Other ANSI QA-related standards designate methods required for specific areas of construction such as high-strength bolting. ANSI Std 45.2.5 (1974) stipulates that "automatic cut-off impact wrenches" must be calibrated at least twice daily by tightening in a device indicating actual bolt tension, using at least three typical bolts of each diameter being installed. Post-installation inspection must confirm that bolts are the correct length, as indicated by at least two threads extending beyond the nut; that the bolt head shows the correct manufacturer's marking; that torque has been applied, as indicated by the burnishing or beening of the corners of the nuts; and that turning elements are on the correct face. The standard also specifies the procedures for calibrating the torque wrenches used to make the inspections.

These examples pertain to installation, inspection, and testing. QA requirements also apply to pre-construction phases such as procurement, fabrication, shipping, and storage. NRC regulations require that each safety-related item be inspected, identified, and traceable — through material heat number, part number, serial number, mill certification, etc. — throughout its fabrication, erection, and installation. In addition, special fabrication and installation processes, including welding, heat treating, and non-destructive testing, must be performed by certified personnel in accordance with applicable codes and standards such as the detailed specifications of the American Society of Testing and Materials.

Even protective coatings (*i.e.*, paints) are addressed by QA requirements. Improperly chosen or applied coatings could come free, clogging flow passages in piping systems or impeding heat transfer in heat exchangers, or they might cause explosive, free hydrogen to form when they are exposed to steam. Two ANSI standards issued in 1972 concern criteria for chemical composition and application of paints to plant equipment and surfaces in different environments.[16]

NRC QA criteria also apply to plant design. Applicable codes, standards and regulations must be "correctly translated into specifications, drawings, procedures, and instructions" for each component.[17] This is to be accomplished by architect-engineer review teams separate from the design group. Where necessary, test programs are to be conducted to check that specific design features can accomplish their intended function.

ANSI Std N45.2.11 further requires that the same degree of review be applied to design changes as to the original design. Sidney Bernsen, Bechtel's QA manager for steam plant construction, contends that "This has severely restricted earlier construction practices that allowed field engineers at the construction site to initiate and approve rather significant departures from the original design."[18] Although field personnel with a demonstrated understanding of the design area may undertake field changes, it has become increasingly common to check changes with the architect-engineer's home office or with authorized architect-engineer representatives on-site.

Documentation: A major part of QA activity is documentation to verify that required procedures have been carried out. This includes drawings, specifications, procurement documents, quality-program manuals, procedures (document-control procedures, installation procedures, inspection procedures, and personnel qualifications), and documentation providing evidence that inspections and tests have been conducted.[19]

The nuclear industry considers the large and growing amounts of paperwork involved in QA documentation to be a major cause of increased nuclear costs. Document control — ensuring that QA documents are available to personnel performing design, construction, or installation of safety-related items — has become especially difficult, particularly with changes in plant designs and equipment. According to Bernsen,

> In general, everyone wants to use the applicable (normally the latest) documents; however, people are reluctant to part with superseded copies of these documents, sometimes because they have recorded important information or notes on a given issue of a drawing. Furthermore, the transient and fluctuating nature of the work force at a construction site and the physical movement of crews from one area of the plant to another make it quite difficult to identify the groups having copies of different issues of documents. Literally tens of thousands of documents that need control are distributed at the jobsite.[20]

One possible solution, deploying crews to distribute documents and mark obsolete information, may lead to labor-relations problems and requires continuous attention in any event.

Applicability of Quality Assurance: QA requirements may appear to be pervasive, reaching into every aspect of nuclear construction, but three caveats should be kept in mind. First, QA applies only to equipment that the licensee designates, subject to NRC review, as "safety-related" (*i.e.*, required to achieve safe reactor shutdown, to maintain core cooling, and to contain radioactivity releases in the event of postulated accidents). Although licensee lists of safety-related equipment have expanded somewhat over time, they still comprise only about one-fifth of the volume of total plant equipment at recently completed reactors. Such equipment does, however, account for a larger share of equipment costs because of the QA process.

Second, new QA-related standards rarely apply to equipment already purchased. Many even apply only to new construction starts, exempting equipment yet to be purchased for plants under construction. Third, virtually all inspections of licensee and contractor work are by the licensees and contractors themselves. NRC auditing consists mostly of reviews of licensees' QA paperwork, except for small samplings of reactor field construction and installation. NRC displeasure over widespread QA inadequacies may lead to more stringent requirements and closer NRC audits, as discussed in Sections 5.1 and 6.3.

Despite these limitations, QA requirements appear to have had a major impact on reactor construction costs and logistics (see box). In addition, procedures for ensuring that safety-related equipment will function under

Quality Assurance Documentation At Diablo Canyon

Pacific Gas & Electric (PG&E) began construction of its two-unit Diablo Canyon station in the late 1960s, but plant start-up has been delayed since approximately 1977 due to unresolved seismic design issues. In the following excerpted testimony, PG&E construction vice-president Donald Brand describes the effect of QA requirements upon construction of Diablo Canyon.

We did not . . .anticipate the detail in documentation and independent inspection of workmanship which would be required by the NRC. For instance, simple field changes to avoid physical interference between components (which would be made in a conventional plant in the normal course of work) had to be documented as an interference, referred to the engineer for evaluation, prepared on a drawing, approved, and then released to the field before the change could be made. Furthermore, the conflict had to be tagged, identified and records maintained during the change process. These change processes took time (days or weeks) and there were thousands of them. In the interim the construction crew must move off of this piece of work, set up on another and then move back and set up on the original piece of work again when the nonconformance was resolved.

Installation of wire must be done according to written procedure and must be documented. Every foot of nuclear safety-related wire purchase is accounted for and its exact location in the plant is recorded. For each circuit we can tell you what kind of wire was used, the names of the installing crew, the reel from which it came, the manufacturing test, and production history. The tension on the wire when it is pulled is recorded and the tensioning device is calibrated on a periodic basis.

None of these requirements were in existence when Diablo Canyon was planned. Hundreds of requirements similar to these give us assurance of the quality of the Diablo Canyon plant. While this assurance is very costly, a precise cost cannot be assigned to this program.[21]

accident conditions have become considerably more demanding, as the next section demonstrates.

Section 4.3:   Equipment Qualification[d]

Equipment qualification refers to the generation of evidence that a manufactured component is capable of performing its intended function under all postulated service conditions. These conditions can encompass a wide range: normal operation; abnormal status such as degraded voltage or frequency; maintenance and testing, both of the component itself and of other parts of the plant which may subject the component to unusual service conditions; and "design basis events" such as earthquakes and accidents.

Demanding Service Conditions: Safety-related equipment may be exposed to two kinds of extreme service conditions. The first is *process conditions.* Valves may have to function under unusual flow conditions such as high or pulsed flow; motors may have to start and run with abnormal supply voltage and frequency if power supplies are degraded; components mounted on piping systems may be subjected to unusual vibrations and shocks due to hydrodynamic forces in the systems; and maintenance or testing can compromise ventilation, power supply, heat removal, lubrication, or other support systems that are ordinarily fully available.

Second, extreme *environmental conditions* may develop due to severe natural phenomena such as floods and tornadoes or accidents within the plant. Safety-related equipment can be shielded to some extent by locating it in suitably designed structures. Consideration of potential tornado forces has led to increased wall thickness at later plants, and safety-related equipment within plants has increasingly been placed in watertight compartments to protect it from water or steam that could issue from ruptured pipes (see Sections 4.4 and 4.5). This approach does not offer protection from earthquakes, however, because seismic disturbances are transmitted by the building to all equipment contained inside. Thus, seismic considerations require not only preserving the plant's structural integrity but also qualifying vital equipment for seismic stress.

Moreover, even sealed compartments cannot shield all safety-related equipment from adverse effects during a postulated loss-of-coolant accident (LOCA). Safety-related equipment located within the containment must function while subjected to an environment of high-pressure steam and radioactive fission products assumed to be released from the reactor and to deluge sprays containing water or caustic solutions designed to scrub radioactive gases. In addition, some equipment located in or near lines carrying radioactive fluids, such as pumps, valves, motors, and valve operators, may be subjected to a high

d. Much of this section draws heavily upon original material provided to the author by William H. Steigelmann, a specialist in nuclear design and engineering, formerly at the Franklin Institute in Philadelphia, now director of energy technologies at Synergic Resources Corp., Bala Cynwyd, PA.

radiation flux from a LOCA or a large pipe break. Equipment could also be exposed to temperatures outside of normal industrial ranges (typically 40-105°F) if the heating, ventilating, and air conditioning (HVAC) system for its location is not energized or fails under accident conditions.

A final environmental consideration for some equipment is that it may be called on to function after standing idle during the first days or even weeks following an accident. Corrosion and insulation degradation promoted by steam, chemical spray, and moisture may be worse in idle equipment than in identical equipment that operates immediately after an accident. This too must be taken into account in qualifying safety-related equipment.

IEEE Standards for Equipment Qualification: Electrical equipment is vital to the performance of safety systems but is potentially sensitve to extreme service conditions. Accordingly, qualification guidelines have been developed for electrical equipment, in consensus standards published by the IEEE and endorsed by AEC/NRC Regulatory Guides. Recently the American Society of Mechanical Engineers has formed committees to write standards for safety-related mechanical equipment — valves, pumps, piping, snubbers, etc. — but there are as yet no published industry standards directly applicable to these specific items.

The IEEE standards permit proof of performance to be established through operating experience, analysis, or test, but testing is the most widely used method. Design basis conditions are frequently so different from those to which equipment has been subjected in the past that qualification cannot be established by citing prior operating experience of identical equipment. Similarly, analytical techniques employing methematical models are generally neither sufficiently accurate nor broad to predict the effects of all possible combinations of conditions during accidents, such as extreme temperatures, humidity, and radiation. (This is not the case when qualifying *structures,* since failure modes are less complex and are generally amenable to an analytical approach.) Thus, testing of equipment specimens is usually necessary for qualification.

IEEE qualification test requirements were first stated in general terms in the 1971 edition of Std 323 (relevant engineering standards and regulatory guides are listed at the end of this chapter). The intent was to provide evidence that equipment required to operate in the severe environment produced by an accident could do so. Samples of electrical cable, motors, sensors, etc. employed in the containment were irradiated and exposed to the same steam and chemical spray environment that was predicted to occur in a LOCA.

Current IEEE standards such as the 1974 edition of Std 323 (endorsed by NRC Regulatory Guide 1.89) include more specific, stringent regimes incorporating potential degrading factors such as equipment aging and seismic stress. They also require a thorough description of the test facility, instrumentation, procedures, and results as well as retention of extensive records. In addition, some minimum test margins are higher in the later standards.

For new reactors, qualification tests now apply to all electrical equipment needed to shut the reactor, control safeguard systems, and operate vital instrumentation and controls. This includes many of the valves, motors, cables, connectors, relays, switches, and transmitters and much of the instrumentation and logic systems in nuclear plants. Specific qualification requirements for some of these component classes were developed in successor standards to IEE Std 323. They include Std 334 for motors (1971, revised 1974), Std 381 for electronic circuit modules (1977), Std 382 for electric valve operators (1972), and Std 383 for electric cables, field splices, and connections (1974). Seismic qualification of safety-related electrical equipment is treated in Std 344 (1971, revised 1975, discussed in Section 4.5).

These standards raised the severity and scope of electrical equipment qualification requirements. The 1971 edition of Std 334 applied only to safety-related motors inside containment, but the 1974 edition extends to critical motors outside containment, such as pumps for the ECCS and the residual heat removal system. The test cycle in the 1974 edition was also expanded to require that motors aged to the end of their expected life be able to function after exposure to severe accident conditions.

Similarly, Std 383 specifies elaborate testing and documentation requirements for electrical connections and splices made at the plant site. For all electrical conductors, the installer must record material type, size, and coating; insulation thickness, material, and application method; types of shielding and covering; and, for instrumentation cables, their capacitance, attenuation, and impedance.

Probably the greatest change in equipment qualification has come from the addition of an *aging* step to qualification programs in the 1974 revision to IEEE std 323. The physical and chemical properties of materials in electrical equipment can deteriorate with time, because of both normal aging and the effect of continuous or occasional thermal cycling, humidity, radiation, etc. Accordingly, equipment is now put in a simulated advanced-life state in evaluating its functionability under postulated environmental conditions.

Unfortunately, the fundamental processes that produce aging degradation (*e.g.*, embrittlement) are only poorly understood, and the possible synergistic influences of multiple environments (temperature, humidity, radiation, and vibration) have not been fully explored. Accordingly, the artificial pre-aging techniques in use, largely thermal stress, are conceded to be of questionable validity for many safety-related materials[22] and are the subject of an extensive NRC research program. Consideration of aging in equipment qualification is nevertheless requiring more complex testing procedures (see box for a complete list of the elements) and is beginning to lead to costly upgradings for some electrical equipment.

Qualification tests and material improvements are the responsibility of

Elements Of Current Equipment Qualification Programs

I. Aging

    A. Environmental
        - Thermal
        - Nuclear Radiation
        - Humidity and Chemical (moisture, oxygen, ozone)
        - Pressurization Cycles*
        - Seismic**

    B. Operational
        - Start/Stop Cycles
        - Total Operating Time
        - Process Conditions Associated with Normal Operation (including testing)***
        - Normal Maintenance

II. Design Basis Events (DBE)

    A. Seismic

    B. Loss-of-Coolant Accident or Pipe Break Outside Containment
        - Nuclear Radiation
        - Steam
        - Deluge Spray (demineralizer water or chemical solution)
        - Flooding

    C. Process Conditions Associated with DBE

* Equipment located within primary containment is subjected to external pressurization during periodic leak rate tests.

** Some number (typically 5) of ''Operating Basis Earthquakes'' (see Section 4.4) during the installed life of the equipment must be assumed.

*** Includes vibration due to fluid flow and/or equipment operation.

---

the equipment manufacturers, and cost increases are passed on to the licensee. Few cost estimates are available, but the increases appear significant. The Washington Public Power Supply System estimates that it has spent $6/kW (apparently in mixed current dollars) for testing and documentation alone for an estimated 4,000 pieces of safety-related equipment at each of its five reactors under construction.[23] Moreover, failures of "qualified" electrical connectors at nuclear plants and NRC's discovery of widespread deficiencies in manufacturers' qualification programs are likely to provoke substantial, costly improvements in the future (see Section 5.1).

Section 4.4:   Protection Against Internally-Induced Accidents

This section discusses changes in design and equipment to prevent damage to safety-related equipment in the event of failure of other equipment. The major failures of concern are the release of objects at high speeds from rapidly rotating machinery (referred to as "missiles"); pipe ruptures causing pipes to move ("whip") violently, release fluids under pressure, or cause flooding; and the release of heavy loads from overhead cranes onto critical equipment.

Consideration of these accidents — referred to here as *internally-induced accidents,* since they are induced when equipment failure in one system adversely affects separate systems — increased significantly during the 1970s. Neither missiles, pipe rupture, nor crane failure figured greatly in AEC review until the early 1970s. By the late 1970s each had become the focus of significant NRC requirements affecting equipment quality (as discussed in the two preceding sections), location, separation, and redundancy. Moreover, a major increase in engineering effort was required to sift the large numbers of potential initiating failures for those that could lead to serious accidents.

Protection Against Missiles: Nuclear plants contain many parts of sufficient mass rotating fast enough to cause significant damage if they should suddenly break free and penetrate their housing (if there is one). A prime example is the flywheels attached to reactor coolant pump motors. They consist of several large, joined discs which store large amounts of rotational energy to drive the pumps for brief periods if the motors lose power. This energy could cause considerable damage if the flywheel should break loose or split apart. The "initiating event" that could produce excessive stress is overspeeding ("windmilling") of pumps due to pressure loss during a loss-of-coolant accident (LOCA).

AEC Regulatory Guide (RG) 1.14, issued in 1971, upgraded the design margins, equipment qualification, and inspectability of reactor coolant pump motor flywheels (see box for description of regulatory guides and other NRC requirements). The guide mandated that flywheel discs be produced only in extremely high-quality fabrication processes (vacuum-melting and degassing or electroslag remelting), with careful checking of the material's fracture toughness and tensile strength, and strict control of welding. The flywheel

Nuclear Regulatory Commission Regulations, Guides And Standards

NRC standards are published in Regulations, in Regulatory Guides, and in the Standard Review Plan.

Regulations, requirements that licensees must meet, are published in different parts of Title 10, Chapter 1 of the Code of Federal Regulations (10 CFR). Part 50, Licensing of Production and Utilization Facilities, includes as its Appendix A a set of *General Design Criteria* establishing, in general terms, the necessary design, fabrication, construction, testing, and performance requirements for equipment and systems important to safety. Among other relevant parts of 10 CFR are Part 20, Standards for Protection Against Radiation, and Part 100, Reactor Site Criteria.

Regulatory Guides describe methods acceptable to NRC staff for implementing specific parts of NRC regulations. They also delineate techniques used by staff to evaluate specific problems or postulated accidents. As described in Chapter 3, most regulatory guides serve as *de facto* regulations since it is usually less costly for licensees to conform with them than to seek approval for alternative approaches.

Regulatory guides are issued in ten different "divisions," covering research reactors, materials transportation, fuel cycle facilities, etc. The division of concern here is Division 1, power reactor guides. All guides and industry consensus standards discussed in Sections 4.3 through 4.10 are listed at the end of this chapter.

The *Standard Review Plan* is a compilation of internal NRC review procedures used in licensing reactors. There is much overlap between the Standard Review Plan and regulatory guides. Both describe acceptable design and construction approaches and both frequently notify applicants of information needed for the staff's licensing reviews.

NRC regulations, regulatory guides, and provisions of the Standard Review Plan are generally referred to interchangeably here as "standards" or "requirements," reflecting their actual, if not legal, status in the regulatory process.

must be designed to fracture at not less than its possible overspeed during a LOCA and at not less than twice its normal operating speed. All high-stress areas of the flywheel and its attachment to the motor must be capable of inspection without removing the flywheel from its shaft. These requirements added significantly to the cost of reactor coolant pump motors.

The steel discs that hold the blades of nuclear steam turbines are also a potential source of damaging missiles due to their high rotational speed (1800 rpm) and large mass. Turbines are equipped with protective devices to prevent overspeeding, but failure of these devices could cause discs to rupture, in which case fragments weighing as much as several tons might leave the rotating shaft at high speeds. The fragments could be contained by the heavy steel turbine casing, but they might also break through, as has happened in several fossil plants and one foreign reactor.[24] The fragment might retain sufficient kinetic energy to penetrate the turbine building and enter the containment building, where it could strike critical equipment such as reactor coolant piping, steam generators, or the control rod drive housing above the reactor vessel (in PWRs).

NRC's approach to low-trajectory turbine missiles (the most likely missiles of concern), set forth in RG 1.115, involves turbine orientation and protective barriers. The simplest defense is to orient the turbine axis so that turbine missiles cannot strike the containment building. This may require separate service cranes for each turbine in a multi-unit plant or other minor adjustments, but the costs are small. For plants already under construciton with a "high-risk" turbine orientation, however, RG 1.115 required considerable analysis. In some cases, moreover, concrete or steel shielding was installed around potentially vulnerable, critical plant systems including the reactor coolant system, emergency and residual cooling systems, the control room, and the gaseous radwaste system.

Other potential missiles among nuclear plant components are the subject of ANSI draft Std N177 which has been incorporated, with little alteration, into NRC's Standard Review Plan. The standard requires consideration of a tremendous number of potential missile accidents. They include overpressure or material fatigue causing devices such as valve stems and bonnets, instrument gauges, and other fragments of pressure-retaining equipment to separate and be propelled by pressurized fluid; seizure or overspeed of pumps, fans, or turbines causing propulsion of blades, bolts, or flywheels; and "indirect" sources of missiles such as chemical explosions, pipe ruptures, and short-circuits in transformers or switchgear causing electrical components to melt and be ejected through electromagnetic repulsion.

Designers must also identify possible critical targets: fluid systems such as the reactor coolant system and emergency and long-term cooling systems; structures such as steam generators and their supports, housing or enclosures for safety equipment, and the containment steel lining; electrical and control equipment and their housing, such as conduits and cable trays; and motors and auxiliary equipment which drive fans, pumps, and valves. This analysis must consider the relative locations of missile sources and targets, the possible energy and angle of ejection, and the likely strike angle and orientation of the missile upon impact.

Not all conceivable accidents require prevention, but protection against missiles has led to considerable equipment changes. Fabrication, equipment qualification, and inspection have been upgraded for much rotating machinery. Barriers and housings have been strengthened, especially for conduits and cable trays, which afford less protection than piping walls. Substantial engineering effort has also been involved.

Protection Against Pipe Failure: Nuclear plant piping systems contain considerable volumes of water and steam maintained under very high pressures. The thrust of high-pressure fluid from a complete pipe rupture could cause the pipe to whip about with great force. Water or steam released in even a partial rupture may strike equipment with high, destructive pressure ("jet impingement") or at least flood or saturate the vicinity. AEC General Design Criterion No. 4 required protection for safety-related equipment from such phenomena, but design requirements to do so were not spelled out until the 1970s.

RG 1.46 issued in 1973, stipulates possible pipe break locations to be analyzed to determine protective measures against pipe whip within the reactor containment. The locations include all *terminal ends* of piping runs — structures such as piping anchors or components such as vessel or equipment nozzles which constrain piping movement or expansion — and *intermediate piping locations* of potential high stress and fatigue, such as pipe fittings (*e.g.,* elbows and tees), valves, flanges, and welded attachments. Conservative stress intensity and usage factors must be employed to calculate stress locations, taking account of possible degradation or errors in design, fabrication, installation, and operation. Two or more intermediate locations must be analyzed for each piping run.

RG 1.46 addresses only "high-energy" piping (above 200°F or 275 psig pressure) in excess of stipulated widths (at least one inch for circumferential pipe breaks, four inches for longitudinal breaks) within containment. Nevertheless, it covers much reactor piping, and later plants contain pipe restraints at most of the examined locations. These are usually concrete or steel encasements that physically separate piping from other components, designed to high seismic standards to ensure that they remain intact during earthquakes. Costs for the restraints and the engineering analysis have been substantial.

Jet impingement and related steam and flooding effects from pipe rupture are addressed in ANSI draft Std N176 which has been incorporated into the Standard Review Plan. It requires designers to postulate complete pipe ruptures, partial breaks, and through-wall cracks that could generate fluid jets at the same locations examined for pipe whipping. This analysis is more complicated than that for pipe whipping, since jet impingement forces, flooding, saturation with steam, and potential pressurization of nearby compartments must be considered.

Although the ANSI standard and the Standard Review Plan recommend rather than require protective measures, plant designers have built in considerable structures to protect safety-related equipment and maintain access to areas required to cope with pipe ruptures. These include deflectors to absorb and shield fluid jets, further enclosure and separation of piping from safety-related structures and equipment, and provision of redundant safety-related equipment where separation is impractical. In addition, much safety-related equipment has been designed to withstand the high pressure, humidity, and temperatures that could result from fluid jets, as discussed in Section 4.3

Overhead Cranes: Nuclear plants contain large overhead cranes for moving and storing fuel assemblies, spent fuel casks, and the reactor vessel head and upper internals. PWRs require two cranes, one inside the containment building over the reactor vessel, the other in the spent fuel storage area outside containment. Most BWRs require only one since the spent fuel pool is inside the reactor building, but two cranes will be needed for the new Mark III design, which has an external spent fuel pool.

Most items handled by the cranes are deemed ''critical loads'' because dropping them could lead to release of radioactivity. For example, the impact of a large load on the spent fuel pool could damage cooling and make-up systems, leading to loss of coolant and heat-up of the spent fuel with subsequent fuel damage and radiation release. The NRC has allowed licensees to meet some safety requirements for critical loads by strengthening vulnerable structures, for example, by adding protective barriers to the spent fuel pool cover. At most recent plants, however, crane handling systems have also been upgraded to reduce the chances of dropping critical loads.

NRC crane reliability requirements are set forth in a 1979 report by the Commission's Office of Standards Development, which superseded a 1976 regulatory guide.[25] The report requires upgrading design and fabrication requirements for crane hoisting and braking systems so that a single failure will not cause critical loads to be dropped. These requirements affect all the structural, mechanical, and electrical systems that enable cranes to lift and move loads, including the overhead bridge and trolley, the braking and control systems, and the hoisting system.

The report's major advance is its application of the *single-failure criterion*[e] to crane systems. The single load block is replaced by a double block system which also requires an equalizer beam to distribute the loads equally. The cable hoisting system must have two holding brakes in addition to the power control braking system, and these must be stronger than previously. A dual braking system is required to guarantee control over bridge and trolley travel. Moreover, the hoisting and trolley systems must be capable of manual operation in the event that control failure immobilizes the crane while it is

e. See p. 156 for a description of the single failure criterion.

holding a critical load. Similarly, the motor controls must be able to accommodate excessive electric current and overspeed or overload conditions.

The NRC report also upgrades crane design margins and material specifications. Since the crane might be operating during an earthquake, the bridge must be strong enough to remain intact and the trolley must be designed to stay in its runway and hold its loads despite seismic stress. Critical bridge and trolley weld joints must be examined prior to installation, materials for critical structural members must be tested extensively prior to fabrication, and the crane manufacturer must conduct a fatigue analysis for critical load-bearing structures and components. The report also mandates 15% greater design margins for component parts subject to wear and exposure. Thicker and stronger cables are also required, which in turn necessitates strengthening other components to handle the higher loads.

These requirements began to be employed in NRC licensing review in 1975, and they subsequently expanded to the level described above. The associated costs are only partially reflected in the later plants in the nuclear data sample and will be more fully experienced by plants still under construction.

Section 4.5: Protection Against Earthquakes And Other Natural Phenomena

AEC and NRC regulations have required that nuclear plants be designed to withstand earthquakes and other natural phenomena without suffering damage that could lead to accidents and without losing safety systems needed to prevent accidents. [26] Interpretation of this requirement became more stringent beginning in the mid-1960s and continuing during the 1970s, leading to substantial plant design and equipment changes and concomitant increased engineering efforts that contributed significantly to increases in nuclear plant capital costs.

The AEC first encountered seismic issues in reviewing proposed reactors at Bodega Bay in 1962 and at Malibu in 1964, both on the California coast. The Commission's modest seismic requirements were criticized in expert testimony sponsored by intervenors, including the Sierra Club (Bodega Bay) and local residents and property owners led by comedian Bob Hope (Malibu). Their testimony established that both sites were potentially subject to greater ground acceleration than AEC staff had initially specified, and that the novel Bodega Bay design could not be guaranteed to attenuate ground movement sufficiently to prevent damage from earthquakes.

Both reviews were prolonged and marked by clashes of opinion between the Advisory Committee on Reactor Safeguards and more conservative factions among regulatory staff. The AEC ultimately rejected both applications, Bodega Bay in late 1964 and Malibu in early 1967. The public uproar and new information (and uncertainties) generated in both battles led the AEC to begin setting detailed seismic protection criteria.

AEC/NRC seismic protection methodology entails a chain of predictions, including:

- the magnitude of possible earthquakes;
- the transmission of earthquake forces through the ground;
- the response of structures to these forces; and
- the behavior of equipment within the structures.[27]

Considerable uncertainties are involved at each level of prediction. Increased information and an expanding reactor population have led AEC/NRC to resolve the uncertainties in the direction of greater conservatism.

Earthquake Magnitude: Following its reviews of Bodega Bay and Mailbu, AEC staff developed the concept of the Operating Basis Earthquake (OBE) and the Safe Shutdown Earthquake (SSE). The OBE is the maximum earthquake ''reasonably'' expected during plant lifetime, based on past earthquakes at the site or in areas with similar geologic characteristics. The SSE is the maximum possible earthquake potential for the site. Reactors must be able to remain operable during an OBE and to shut down without losing any safety systems during an SSE. Both are specified by the applicant, subject to AEC/NRC review. The OBE is generally at least half the SSE. The SSE varies considerably among reactor sites, ranging from 0.10g — equivalent to one-tenth the acceleration of gravity — in the least seismically active regions, to 0.75g for several West Coast reactors. The 0.10g floor was established informally by AEC staff during 1965-67 and codified as a regulation in 1973.[28]

The magnitudes of reactor SSEs have increased only slightly over time, with several notable exceptions. When an assertedly minor fault 3½ miles form the Diablo Canyon site in California was later identified as an extension of the more potent Hosgri fault, the NRC required the utility to raise the SSE from 0.40g to 0.75g. This led to extensive re-analysis and design modifications which have prevented start-up since 1977. Similarly, the 0.67g SSE for San Onofre 2 and 3 in California is higher than the 0.50g value for Unit 1.

Architect-engineers have conservatively specified the highest SSE values outside the West Coast, 0.25g to 0.30g, for the several standardized plants under construction in an effort to avert the possible later need for recalculations and the resulting expensive re-design during construction. Nevertheless, there has been little general move toward increases in earthquake design bases since the early commercial-size reactors. Seismic criteria have changed primarily in projecting the effects of earthquakes of given magnitudes and in specifying necessary protective standards and systems.

Earthquake Forces and Structural Response: Four regulatory guides issued between 1973 and 1976 (but reflected in earlier regulatory review) have altered the methods used by nuclear designers to translate the intensities of postulated earthquakes into potential effects on plant structures. RG 1.60 and 1.61 defined calculational procedures for estimating the maximum responses of hypothetical structures to a given earthquake's motion. The guides specified the degrees of horizontal and vertical acceleration (response spectra) to be assumed for different earthquakes, as well as the "damping" capabilities of various plant structures — their ability to dissipate vibrational energy rather than transmitting it to connected structures. Most specifications in the guides were more conservative than past practice.

RG 1.92, issued in 1974, presented a new method for combining loads to estimate the total response of a structure to the components of earthquake-induced forces, *e.g.*, east-west vibratory motion concurrent with north-south motion. The guide essentially replaced "static" models, which assumed that forces would act on the center of gravity of structures, with "dynamic models" which effectively divide structures into sections connected by springs representing stiffness. Greater deformation of structures is predicted by the dynamic models.

Finally, RG 1.122, issued in 1976, provided a means of estimating the responses of equipment or structures supported at various levels above the main foundations of the plant buildings. The previous method had neglected most of the amplification of vibratory forces from the flexibility of supporting structures.

An earlier regulatory guide, RG 1.12, required that reactors include an instrumentation system to measure the input vibratory ground motion from actual earthquakes and the resultant vibratory response of critical plant structures. Its purpose is to provide immediate indication of the plant's response to earthquakes and also to compare the actual vibratory response to that predicted in developing the plant design. The system includes triaxial peak accelerographs and triaxial response-spectrum recorders installed on reactor equipment, reactor piping, the containment foundation, and safety-related auxiliary structures outside containment.

Seismic Protection and Equipment Standards: The most significant cost impacts from seismic protection have been felt directly in equipment requirements and quality standards. Since the early 1970s, AEC/NRC has maintained a roster of plant equipment which must be designed, manufactured, and installed to exacting requirements to ensure that it can remain functional despite the effects of the SSE. The "Seismic Category I" list has been expanded through four editions of RG 1.29. Originally it comprised only the reactor coolant pressure boundary (essentially the reactor vessel, primary cooling loops, and primary reactor piping). It now also includes the ECCS, systems to remove heat and radionuclides from containment, normal and auxiliary electric power systems and circuitry, and the control room, among other equipment. The quality assurance requirements for these systems have led to more exacting, hence costly, manufacturing standards and verification procedures, as discussed in Section 4.2.

Procedures to qualify (*i.e.*, demonstrate the adequacy throughout service life) electrical equipment needed to ensure safe shutdown during an earthquake were especially upgraded during the 1970s. They are delineated in IEEE Standards 323 and 344, first published in 1971 and revised in 1974 and 1975. These standards were applied voluntarily in nuclear construction until they were endorsed by NRC Regulatory Guides 1.89 and 1.100 in 1974 and 1976, respectively.

The IEEE standards require that electrical equipment (including batteries, switchgear, cables, relays, motors, and electronic sensors and indicators of pressure, temperature, and flow) used in safety-related systems have demonstrated capability to withstand the vibratory effects of earthquakes throughout their installed life in the plant. Although proof of seismic qualification may be provided by analysis, in which seismic effects are calculated using elaborate computer models, this approach is generally applied only to large pieces of equipment that are difficult to test due to limitations in the loading capability of vibration equipment. The preferred method is direct testing of the equipment by subjecting it to a simulated earthquake while measurements are made of its mechanical strength, alignment, electrical performance, and non-interruption of function.

Elaborate tests are required to ensure accurate mounting, monitoring, assessment, and documentation. They must be geared to the ''response spectrum'' that the SSE would generate in the equipment, as predicted by the structural response analysis discussed earlier. A minimum of five OBEs and one SSE are simulated. Current tests are more complex than earlier ones, applying simultaneous excitation along two axes of the specimen (versus one previously) and employing random frequencies and amplitudes — much like the time history of an actual earthquake (versus only one frequency component previously). In addition, the number of components required to be tested has grown with the expanding designation of Seismic Category I equipment.

Another series of regulatory guides significantly upgraded requirements for fluid systems, containment components, and component supports. Fluid systems are vessels, pipes, pumps, and valves containing or controlling water or steam. Prior to 1973, licensees designed the fluid system components of their plants by consulting the Boiler and Pressure Vessel Code of the American Society of Mechanical Engineers (ASME), a professional society that sets standards for industrial fluid-bearing materials and equipment. The code specifies the design requirements (*i.e.*, the anticipated mechanical, pressure, and thermal loads) for various nuclear plant components. But because different licensees faced with identical seismic and other service conditions frequently assumed different loading combinations and associated design limits, the AEC issued RG 1.48 in May 1973.

This guide chose the particular ASME code requirements for each equipment category. Its requirements were generally more conservative than previous practice, with the result that valves, pipes, etc. were designed to accommodate higher pressures and temperatures and greater combinations of mechanical loads (*e.g.*, pipe reactions or pressure loads generated by accidents). Another aspect of these changes was a substantial increase in the numbers of supports and restraints provided for nuclear piping, both inside and outside containment, with a concomitant significant increase in costs.

RG 1.48 also specified more rigorous design limits and performance specifications (*e.g.*, response times) for *active* pumps and valves — those which must perform mechanical motion to accomplish their safety function. It superseded the ASME code requirement that pumps and valves merely maintain their presssure-retaining integrity without assurance of operability.

RG 1.57 defined analogous requirements for components of the metal primary reactor containment. These include the steel containment lining, containment penetration assemblies[f] and access openings, and piping systems attached to containment vessel nozzles or penetration assemblies. As with fluid systems, reactor designers had not been applying the ASME Code uniformly in specifying the capabilities required for these components.

The guide primarily lists the applicable ASME Code sections for containment components, with certain key modifications. Vibratory motion from earthquakes must be included in calculating material strengths required to withstand the long-term strain induced by varying plant power levels. Similarly, seismic stress must be considered in tandem with the effects of water jets from possible pipe ruptures in specifying the design requirements of metal components of the containment structure.

Finally, two later regulatory guides, RG 1.124 and RG 1.130, set similar requirements for critical component supports. These are structural elements such as beams, columns, trusses, and vessel skirts which carry the weight of components or give them structural stability. The guides mandate considerable increases in the physical attributes, quality assurance steps, and engineering analysis for much of the support structure of nuclear plants.

Both guides direct that the same QA requirements and design limits be applied to component supports for safety-related equipment as to the components themselves. They also require that seismic loads be included in the loading calculations used to determine the design strengths of component supports. They surpass the ASME Code in some respects, requiring, for example, that structures have adequate tensile strength over the possible range of accident temperatures rather than merely at room temperature as some sections of the Code require.

---

f. RG 1.57 defines penetration assemblies as parts or appurtenances required to permit piping, mechanical devices, and electrical equipment to pass through the containment vessel shell or head and maintain leaktight integrity while compensating for such phenomena as temperature and pressure fluctuations and earthquake movement.

Protection Against Floods and Tornadoes: Nuclear plant protection from non-seismic natural phenomena also increased during the 1970s. New geographical data have led to higher-strength "design basis" floods and tornadoes at some sites, and equipment design margins have been increased.

Nuclear plant structures are potentially vulnerable to tornado-generated pressure and to tornado-induced missiles such as telephone poles. Protective measures were first seriously considered in licensing reviews in the late 1960s and were applied in the early 1970s by new requirements subsequently issued as regulatory guides.

RG 1.76 requires plants east of the Rockies to be able to withstand 360 mile-per-hour tornado wind speeds. Western plants are designed to 240-300 mph speeds and to commensurately lower pressure drops. RG 1.117 names plant structures and systems requiring special protection. It has led to increases in wall thicknesses, typically from 18 inches to 24-28 inches, in auxiliary buildings, housings for diesel generators and their fuel supply, and condensate storage tank structures. (containment buildings were generally already designed to tornado strength for reactor accident considerations.) Doors, windows, and safety-related cable and pipe penetrations have also been strengthened against tornado-generated missiles.

A recent revision of RG 1.117 requires tornado protection for systems that provide long-term post-accident cooling and limit radiation releases. These include the residual heat removal system, the gaseous radwaste treatment system, the control room, and supporting instrumentation and controls. Although this requirement applies only to recent construction starts, it may also lead to increased shielding and wall strengthening at plants in advanced stages of construction.

Flooding is also of concern for nuclear plants, since most are located near water sources which have potential for flooding. As with tornadoes, requirements for protection against floods grew in the late 1960s and early 1970s and were then codified in two regulatory guides.

RG 1.59 spelled out methods for determining flood conditions which a plant must be designed to withstand without losing safety system functioning. These conditions could result from the "probable maximum flood" for the site (as estimated by the Army Corps of Engineers), from seismically-induced floods or dam failures, from other dam failures, or from hurricanes, tides, snowmelt, wind-generated wave activity, or severe local precipitation.

RG 1.102 specifies flood protection design alternatives. A plant may be constructed at a *dry site* with the natural terrain or engineered fill raising it above the design flood level. It may be protected by *exterior barriers* — levees to prevent inundation, seawalls or bulkheads to defend against wave erosion, breakwaters to deter wave attack. Or important equipment may be shielded by *incorporated barriers*, essentially heavy wall reinforcing and water-tight sealing of chambers, piping penetrations, and equipment hatches. Although design approaches and equipment for flood protection are site-specific, most later plants have incorporated greater protective measures. The increased engineering and site analysis have also added to costs.

Section 4.6:   Fire Protection

Fire protection requirements for nuclear plants have become more stringent in recent years, largely because of the very serious fire at the Browns Ferry plant in March 1975. Plants starting construction after 1976 have had to incorporate extensive design features intended to prevent, contain, and mitigate fires that might disable safety-related equipment. Plants already operating or in construction have been exempted from many of the requirements but have been required to take substantial alternative protective measures. All generations of plants have incurred significant costs, and further requirements lie ahead.

Federal fire protection regulations were limited before Browns Ferry, despite the occurrence of several dozen fires at operating reactors.[29] AEC General Design Criterion No. 3 directed only that noncombustible and heat-resistant construction materials be used ''wherever practical,'' and that fire detection and control systems be provided to prevent fires from affecting safety-related equipment. Regulatory Guide 1.75, published in early 1974, specified that electric power supplies and cables important to safety be physically separated to prevent fires from disabling redundant safety systems. The guide applied only to new construction starts, however, and did not address operating plants or those in construction.

The Browns Ferry Fire: The fire on March 22, 1975 at the Tennessee Valley Authority's Browns Ferry nuclear plant in Alabama provided a powerful impetus for change. The fire was initiated by an electrician's use of a candle to check for air leaks in a cable-spreading area beneath the control room. It spread rapidly and lasted for six hours, consuming 1600 cables, including 618 related to safety systems, and disabling much of the instrumentation and control systems for the plant's two operating units. For several hours, one unit was without its normal feedwater system, all three emergency core cooling mechanisms, and most of its reactor monitors. The reactor cooling water level fell dangerously close to the top of the fuel and was maintained only by a condensate booster pump that was barely able to deliver the minimum water required. Much ventilating and firefighting equipment either malfunctioned or was inaccessible. Thick smoke and fumes filled the control room during much of the accident, adding to the problems created by equipment and instrument failures.[30]

The fire was publicized around the world and was considered the most perilous accident at a commercial nuclear plant until Three Mile Island. The NRC's regulatory response is stated primarily in a section of the Standard

Review Plan issued in late 1976.

Fire Protection For New Plants: The Standard Review Plan requires that new plants — those whose construction permit applications were docketed after June 1976 — contain a comprehensive design for fire protection. Redundant divisions ("trains") of safety-related systems must be separated so that both are not subject to damage from a single fire hazard. Similarly, fire barriers (fire-resistant walls, doors, floors, etc.) must be provided that can isolate safety-related systems from fires in non-safety-related areas for three hours. Cable-spreading rooms may not contain both redundant safety divisions, and adjoining reactors may not share the same cable-spreading room. To prevent fires from propagating in electrical cables, cable trays must be made of metal and equipped with continuous heat detectors, cable tray penetrations through fire barriers must be sealed and fireproofed, and fire stops must be installed along cable routings at frequent intervals.

The plan also stipulates noncombustible material requirements for much of the plant, including interior walls, structures, and finishes, and radiation shielding. In addition, ventilation systems must be capable of removing heat and smoke from a fire vicinity without conveying combustion products to critical areas such as the control room. Fire-fighting systems must include automatic systems in inaccessible spaces such as primary containment or in areas where safety considerations preclude fire barriers.

Drains must be provided to remove firefighting waterflow from the vicinity of safety-related equipment, but they must not provide fires with a path to spread. The capability to collect, sample, and analyze drainage from potentially radioactive areas is also required. Guaranteeing a water supply for firefighting requires a system of pumps, valves, pipes, storage tanks, and hoses.

New plants must also have fire detectors in all areas in which safety-related equipment might be exposed to fire. Fixed and portable emergency lighting and communication systems must be provided. Finally, since earthquakes can induce electrical fires, NRC Staff now considers the need to qualify new plants' fire detection and fighting equipment to be functional following the designated Safe Shutdown Earthquake (see definition in preceding section).

These are merely the highlights of the extensive, detailed fire protection rules in the Standard Review Plan. They affect a large portion of plant design, not only by calling for additional equipment but also by requiring redundant safety systems to be separated by distance and/or physical barriers. Separation is costly because the increased space necessitates larger structures to house cables and other equipment as well as more cable and piping to span the longer distances. Considerable engineering, moreover, is involved in both the fire hazards analysis and the actual design effort — in planning cable routing, for example.

Fire Protection for Plants in Progress:　Many of the 1976 fire protection provisions of the Standard Review Plan could be difficult and expensive to implement in plants already built. Accordingly, the plan permits deviations for plants in operation or still in construction where strict adherence would be impracticable. Plants in progress are not required to meet all physical separation requirements for redundant safety-related systems. Instead they may rely on fire-retardant coatings and fire detection and fighting systems to prevent or extinguish fires that might otherwise propagate from control cables of one safety system to those of its back-up.

Similarly, cable-spreading rooms may be shared between reactors operating or under construction and need not be separated from other plant areas by three-hour fire barriers, provided that fire-retardant materials are used and fire detection equipment and automatic suppression systems are installed near electrical cables. There are similar exemptions for separation criteria for pumps and other safety-related equipment. Fire-retardant coatings are permitted as an alternative to requiring safety-related cables to be enclosed in covered conduits or trays and requiring highly fire-resistant seals for cable penetrations through walls and structures. Cables that fail standard fire-propagation tests need not be replaced if they are retardant-coated and derated (the thermally-insulating coating could build up excessive heat at high amperages).

Plants in progress also need not be able to ventilate smoke and corrosive gases to safe locations or monitor the content and pathways of radioactive combustion products. Similarly, floor drains are not required near fire-fighting fixtures or near sensitive equipment unless water accumulating from their operation could create "unacceptable consequences." Some deviations from fire detection guidelines are permitted, and most fire-fighting equipment need not be operable following a severe earthquake.

Notwithstanding these liberal allowances, considerable expenditures were required for most recently completed plants to comply with the plan. They have installed fire barriers and suppression systems; applied retardant coatings to cables and cable penetrations and added compensating cable capacity; and performed fire hazards analyses to satisfy NRC review. Some plants under construction that could not meet the physical separation criteria of RG 1.75 have had to add auxiliary reactor shutdown systems whose cabling is routed outside the cable-spreading room. The logistical problems in treating cables already installed and erecting walls in previously constructed areas have added further to costs.

Existing plants will be affected further by a new fire protection regulation adopted by the NRC in October 1980.[31] Most plants must install auxiliary shutdown systems as described above; bolster automatic fire-fighting capability with manual capability throughout the plant; install sectional control valves on fire mains to allow local fire-fighting system maintenance; provide a dedicated fire-fighting water distribution system with two separate trains; install emergency lighting near reactor shutdown equipment; and upgrade

Chapter 4

automatic fire detection systems. Licensees have until early 1983 to make most improvements.

Fire protection costs would probably be still higher for any plants making construction permit applications after July 1976, for which the foregoing deviations would not be permitted. The re-design and re-construction problems avoided in those cases would probably be outweighed by the extensive requirements for physical separation, ventilation, and electric cable verification, as discussed in Section 5.1.

Section 4.7:   The Reactor Core And Cooling Systems

The heart of a nuclear power plant is the reactor pressure vessel, its fuel and control rod assemblies, and the pipes, pumps, valves, and tanks that circulate water through the reactor "core" under both operating and accident conditions. Core designs and standards have not changed greatly since the first commercial plants. Cooling water circulating equipment has evolved somewhat, however, with a noticeable effect on costs.

Emergency Core Cooling System: The Emergency Core Cooling System (ECCS) comprises equipment intended to replenish reactor cooling water to prevent fuel overheating and melting following a loss-of-coolant accident (LOCA). It encompasses several equipment "trains" designed to deliver water to the core under a wide range of pressures corresponding to different types of LOCAs (*e.g.*, breaks in coolant pipes of varying sizes and at varying locations). PWRs have pump-actuated high- and low-pressure coolant injection systems and "accumulators" containing water and nitrogen under pressure which automatically release water if core pressure falls below a set level. BWRs have low- and high-pressure core spray systems, a low-pressure core flooding system, and a system to automatically depressurize the reactor coolant to enable the low-pressure systems to be effective.

ECCS changes have been primarily incremental, directed at improving early designs rather than developing radically new configurations. Changes have come in response to pressure from the Advisory Committee on Reactor Safeguards (see preceding chapter) and as a result of the AEC's ECCS rule-making hearings from mid-1971 to early 1973. The hearings led to new ECCS performance criteria including closer limits on fuel temperatures and rates of cladding oxidation during accidents to improve control of LOCAs.[32] Although the new criteria did not directly require significant equipment changes, they made clear the AEC's intent to upgrade ECCS capacity and reliability. They also required the reactor vendors to improve their analytical models to demonstrate that the ECCS could meet the new criteria — a major effect that has been reflected in nuclear steam supply system costs.

The capacities of ECCS injection and flooding systems have increased over time, requiring expansion of circulating pumps and pipes, holding tanks, and vessel inlets that deliver water to the core. The delivery conduit for BWR high-pressure coolant systems has evolved from a simple pipe to a more complex "header" device — a ring with spray inlets on top of the core. Recent PWRs have supplemented their small-break response capability with "charging pumps" that can operate at higher pressures than the high-pressure coolant injection system.

The ECCS in PWRs is served by sumps within the primary containment designed to collect the reactor coolant lost from the primary system in the event of a LOCA so it may be recirculated into the core. RG 1.82, issued in 1974, specified that the containment have two sumps, one for each of the redundant halves of the ECCS. They must be physically separated from each other and from high-energy piping systems to prevent a pipe break from initiating a LOCA and also disabling part of the ECCS. The guide also included design specifications to ensure that sump intake systems effectively collect water without introducing air through the suction intakes of the recirculating pumps. Trash racks and fine-mesh screens were required to protect sumps from LOCA-generated missiles and prevent clogging with debris such as insulation that could be ripped off by a pipe rupture. The vendors have also had to build elaborate scale models to test the ability of different sump configurations to recirculate cooling water to the ECCS.

Residual Heat Removal: Reactor fission products continue to generate "decay heat" after insertion of control rods has terminated the nuclear chain reaction. Reactors therefore contain a residual heat removal (RHR) system designed to operate at low pressures for days or even months after an accident, when the higher-pressure ECCS systems may not be able to remove the substantial decay heat still being produced. It comprises steam lines, feedwater systems, heat exchangers, and parts of the ECCS.

The NRC's Reactor Safety Study (WASH-1400) concluded that a meltdown was more likely to result from inability to remove decay heat than from failure of the ECCS to replace primary coolant directly after a large pipe break. RG 1.139 upgrades the RHR system at new plants by requiring that it be built to "safety-grade" quality (subject to quality assurance) and with sufficient redundancy to be operable despite a single failure (*e.g.*, to power supply). Interlocks, valves, and controls must be provided to ensure that the system is activated only on low reactor coolant system pressure to guarantee that it is available *after* an accident. Pressure relief capacity is required to prevent damage from accidental overpressurization, and any fluid discharged through relief valves must be collected to avoid flooding safety-related equipment or interfering with the ECCS.

Although the guide is directed only at construction permit applications docketed after 1977, some of its provisions have been carried out at currently operating plants, for example, interlocks to isolate the RHR system from the reactor coolant system. The problems encountered in maintaining long-term cooling at Three Mile Island and the more recent loss of decay heat removal capability for several hours at Davis-Besse may cause many of the provisions of RG 1.139 to be applied to plants under construction, as discussed in Section 6.2.

The Reactor Coolant System: The pumps, pipes, valves, tanks, and related equipment which continuously circulate primary coolant water during normal operation are referred to as the reactor coolant system (RCS). This system was refined during the 1970s both to improve plant performance reliability and to reduce accident risks.

Cracking of RCS pipes and other metal components has been the single largest cause of nuclear plant shutdowns. BWRs have developed cracks in ECCS core spray piping and in "recirculation" pipes in the normal circulating water system. PWRs have developed cracks in steam generator tubes and, more recently, support structures. Most of the cracks require prompt repair since they affect the reactor coolant pressure boundary and thus are safety-related. High radiation fields sometimes make the cracks difficult to diagnose and repair.

Design changes have been made at newer plants in an effort to reduce cracking incidence by keeping corrosion-inducing impurities out of the coolant system. These are primarily metals and salts that may leak through the condenser and flow directly to the "secondary" side of PWR steam generators and into the BWR primary coolant system. Both reactor types have upgraded their "condensate demineralizers" — ion-exchanging resins which filter out dissolved solids. RG 1.56 has further required that BWRs, which have a direct cycle that makes the reactor core especially susceptible to solids deposition, install meters at the condenser and demineralizer inlet and outlet to record flow and conductivity. The meters trigger control room alarms when the chemical content of the water reaches high levels.

The cracking problem has also led to costly materials changes in circulating water systems at some new plants. Copper is increasingly being supplanted in condensers by titanium, which is more resistant to chemical attack, particularly by chlorides in seawater. Stainless steel is replacing carbon steel in steam generator support plates, and new techniques of heat-treating stainless steel and of cladding welds are being applied to BWR piping to improve resistance to stress. Funds have also been expended in analysis and testing. Utilities in the PWR "Steam Generator Owners Group" have spent several tens of millions of dollars since 1977 on EPRI programs, and the vendors have matched this with outlays on their own steam generator research facilities.

A critical RCS component at BWRs is the main steam isolation valves (MSIVs) on the steam lines connecting the reactor vessel to the steam turbine. They are designed to isolate the RCS in the event of a break in a coolant pipe or in a steam line outside containment. Chronic MSIV leakage led the NRC to issue RG 1.96 in 1975, requiring leakage control systems for all BWRs with construction permits granted after February 1970. (This includes the last two BWRs in the data base employed to calculate nuclear capital cost increases — Fitzpatrick and Duane Arnold.) The systems consist of supplementary valves, vents, ducts, interlocks, and associated instrumentation and circuitry. The components must meet seismic and quality assurance criteria and must function despite LOCAs, missiles, pipe whipping, or water jets. Substantial costs appear to have been involved for the high-quality equipment and supporting engineering analysis.

An earlier regulatory guide, RG 1.45, required equipment to detect reactor coolant leakage into the containment in order to reduce radioactivity in the plant and to abort potential LOCAs. Three or more monitoring systems must be employed, for sump level and flow, for airborne particulate radioactivity, and for condensate flow rate or airborne gaseous radioactivity. They must be able to distinguish normal, small leakage in equipment that can't be made leaktight (*e.g.,* valve stem packing glands, pump shaft seals) from abnormal leakage with potential safety significance. The detection systems must also be able to locate leaks and report them to the control room through indicators and alarms. The guide also urged use of sensors to detect changes in temperature and humidity.

Ultimate Heat Sink: Piping and intake requirements outside of the reactor coolant system were upgraded by RG 1.27 to ensure that the plant's water source — river, pond, cooling tower, etc. — can supply water to operate the residual heat removal system at sufficiently cool temperatures for at least a month after an accident. The guide's most recent revision in 1976 requires two separate sets of intake structures and piping routes, both designed to withstand an Operating Basis Earthquake (see Section 4.5) and sized according to severe historical climate conditions.

Reactor Core Equipment: Designs of reactor vessels, fuel and control rod assemblies, and supporting structures have changed little since the earliest commercial plants. Several changes have added slightly to costs, however. Newer plants and some older ones have added more neutron detectors to monitor reactor power fluxes. These reduce uncertainties in fuel "peaking factors" so that reactors can be operated closer to design capacity without violating the fuel temperature restrictions imposed by the revised ECCS criteria. Similarly, RG 1.20 has required utilities and vendors to develop baseline core vibration data prior to startup to validate analytical models which estimate vibration during plant operation — a response to early instances of flow-induced vibration affecting core support structures and control rod assemblies at many plants. Third, General Electric has divided the hydraulic drive system for control rods into separated banks, each of which covers enough of the core geometry to reduce the likelihood that a single equipment failure could prevent reactor "scram" (rapid insertion of control rods).

Tests and Monitoring: Testing and inspection of reactor core and cooling systems also grew in the 1970s. RG 1.133 mandated that recent plants include automatic systems to detect loose parts that may be left in the reactor during construction or maintenance or that may be deposited in the primary coolant system from component failure. Disengaged parts can damage or wear out critical, hard-to-repair equipment, block coolant flow, or jam the control rods. Most detection systems place up to a dozen pairs of redundant sensors in the reactor vessel and PWR steam generators and include amplifiers to magnify signals and cabling to transmit them to the control room. The members of each pair of instrumentation channels must be physically separated, and the system's operability must be verifiable through in-service inspection.

Two other regulatory guides issued in 1974 mandate special test and inspection programs for PWRs. RG 1.79 requires that PWR operators run through tests of ECCS components and system response prior to fuel loading under both cold and (simulated) hot operating conditions. This necessitates filling the containment sump with water and draining it before start-up. (Preoperational ECCS testing is less eventful for BWRS because the suppression pool sump is kept filled.) RG 1.83 specifies inservice inspection requirements to detect corrosion and cracking in PWR steam generator tubes. The guide mandates pre-service inspection of all tubes to establish baseline conditions as well as regular inspections during plant operation.

Equipment Quality: NRC regulations require that the reactor core and its cooling systems be designed, fabricated, and installed in accordance with the ASME Boiler and Pressure Vessel Code (see Section 4.5). It is beyond the scope of this study to trace the evolution of ASME nuclear plant standards, but several AEC/NRC regulatory guides in particular have extended the ASME requirements, adding to fabrication and installation costs.

Section 50.55a of 10 CFR 50, adopted in 1972, applied the stringent "Class 1" quality standards of Section III of the ASME Code to the reactor pressure vessel and to components of the reactor coolant pressure boundary (see definition in Section 4.5) whose failure could prevent reactor shutdown or cooldown. Starting the same year, RG 1.26 applied the substantial (although less stringent) "Class 2" ASME standards to other significant cooling system components: pressure vessels, piping, pumps, and valves of the reactor coolant pressure boundary that were excluded from Section 50.55a, and much of the ECCS, the RHR system, and equipment to remove fission products and heat from the containment after accidents.

In the following year, RG 1.65 strengthened the ASME standards for bolts and studs which fasten the pressure vessel head, requiring them to be forged from particularly tough and durable steels that are quenched and tempered under closely controlled procedures. RG 1.67 applied the ASME

Code to piping subjected to large reaction forces from pressure discharges from safety valves and relief valves. The guide required that the maximum stress-inducing sequence of valve openings be assumed in determining possible piping stresses, thus contributing further to pipe strength requirements.

Several early regulatory guides upgraded the steel welding and fabrication techniques used in Class 1 and 2 cooling system components. RG 1.31 increased testing of weld materials used to fabricate and join components made of austenitic stainless steel — a nickel-chromium alloy steel used in much primary system piping and components. RG 1.43 required closer control of welding techniques for applying austenitic steel claddings to surfaces of forgings and platings made of stronger but less ductile ferritic steel. RG 1.44 required carefully controlled solution heat treating and similar measures to be used in manufacturing Class 1 and 2 component stainless steels in order to reduce contaminants that could induce stress corrosion cracking. RG 1.50 specified close monitoring of welding for low-alloy structural steel components to avoid heat variations that can enhance cracking.

Section 4.8: Containment Structures And Systems

The containment structure of a nuclear power plant is generally a steel-and-concrete shell enclosing the reactor coolant system. It contains machinery to remove heat, radioactive particles, and combustible gases from the containment atmosphere to minimize radiation releases and ensure that the containment structure remains intact despite high pressures and temperatures generated during accidents. Together, the containment and its systems are major "engineered safeguards" intended to prevent the release of radioactivity from reactor accidents and to protect the reactor coolant system from natural phenomena.

Containment structures and systems have evolved gradually since the first commercial-size reactors. Containment shells have generally been made thicker to accommodate increases in design-basis "loads" resulting from asymmetric (non-uniform) pressures and earthquakes. Design and environmental requirements for "penetrations" which convey piping, electrical cables, and other equipment through the containment wall have been upgraded to prevent leakage throughout the course of postulated accidents. Post-accident "air cleanup" systems to control radioactivity have been designed to more stringent standards. And quality assurance and testing requirements have been toughened for construction materials in the containment shell and for some containment systems. These changes had a modest but noticeable impact on nuclear plant costs in the 1970s.

Containment Designs:[g] Containment structures vary in design configuration and material composition. The ''full-pressure'' containment employed by a majority of PWRs has a several-foot-thick wall of reinforced or prestressed concrete to withstand internally generated pressures and an inner lining of one-quarter to one-half inch-thick welded steel to provide a leaktight membrane. A prominent variant of this design is a double containment structure in which a thicker, pressure-withstanding steel shell is surrounded by a concrete shield building, with filtering equipment in the several-foot space in between.

A different design, the ''ice-condenser'' containment used in some large Westinghouse plants, maintains large stores of ice in an annular region between the reactor coolant system and the inner containment wall. The ice provides a heat sink to significantly reduce pressure build-up in case of accidents. This permits a 10-20% reduction in the containment diameter.[h]

The BWR ''pressure-suppression'' containment design is similar in general concept to the ice-condenser, but water is used instead of ice as the steam-condensing heat sink. The reactor vessel and coolant system are housed in a ''dry-well'' structure that is connected by large vent pipes to a surrounding, water-filled pressure-suppression chamber. In the event of an accident, steam from the drywell would flow through the partly submerged vents and condense, reducing the maximum pressure reached. The ''Mark I'' containments in operating BWRs employ steel walls for both the drywell and the donut-shaped pressure-suppression chamber. Concrete, steel-lined vessels are used in later BWR containments: the Mark II design used in eleven BWRs now in construction, in which the suppression chamber is under the drywell; and the later Mark III now in design and construction, in which the suppression chamber surrounds the drywell.

Changes in Containment Structures: Although the diversity of containment designs precludes quantification, containment structures were generally made thicker and stronger during the 1970s in response to increases in postulated dynamic loads during accidents. Seismic loads were generally higher for later plants, primarily because of changes in calculational techniques for translating anticipated earthquake motion into predicted structural responses (see discussion in Section 4.5). Tornado loads were similarly higher

g. Much of the discussion of containment design and materials is drawn from W.H. Stiegelmann and C.P. Tan, ''Containment-System Design and Construction Practices in the United States,'' *Reactor and Fuel-Processing Technology, 12* (No. 2), 151-172 (1969).

h. Cook 1 and 2 are the only ice-condenser plants in the nuclear data base. Their costs were several percent higher than would have been expected for plants of their size, chronology, etc. — a statistically insignificant difference. Tennessee Valley Authority's Sequoyah 1 and 2 and Duke Power's McGuire 1 and 2 are ice-condenser plants scheduled for 1981 start-up. The low design pressure of ice-condenser plants has been a source of concern since Three Mile Island, as discussed in Chapter 6.

for some later plants. The result was an increase in the quantity of steel reinforcing bars and tensioned steel cables ("tendons") required for concrete containments using reinforced and pre-stressed concrete, respectively.

Structural improvements also arose from more conservative loading criteria. Regulatory Guide 1.57 increased the design requirements for steel in containment lining, containment penetration assemblies, and piping systems attached to the containment or to its penetrations (see Section 4.5). A new "Code for Concrete Reactor Vessels and Containments" added to the ASME Boiler and Pressure Vessel Code in 1975 (applied in draft form in the early 1970s) increased the volume of concrete and steel considered necessary to maintain structural integrity under postulated loads. The code was subsequently endorsed by RG 1.136.

In addition, AEC/NRC- and vendor-sponsored accident research identified situations such as "asymmetric blowdown loads" that had previously been overlooked and which required increased containment wall thicknesses (see discussion in Chapter 3). Similarly, General Electric discovered that accidents could generate severe vibratory pressures on its pressure-suppression containments, such as "steam-chugging" — periodic rushes of water up the discharge vents causing pressure spikes.[33] This led to increases in BWR steel containment wall thicknesses, greater steel placement in reinforced concrete containments, and stronger supports for pipes and other equipment in the suppression-pool area.[34] These changes affected several later plants in the data base but will be felt primarily at plants under construction. The vendors have already expended significant funds in developing and operating test facilities and computer simulation models to analyze the effects of the newly identified loads.

Containment vessel construction techniques also came under closer AEC/NRC scrutiny in the 1970s and were upgraded in regulatory guides. In addition, many of the guides apply to interior concrete and steel structures which support, house, and separate reactor equipment and engineered safety systems, such as reactor vessels, PWR steam generators, and reactor coolant pumps. Regulatory Guides 1.10 and 1.15 upgraded the testing and sampling of "rebars" — steel reinforcing bars which add strength to concrete — and of mechanical splices that join rebars together. RG 1.18 mandated testing of containment leak-tightness prior to start-up, and RG 1.19 required radiographic examination of welds in containment steel liners and penetrations. RG 1.35 strengthened inservice inspection requirements for tendons in prestressed concrete containments, modifying anchoring hardware so it would be accessible for periodic post-construction examinations.

Two later regulatory guides had more sweeping impacts. RG 1.55 required upgraded methods of placing concrete to solve "recurring problems of voids, cracks, and bulges" in nuclear plant concrete structures, especially in critical but hard-to-inspect areas such as foundation mats and containment walls. It directed licensees to monitor shop drawings and construction and to upgrade concrete-placing equipment to ensure that rebar and concrete installations satisfied design pressure specifications and minimized voids. ANSI Std N45.2.5, endorsed by RG 1.94, strengthened quality assurance requirements for the major construction uses of concrete and structural steel: foundation preparation, formwork, steel reinforcement, and embedded items such as anchor bolts. The standard upgraded specifications for cement and aggregates used in making concrete and reduced permissible variations in moisture content, temperature, and surface characteristics in curing and finishing concrete. The ANSI standard also required that more qualification splices be inspected and tested prior to splicing rebars and toughened inspection of most major steel construction including erection, fastening and bolting, welding, protective painting, and cleaning.

Containment Penetrations: Containment vessels cannot be built completely sealed. They are equipped with hundreds of ''penetrations'' — sleeves welded to the steel containment plate or liner and embedded in the concrete wall — to enable piping and electrical cables to convey water, steam, power, and electrical signals between the containment interior and buildings housing key equipment such as the steam turbine, the control room, and auxiliary and safety system pumps. A containment vessel must also contain access hatches for equipment and personnel, and it is joined with thousands of construction welds. Equipment standards and testing requirements to reduce radiation leakage through these penetrations, hatches, and welds were upgraded during the 1970s.

RG1.11, issued in 1971, required that isolation valves be added to reactor *instrument lines*. These are small-diameter conduits that continuously convey primary coolant samples outside containment to measure the status of key reactor parameters. In the event of a rupture in instrument lines outside containment, the isolation valves must be capable of closing on both automatic signals and control room initiation to minimize primary coolant leakage. Otherwise they must remain open to maintain coolant sample monitoring. RG 1.11 also required instrumentation to indicate closed or open valve position in the control room and specified inspection and separation criteria to prevent failure of redundant lines.

Main steam lines, feedwater piping, and auxiliary and safety system piping convey fluids through the containment as well. Isolation valves for these piping systems must also be capable of preserving containment boundary integrity to prevent the escape of radioactivity, while allowing normal or emergency passage of fluids. ANSI Std N271, endorsed by RG 1.143, applied the RG 1.11 requirements for valve actuation, status indication, and physical separation to these isolation valves. The standard also specified that isolation valves in series be actuated by independent power sources, stipulated rapid valve closing times, required that valves be protected against missiles and the effects of pipe ruptures, and applied Seismic Category I requirements to ensure that they can close and re-open as necessary during earthquakes.

Electric cables are conveyed into containment through *electric penetration assemblies* — insulated electric conductors and conductor and aperture seals which together maintain a leak-tight pressure barrier between the inside and outside of the containment structure. Detailed design, material, and qualification requirements for electric penetration assemblies were developed in IEEE Std 317 (twice revised and endorsed by RG 1.63). They include grounded barriers to separate different-voltage conductors; ability to withstand fires, high radiation, and current surges; and testing under a variety of adverse conditions including high temperature, pressure, humidity, chemical spray, and simulated aging (see Section 4.3).

General Design Criterion No. 53 stipulated general testing requirements for measuring containment leakage through both individual penetrations and the entire containment structure prior to start-up. Appendix J to 10 CFR 50 added specific testing requirements in 1973.[35] It expanded the tests to cover a range of pressure differentials approximating potential accident conditions and upgraded them to employ advanced leak-detection methods and to include detailed reporting of test conditions.

Containment Systems: Systems for removing heat, radioactive gases, and combustible gases from the containment atmosphere following accidents changed little in basic concept but underwent extensive refinement during the 1970s. Nuclear plants contain multi-purpose *containment spray systems.* In the event of a LOCA they would condense steam, reducing pressure and thereby reducing leakage; provide long-term cooling of the containment and its contents; and precipitate and "scrub" radioactive iodine particles from the containment atmosphere. Later plants have higher-capacity spray systems with an increased number of spray ring headers around the containment dome to ensure coverage of the containment volume, and they are also designed for automatic initiation on indication of high containment pressure. RG 1.82 also mandated improvements in the sumps which collect the spray and spilled coolant at the bottom of PWR containments for recirculation: redundant, separated sumps; screens and racks to prevent clogging and to protect sumps from missiles; and design improvements to prevent vortexing.

Reactors also contain a post-accident *air-cleaning system* to remove accident-generated fission products from the containment atmosphere. It consists of a multiplicity of filters, adsorbers, fans, cooling coils, etc., which, in addition to removing airborne radioactivity, also cools the atmosphere and thereby supplements the containment spray system. To ensure that the system can function under post-accident conditions, RG 1.52 has required that at later plants it include redundant, physically separated equipment trains, be protected from pressure surges by relief valves, and be built to more stringent seismic and quality assurance standards.

A LOCA can cause hydrogen gas to form and accumulate within the containment through chemical reactions between overheated fuel cladding and the reactor coolant or by releasing radiolytic hydrogen generated by the decomposition of the cooling water. RG 1.7, issued in 1971 and twice revised, required the capability to measure and reduce hydrogen concentrations ranging up to 5% of the maximum potential from the cladding-coolant reaction. The installed equipment generally includes *recombiners* consisting of a reaction chamber, fans, and valves to collect and gradually heat the hydrogen to react (oxidize) it harmlessly to form water. Because the cladding-coolant reaction reached at least a 25% level at Three Mile Island, future and even existing plants may be required to increase their recombiner capability.

Section 4.9: Systems For Electric Power, Instrumentation, And Control

Reactor functions and mechanical systems in nuclear plants are woven together and regulated by networks of sensors, controls, actuators, circuits, and power sources which monitor plant conditions, transmit operator commands, and activate equipment. This equipment comprises two systems: the *electric power system* and the *instrumentation and control system*. Although the two systems function separately, they consist of equipment that is somewhat similar (*e.g.*, cables, relays, electrical penetrations) or is manufactured from very similar materials. As a consequence, both systems have been affected similarly by increased design and quality requirements; hence they are discussed together here.

These requirements fall into three broad areas: (i) increased system requirements, such as the number of plant variables to be monitored or the power requirements of safety equipment; (ii) increased redundancy and separation to ensure that electrical and control systems function despite equipment failure; and (iii) upgrading of standards and tests for individual components and equipment items to provide increased confidence that they will perform their needed functions. Major changes added significantly to costs in each area. The third area, standards and tests, was discussed in Section 4.3; this section treats increased system requirements and redundancy and separation criteria.

Many of the changes were stipulated in standards (detailed rules of practice) issued by the Institute of Electrical and Electronics Engineers (IEEE). Over two dozen IEEE standards pertaining to nuclear plant electrical systems and equipment were issued in the 1970s and subsequently adopted by AEC/NRC (with modest amendments) as regulatory guides. Due to the complexity of the standards and because many of them were applied as drafts to reactor design and licensing reviews, it is not possible here to identify which generation of plants was (or will be) the first affected by each provision of each standard. The discussion of IEEE standards is therefore offered as a general account of the major changes in electrical equipment and instrumentation and control systems.

Equipment Redundancy and Separation: Electrical equipment and systems that are essential to shut down the reactor and to operate engineered safeguards such as the ECCS are designated as *Class 1E* equipment. The General Design Criteria stipulate that Class 1E equipment must satisfy the "single failure criterion," *i.e.*, the failure of any single "active" component or piece of equipment — the failure of a motor or sensor, for example — must not prevent accomplishment of any vital safety "functions." The primary means of complying with this criterion — providing redundant, mutually separated and independently operated "divisions" of Class 1E equipment — was applied with increased stringency during the 1970s.

The major statement of criteria for *power supplies* for electrical equipment is IEEE Std 308. It establishes design criteria to ensure that power is available for Class 1E instrumentation and control systems and for electrical components of engineered safeguards. It also establishes criteria for the normal and emergency power systems themselves, which are needed in the event of failure in the circuits connecting the plant to the utility's transmission network. They include: (i) batteries to supply d/c power for vital instrumentation and controls; (ii) a back-up source of a/c power — on-site standby diesel generators — to supply power to motors driving pumps, fans, and valve actuators in systems needed to mitigate the consequences of accidents (*e.g.*, ECCS, hydrogen controls, and containment cooling) and bring the plant to a safe shutdown condition; and (iii) power distribution systems to connect these power sources to the equipment that requires it.

IEEE Std 308 essentially stipulates that all electrical equipment in the vital systems named above must be provided in two or more independent, redundant divisions. For example, there must be two or more separate feeder lines from the plant to the transmission network, and two or more diesel generators and battery supplies, each connected separately to independent sets of safeguard equipment which are individually capable of shutting down and cooling the reactor. Vital instruments must also have two or more independent power supplies and two or more separate circuits to transmit data to the control room.

Although some of these requirements were specified prior to issuance of IEEE Std 308 in 1970 (in draft editions of the General Design Criteria), that and later editions have applied them more firmly and to a greater range of equipment. For example, the standard required that auxiliary devices (providing cooling, lubrication, etc.) needed to operate electrical equipment receive their power supply from the same bus section so that loss of power in one load group would not disable the other load group. The 1978 edition of the standard required physical separation of redundant circuits to the off-site grid (not required in the General Design Criteria). That edition also barred automatic transfer of electric loads between redundant power supply buses and added electrical equipment in engineered safeguards systems to the group requiring separation. These requirements have effected a significant increase in the amount of electrical equipment required to be installed in nuclear plants, and the items added must satisfy stringent performance requirements and quality standards.

IEEE Std 308 was endorsed in several regulatory guides, particularly RG 1.6 and RG 1.32, which were issued in the early 1970s and appear to have influenced revisions of the standard. RG 1.41 required pre-operational testing to verify that redundant load groups (*e.g.*, engineered safeguard systems) were assigned to independent power sources. RG 1.81 prohibits sharing of a/c and d/c power sources and distribution systems at multi-unit plants, such as providing only one "swing diesel" to simultaneously back up both units' diesel generators. This requirement stems from concern that a short circuit could keep the swing diesel tied to one plant and unavailable for the other. It is primarily aimed at future plants, where it will add a diesel generator and battery capability to the emergency power supply.

Requirements governing separation of electric power sources, electric circuits, and other Class 1E electric equipment were also significantly upgraded in the 1970s. IEEE Std 279 first stated general separation criteria in 1968. Specific guidelines were subsequently delineated in IEEE Std 384 and endorsed in Regulatory Guide 1.75. IEEE Std 384 extended separation criteria to major new classes of electrical equipment: redundant Class 1E *instruments* must be located in separate cabinets or compartments; *electric penetrations* conveying redundant Class 1E circuits through the containment must be widely dispersed around its circumference; *auxiliary systems* essential to functioning of Class 1E equipment, such as ventilation systems for switchgear, cooling water systems for motors, and fuel oil supply systems for emergency diesel generators, were required to be provided in redundant, physically separated pairs. Std 384 also barred power cables from the cable spreading area, where instrumentation and control cables converge prior to entering the control room, unless they are contained in embedded conduits or similar enclosed structures.

The minimum separation distances specified in IEEE Std 384 frequently exceeded past practice. In addition, the standard required greater protection of Class 1E equipment from external hazards. Although fires, missiles, vibration, pipe whip. water sprays, and high-energy electrical switchgear were all mentioned as important design considerations in Std 279, they received much greater emphasis in Std 384. The standard also mandated protection of Class 1E equipment from possible high radiation, pressure, temperature, and humidity due to failure of operation of plant mechanical systems.

Finally, IEEE Std 384 supplemented physical separation with *electrical isolation* to maintain independence of redundant circuits and equipment. Class 1E power circuits must be protected from their redundant counterparts and from non-Class 1E circuits by circuit breakers, input current limiters, and other isolation devices to ensure that they are not degraded by current transients or other high voltage sources. Similarly, instrumentation and control circuits must be protected from electrical interference (*e.g.*, electromagnetic induction) by filtering circuits, grounding, or shielding or by isolation devices such as amplifiers, fuses, or transducers.

The increased separation requirements substantially increased the length of cables needed for circuitry and the volume of structures required to support, shield, and separate the increased quantity of equipment. The increases have been especially marked for instrumentation and control circuits because of the expanding number of variables requiring monitoring. More engineering effort has also been expended to analyze the vulnerability of greater amounts of electrical equipment to accidents and environmental hazards.

Emergency Power Sources: The standby a/c power system, powered by large diesel generators, provides the energy to drive safeguard systems in the event of an accident and simultaneous loss of off-site power. The standby d/c power system, powered by lead-acid batteries, provides energy for plant instrumentation and controls, including the starting controls of the diesels. Both systems are considered vital to safety, and so are subsumed in the preceding discussions of equipment redundancy, separation, and qualification (in Section 4.3) for Class 1E equipment. This section treats changes in design and equipment requirements specific to the diesels and the batteries.

Diesel and battery power capacities grew relative to plant generating capacity in the 1970s. This was due to growth in emergency electric loads (*e.g.*, ECCS pumps) and to performance problems demonstrating a need for greater standby power design margins. IEEE Std 387 and AEC Regulatory Guide 1.9, both issued in the early 1970s, reduced permissible frequency and voltage fluctuations by diesel generators during starting of emergency loads. Together with the increased load sizes, this doubled average diesel generator capacities from 2-2.5 MW in the early 1970s to 4.5-5 MW for recently completed plants. Battery capacities also increased because of the expanding number of vital instrumentation and control functions. RG 1.32 required battery chargers to be sized according to the largest possible total loads and thus added to their capacity requirements.

Improvements in quality assurance, installation-design, and auxiliary systems also added to diesel and battery equipment and installation costs. Recent revisions of IEEE Std 387 and RG 1.9 required elaborate instrumentation of systems for start-up, lubrication, fuel supply, and cooling as well as of the generator itself. The diesels must also be designed with controls, bypasses, and instrumentation to allow testing while the plant is operating. IEEE Std 484 and RG 1.128 specified criteria for battery location, mounting, ventilation, and instrumentation to provide protection from adverse conditions such as fire and pipe whipping. Temperature differentials between battery cells must be minimized, and the batteries must be mounted with embedded anchor bolts or racks welded to structural steel plates for seismic restraint.

Instrumentation and Control Systems: Monitoring and control devices necessary for safe reactor shutdown are also subject to the Class 1E equipment requirements discussed earlier. This section treats other requirements specific to instrumentation and control functions: in-service testing, bypassed status indication, control rooms and panels, and instrumentation to follow accidents.

Much instrumentation and control equipment is intended to operate only in emergencies. Since such equipment ordinarily does not generate enough performance data to measure its reliability, it has been subjected to expanded in-service testing. IEEE Stds 279 and 338 and RGs 1.22 and 1.118 require periodic testing of vital instrumentation and control equipment, including testing electrical channels independently to detect loss of redundancy.

To determine equipment functionality without shutting the plant, many utilities have installed additional signal-generating devices, relays, and circuitry allowing testing of component reliability without activating the system functions themselves. This has increased the number of channels in the logic systems which activate plant equipment. Conventional "two out of three" logic systems were activated if a majority of the three channels registered a positive signal; these have increasingly been supplanted by "two out of five" systems which can activate equipment despite failure of one channel while another is down for repair and a third is out for testing.

IEEE Std 279 mandated that the control room be equipped with indicators to signal the unavailability — due to test or maintenance — of electrical or mechanical equipment needed for safe shutdown. Operating incidents in which *tagging* of bypassed valves or switches failed to make operators aware of equipment unavailability led the AEC to stipulate in RG 1.47 that *automatic* indication of the bypassing or inoperability of safety-related equipment be provided in the control room.[i] This has required additional signal-transmitting circuitry from each safety-related component or equipment item to the control room — a major source of the additional conduits into the control room that have required buffering and spacing to satisfy physical and electrical isolation criteria.

RGs 1.78 and 1.95 require that control rooms at newer plants be kept habitable and functional despite spillage of hazardous chemicals. New plants have installed redundant, physically separated instrumentation divisions to detect chlorine (used in circulating water systems and other auxiliary systems) and other chemicals. Equipment to isolate the control room from its vicinity, to filter outside air, and to provide a breathable air supply has also been upgraded

i. In the Three Mile Island accident, discovery of blocked valves closing off emergency feedwater lines was delayed because maintenance tags on the console (the old method of bypassed indication) covered the indicator lights (the new method) — a truly mischievous common-mode failure.

to meet the single-failure criterion.

The General Design Criteria require that reactors be designed so that operators can shut down the plant from outside the control room. It had been standard practice to provide "control panels" beside major equipment, but fires at Indian Point 1 in 1972 (initiated by Unit 2 construction) and at Browns Ferry in 1975 (in which smoke hindered control room action) pointed up the need for alternate *central* control areas. Accordingly, most new plants are being built with emergency shutdown rooms in auxiliary buildings. These have separate ventilation systems for habitability during accidents and are supplied with independent control cables routed outside the primary cable spreading room. Prospective changes in control room instrumentation are discussed in Section 6.2.

Reactor instrumentation designed to monitor plant conditions during accidents grew in scope in the 1970s and will almost certainly expand markedly in the future. Its purpose is to help operators determine the nature of an accident, the response of plant safety features to automatic and manual commands, and the response of the plant to the safety measures. Systems requiring monitoring include the reactor core, the reactor coolant system, containment systems, secondary systems such as steam generators, auxiliary systems, and power supplies and distribution. Variables measured include temperatures, pressures, flow rates, water levels, valve positions, gas concentrations, and radioactivity.

RG 1.97, issued in 1975 and significantly toughened two years later, increased instrumentation requirements by specifying the plant capabilities requiring monitoring and adding design criteria such as physical separation of redundant channels and seismic protection. It applied mostly to future construction starts, however, and largely permitted the licensee to determine the specific instrumentation to be installed. A pending revision of RG 1.97[36] identifies each of the hundreds of instruments required and their measurement ranges. Many of its provisions are likely to be applied to plants under construction and, perhaps, to operating plants, as discussed in Section 6.2.

The instrumentation systems are extensive and costly, especially as provided by the proposed revision. The instruments and their circuitry to the control room must be capable of operating under extreme conditions for the anticipated duration of postulated accidents. Containment-pressure instruments at future plants, for example, must be able to register at least three times the design pressure for concrete and four times that for steel. Many instruments must also include recording capability for diagnoses during and after accidents.

Section 4.10:   Radiation Control Systems

Radionuclides produced by nuclear plants can affect public health in three broad ways:

- through continuous release of small quantities to the environment during routine operation;
- through large releases during possible reactor accidents; and
- through exposure of plant workers to radiation during maintenance, repair, and normal operation.

Equipment for controlling radiation releases from accidents was discussed in Section 4.8. This section considers radiation waste ("radwaste") control for routine emissions, which was significantly upgraded in the 1970s, and design and equipment changes to reduce occupational exposures, which will be reflected primarily in future capital costs.

Releases During Routine Operation: In 1969, with only a handful of operating reactors but with dozens under construction and many hundreds projected, public concern over "routine" reactor emissions was aroused by charges by two senior AEC scientists that subjecting all Americans to legally permissible doses of radioactivity could cause up to 32,000 cancer deaths per year. Although actual radiation exposures from operating reactors were considerably less than the allowable doses, in June 1971 the AEC proposed to reduce the permissible release levels approximately a hundred-fold at both operating and new reactors. The proposal was formally adopted as a regulation in 1975 (Appendix I to 10 CFR 50), consistent with the basic objective that "exposures should be as low as practicable." It required virtually all licensees to augment their waste-treatment systems and procedures. Although the improvements did not need to achieve a full hundred-fold reduction, radwaste treatment systems on both old and new plants are now considerably more elaborate than those originally installed on early plants.

Reactors produce radionuclides directly through nuclear fission and indirectly through neutron bombardment of impurities dissolved in reactor primary coolant. The primary radwaste source is the gaseous waste stream, consisting of gaseous radionuclides and particles such as iodine which are carried through ventilation systems. It is processed by several types of equipment: "hold-up" tanks detain gases with short half-lives until they have lost most of their radioactivity, and impregnated activated carbon adsorbers and "high efficiency particulate air" (HEPA) filters remove iodine and other radioactive particles from the gas stream. Associated with the tanks and filters are heating and cooling coils to regulate humidity before the gas stream reaches the filters and adsorbers, fans and ductwork to route the gas stream through the plant, and instrumentation to monitor radionuclide concentrations. Finally, "recombiners" mix potentially explosive hydrogen gas (generated through radiolytic decomposition of primary coolant water) with oxygen in a controlled catalytic reaction to yield water again and reduce the volume of the contaminated gaseous waste stream. This permits a relatively long "hold-up" of gases in the plant to allow radioactive decay to reduce the radioactivity levels.

All of these systems were upgraded at operating and new reactors in the 1970s, both to comply with the Appendix I criteria and to satisfy RG 1.140, issued in 1978 but employed as an NRC "branch technical position" in 1975. BWRs produce more radioactive gases than PWRs and have been more affected. In addition to improvements provided by recombiners, hold-up tanks and "delay lines" (large buried ducts) were enlarged to expand their storage capacities from as little as half-an-hour in some early plants to several weeks or months. The efficiency and reliability of HEPA filters have also been improved, in part by increasing the number or filters to reduce the load on each.

Liquid radwaste control systems have also been augmented. Licensees have strengthened and expanded demineralizers — systems consisting of ion-exchange resins which chemically remove solids dissolved in liquid waste as well as remove particulate solids by filtration. "Full-flow" condensate demineralizer systems have been added at many new PWRs to protect the steam generators and relieve loading on the radwaste demineralizers from the occasional flushing of corrosive steam generator impurities. This has required additional pressure vessels, piping, instrumentation, and wiring. New plants have also added level monitors, floor drains, and extra piping to reduce leakage and overflow from pipes and tanks containing radioactive liquids.

In addition, RG 1.143, also adopted in 1978 but employed as a branch technical position since 1975, applied more stringent design criteria to gaseous and liquid radwaste systems for seismic protection, pump and valve reliability, and integrity of waste-containing pressure valves and piping. RG 1.21, adopted in 1971 and later revised, has increased the number of radwaste samplers, the frequency of their use, and their capability for measuring smaller quantities of a greater variety of radionuclides. Regulatory Guides 1.21 and 1.112 have also required that licensees employ more elaborate calculational models to quantify radiation releases and translate them to estimated doses received by humans. This has added to the scope of programs to trace the pathways of radionuclides through the biosphere.

Solid radwaste systems at nuclear plants consist of centrifuges, settling tanks, and waste collection and storage tanks to solidify liquid wastes for shipment to offsite burial grounds. These systems underwent few changes in the 1970s. For example, they were exempted from the seismic criteria for liquid and gaseous radwaste systems in RG 1.143. Most current solidification systems leave small amounts of free-standing liquids, however, which render the waste containers more susceptible to corrosion and leakage. Costly cement solidification systems may be required in the future for complete solidification, as discussed in Section 5.1.

Occupational Exposure to Radiation: Utilities have begun making design and equipment changes to reduce levels of worker exposure to radiation, which at most reactors have exceeded early expectations. Total occupational exposures initially averaged approximately 300 person-rems per reactor-year and have generally increased with reactor age to a current average of around 500 person-rems.[37]

Exposures have increased in spite of Regulatory Guide 8.8, which the AEC issued in 1973 because of concern with worker exposure "in view of the anticipated growth of nuclear power stations over the next few decades." The guide did not alter the allowable exposure levels set in 1960 — three rems per quarter-year, five per year.[j] Rather, it instructed licensees to design, construct, and operate reactors so that occupational exposures would be "as low as is reasonably achievable" (ALARA). This meant, in part, that utilities should not satisfy per-worker standards simply by hiring more workers to share the total dose.

Utilities themselves have increasingly identified a self-interest in reducing radiation fields and exposures, primarily to avoid using up large numbers of workers' exposure allowances in brief stints in high-radiation areas — a phenomenon that might eventually lead to crippling increases in shutdown times to maintain and repair aging reactors.[k] Reducing exposures requires improvements to plant designs, equipment, monitoring, and management control. Most design changes have been reserved for reactors under construction, although some recently installed reactors have made equipment fixes. Most changes instituted at operating plants have been less effective administrative changes.

Exposure-reducing equipment includes fixed monitors to measure radiation fields, personnel monitors to measure exposures, and sampling rooms to measure radiation concentrations in process equipment. Radiation absorbed in servicing radwaste systems can be reduced by employing backflushable filters and remote means of changing charcoal adsorbers to reduce radiation concentrations.

Design features for reducing worker exposure focus on reducing radiation fields and speeding maintenance and repair of the reactor coolant system, especially BWR primary system piping and components and PWR steam generators — the greatest sources of worker exposure. This includes shielded cubicles for high-radiation equipment, providing space for deploying portable local shielding, and designing frequently inspected or serviced equipment for rapid removal and reassembly. Deposition of radioactive airborne particles

j. NRC regulations permit 12 rems per year for workers whose lifetime occupational dose does not exceed five times the difference between their current age and 18. The ALARA requirement, however, serves to make the five-rem annual limit the practical upper bound.

k. Welding of Indian Point 1 primary system piping in 1970, for example, when the reactor was eight years old, required over eight months instead of the several weeks needed for comparable but non-radioactive work in a fossil plant. The high radiation field used up Con Edison's 60 expert welders and required 50 health physicists and 600 additional personnel, complicating outage logistics and raising personnel costs.[38]

and gases can also be reduced by careful design of ventilation systems (*e.g.*, to avoid pressure gradients from high-radiation areas to frequently visited areas). Primary coolant contamination may be diminished by minimizing use of materials in primary piping that become highly radioactive under neutron bombardment.

These measures are limited to differing degrees by design trade-offs, however. Physical separation and greater space add to length and penetration requirements for piping and cables. Ventilation systems are governed by other design criteria such as fire-spread prevention and post-accident cleanup. Radioactivity-engendering materials such as cobalt and nickel are valued for their high corrosion resistance and ductility. Most design changes to reduce worker radiation must be analyzed on a whole-systems basis, adding to the amount of engineering analysis. Similarly, frequent decontamination to reduce radiation fields can add to reactor downtime.

Measures to reduce occuptional exposure have varied among plants. Although no definitive analysis of occupational exposure data trends is available, new plants appear to show little, if any, reduction in total dose per unit of electric output. This suggests that relatively few exposure-reducing design changes have been made to date.[1] Greater expenditures will be made at future plants to effect the equipment improvements and design changes described above.

References

1. Atomic Industrial Forum, "Licensing, Design and Construction Problems: Priorities for Solution" (Washington, D.C., January 1978), p. 8.

2. Gordon MacKerron, "Capital Costs of Light Water Reactors: The USA" (Science Policy Research Unit, University of Sussex, U.K., October 1979), p. 33 (emphasis added).

3. Reference 1, p. 8.

4. R.R. Bennett and D.J. Kettler, "Dramatic Changes in the Costs of Nuclear and Fossil-Fueled Plants," Ebasco Services (New York, September 1978), p. 6.

5. Reference 1, Exhibits 1 and 9; and Reference 2, p. 28.

6. Reference 1, p. 1.

7. John H. Crowley, Manager of Advanced Engineering, United Engineers and Constructors, Inc., "Trends Influencing the Continuing Rise in Costs for Nuclear and Coal-Fired Electric Generating Stations," presented to the Washington State Senate Energy Committee, September

1. For example, Carolina Power & Light (CP&L) has reportedly upgraded radwaste systems and increased shielding to reduce occupational exposure at its Harris plant under construction, but exposure data suggest little backfitting at CP&L's existing reactors. Total worker exposure at its veteran (1971) Robinson 2 PWR was third highest among PWRs during 1973-78 and has been rising, while exposure at CP&L's Brunswick BWRs (1975, 1977) has been no lower than the BWR average.[39]

1980, Figures 13 and 14. The estimates of labor and engineering hours therein from the "WASH-1230" and "EEDB" reports correspond approximately to early and late seventies plant completions, respectively (telecom, 16 January 1981). Applying 1979 wage rates of about $13/hr and $25-30/hr, respectively (same telecom), the increased labor and engineering costs were approximately $90/kW and $60/kW, respectively, in 1979 dollars.

8. Donald A. Brand, Vice President for General Construction, Pacific Gas & Electric Company, testimony before the California Public Utilities Commission, Application Numbers 58911 and 58912, 6 June 1979, p. 29.

9. Christopher Bassett, "The High Cost of Nuclear Power Plants," *Public Utilities Fortnightly*, 27 April 1978.

10. Reference 7, p. 22.

11. Reference 8, pp. 23-24.

12. Reference 1, p. 7.

13. This definition was first presented by AEC Commissioner J.T. Ramey in an address to the November 1966 winter meeting of the American Nuclear Society in Pittsburgh. It is contained in all AEC/NRC regulations and American National Standards Institute standards related to QA.

14. 10 CFR 50, Appendix A, General Design Criterion No. 1.

15. Sidney A. Bernsen, *Nuclear Safety*, *16* (No. 2), 127 (1975).

16. ANSI Standards N101.2-1972, "Protective Coatings (Paints) for Light Water Nuclear Reactor Containment Facilities," and N101.4-1972, "Quality Assurance for Protective Coatings Applied to Nuclear Facilities."

17. 10 CFR 50, Appendix B, Section III, "Design Control."

18. Reference 15, p. 135.

19. Reference 15, p. 133.

20. Reference 15, p. 135.

21. Reference 8, pp. 17-18.

22. See, for example, G.L. Bennett, *Summary of NRC LWR Safety Research Programs on Fuel Behavior, Metallurgy/Materials and Operational Safety*, NRC, NUREG-0581 (1979), p. 44.

23. M.D. Sulouff, "Equipment Qualification," WPPSS-0026-SA (June 1979).

24. *EPRI Journal*, July/August 1980, p. 48.

25. NUREG-0554, *Single-Failure-Proof Cranes for Nuclear Power Plants* (May 1979), replacing RG 1.104, "Overhead Crane Handling Systems for Nuclear Power Plants," issued for comment in February 1976.

26. 10 CFR 50, Appendix A, General Design Criterion No. 2.

27. This description paraphrases *Inside N.R.C.*, 14 January 1980, p. 16.

28. 10 CFR 100, Part A.

29. See Regulatory Guide 1.120, "Fire Protection Guidelines for Nuclear Power Plants," p. 1.

30. An excellent brief account is David Comey's "The Incident at Browns Ferry" in *Not Man Apart*, 5 (No. 18) (mid-September, 1975). The major NRC post-mortem is NUREG-0050,

Recommendations Related to Browns Ferry Fire (1976). See also Union of Concerned Scientists, *Browns Ferry: The Regulatory Failure* (1976).

31. *Federal Register, 45* (No. 225), 76602-76615 (19 November 1980). This contains Appendix R to 10 CFR 50, "Fire Protection Program for Nuclear Power Facilities Operating Prior to January 1, 1979."

32. The new criteria were published in 1973 as Appendix K to 10 CFR 50. They replaced interim ECCS rules issued in 1971.

33. C.K.B. Lee and C.K. Chan, *Steam Chugging in Pressure Suppression Containment,* NUREG/CR-1562 (July 1980).

34. See I. Goozman and C.M. Jan, *Power Engineering, 83* (No. 11), 64-67 (1979).

35. 10 CFR 50, Appendix J, "Primary Reactor Containment Leakage Testing for Water-Cooled Power Reactors."

36. Revision 2 of RG 1.97 endorses, with exceptions, Draft 4 of ANS-4.5, "Functional Requirements for Accident Monitoring in a Nuclear Power Generating Station," November 1979.

37. Robert O. Pohl, "Radiation Exposure in LWRs Higher Than Expected," *Nuclear Engineering International, 24* (No. 2), 36-38 (1979). See also NUREG-0594, *Occupational Radiation Exposure at Commercial Nuclear Power Reactors, 1978* (1979).

38. C. Komanoff, *Power Plant Performance* (Council on Economic Priorities, 1976), p. 55.

39. NUREG-0594 (Reference 37), Tables 4 through 6.

Regulatory Guides And Industry Standards

Section 4.3

RG 1.89, "Qualification of Class 1E Equipment for Nuclear Power Plants," November 1974.

IEEE Std 323, "Qualifying Class 1E Equipment for Nuclear Power Generating Stations," 1971, revised 1974.

IEEE Std 334, "Type Tests of Continuous Duty Class 1E Motors for Nuclear Power Generating Stations," 1971, revised 1974.

IEEE Std 381, "Type Tests of Class 1E Modules Used in Nuclear Power Generating Stations," 1977.

IEEE Std 382, "Type Test of Class 1 Electric Valve Operators for Nuclear Power Generating Stations," 1972.

IEEE Std 383, "Class 1E Electric Cables, Field Splices, and Connections for Nuclear Power Generating Stations," 1974.

Section 4.4

RG 1.14, "Reactor Coolant Pump Flywheel Integrity," October 1971, revised August 1975.

RG 1.46, "Protection Against Pipe Whip Inside Containment," May 1973.

RG 1.115, "Protection Against Low-Trajectory Turbine Missiles," March 1976, revised July 1977.

ANSI N176, "Design Basis for Protection of Nuclear Power Plants Against Effects of Postulated Pipe Rupture" (draft), June 1974.

ANSI N177, "Plant Design Against Missiles" (draft), April 1974.

Section 4.5

RG 1.12, "Instrumentation for Earthquakes," March 1971, revised April 1974.

RG 1.29, "Seismic Design Classification," June 1972, revised August 1973, February 1976, September 1978.

RG 1.48, "Design Limits and Loading Combinations for Seismic Category I Fluid System Components," May 1973.

RG 1.57, "Design Limits and Loading Combinations for Metal Primary Reactor Containment System Components," June 1973.

RG 1.59, "Design Basis Floods for Nuclear Power Plants," August 1973, revised April 1976, August 1977.

RG 1.60, "Design Response Spectra for Seismic Design of Nuclear Power Plants," October 1973, revised December 1973.

RG 1.61, "Damping Values for Seismic Design of Nuclear Power Plants," October 1973.

RG 1.76, "Design Basis Tornado for Nuclear Power Plants," April 1974.

RG 1.92, "Combining Modal Responses and Spatial Components in Seismic Response Analysis," December 1974, revised February 1976.

RG 1.100, "Seismic Qualification of Electric Equipment for Nuclear Power Plants," March 1976, revised August 1977.

RG 1.102, "Flood Protection for Nuclear Power Plants," October 1975, revised September 1976.

RG 1.117, "Tornado Design Classification," June 1976, revised March 1978.

RG 1.122, "Development of Floor Design Response Spectra for Seismic Design of Floor-Supported Equipment or Components," September 1976, revised February 1978.

RG 1.124, "Service Limits and Loading Combinations for Class 1 Linear-Type Component Supports," November 1976, revised January 1978.

RG 1.130, "Service Limits and Loading Combinations for Class 1 Plate-and-Shell-Type Component Supports," July 1977, revised October 1978.

IEEE Std 344, "Seismic Qualification of Class 1E Equipment for Nuclear Power Generating Stations," 1971, revised 1975.

Section 4.6

RG 1.75, "Physical Independence of Electric Systems," February 1974, revised January 1975, September 1978.

RG 1.120, "Fire Protection Guidelines for Nuclear Power Plants," June 1976, revised November 1977.

Standard Review Plan, Branch Technical Position APSCB 9.5-1, "Fire Protection for Nuclear Power Plants," and Appendix A.

Section 4.7

RG 1.20, "Comprehensive Vibration Assessment Program for Reactor Internals During Preoperational and Initial Startup Testing," December 1971, revised June 1975, May 1976.

RG 1.26, "Quality Group Classifications and Standards for Water-, Steam-, and Radioactive-Waste-Containing Components of Nuclear Power Plants," March 1972, revised December 1974, June 1975, February 1976.

RG 1.27, "Ultimate Heat Sink for Nuclear Power Plants," March 1972, revised March 1974, January 1976.

RG 1.31, "Control of Ferrite Content in Stainless Steel Weld Metal," August 1972, revised June 1973, May 1977, April 1978.

RG 1.43, "Control of Stainless Steel Weld Cladding of Low-Alloy Steel Components," May 1973.

RG 1.44, "Control of the Use of Sensitized Stainless Steel," May 1973.

RG 1.45, "Reactor Coolant Pressure Boundary Leakage Detection Systems," May 1973.

RG 1.50, "Control of Preheat Temperature for Welding of Low-Alloy Steel," May 1973.

RG 1.56, "Maintenance of Water Purity in Boiling Water Reactors," June 1973, revised July 1978.

RG 1.65, "Materials and Inspections for Reactor Vessel Closure Studs," October 1973.

RG 1.67, "Installation of Overpressure Protection Devices," October 1973.

RG 1.79, "Preoperational Testing of Emergency Core Cooling Systems for Pressurized Water Reactors," June 1974, revised September 1975.

RG 1.82, "Sumps for Emergency Core Cooling and Containment Spray Systems," June 1974.

RG 1.83, "Inservice Inspection of Pressurized Water Reactor Steam Generator Tubes," June 1974, revised July 1975.

RG 1.96, "Design of Main Steam Isolation Valve Leakage Control Systems for Boiling Water Reactor Nuclear Power Plants," May 1975, revised January 1977.

RG 1.133, "Loose-Part Detection Program for the Primary System of Light-Water-Cooled Reactors," September 1977.

RG 1.139, "Guidance for Residual Heat Removal," May 1978.

Section 4.8

RG 1.7, "Control of Combustible Gas Concentrations in Containment Following a Loss-of-Coolant Accident," March 1971, revised September 1976, November 1978.

RG 1.10, "Mechanical (Cadweld) Splices in Reinforcing Bars of Category I Concrete Structures," March 1971, revised January 1973.

RG 1.11, "Instrument Lines Penetrating Primary Reactor Containment," March 1971, revised February 1972.

RG 1.15, "Testing of Reinforcing Bars for Category I Concrete Structures," October 1971, revised December 1972.

RG 1.18, "Structural Acceptance Test for Concrete Primary Reactor Containments," October 1971, revised August 1972.

RG 1.19, "Nondestructive Examination of Primary Containment Liner Welds," December 1971, revised August 1972.

RG 1.35, "Inservice Inspection of Ungrouted Tendons in Prestressed Concrete Containment Structures," February 1973, revised June 1974, January 1976.

RG 1.52, "Design, Testing, and Maintenance Criteria for Post Accident Engineered-Safety-Feature Atmosphere Cleanup System Air Filtration and Absorption Units of Light-Water-Cooled Nuclear Power Plants," June 1973, July 1976, March 1978.

RG 1.55, "Concrete Placement in Category I Structures," June 1973.

RG 1.57, "Design Limits and Loading Combinations for Metal Primary Reactor Containment System Components," June 1973.

RG 1.63, "Electric Penetration Assemblies in Containment Structures for Light-Water-Cooled Nuclear Power Plants," October 1973, revised May 1977, July 1978.

RG 1.82, "Sumps for Emergency Core Cooling and Containment Spray Systems," June 1974.

RG 1.94, "Quality Assurance Requirements for Installation, Inspection, and Testing of Structural Concrete and Structural Steel During the Construction Phase of Nuclear Power Plants," April 1975, revised April 1976.

RG 1.136, "Material for Concrete Containments," November 1977, revised October 1978.

RG 1.143, "Design Guidance for Radioactive Waste Management Systems, Structures, and Components Installed in Light-Water-Cooled Nuclear Power Plants," July 1978.

ANSI N45.2.5, "Supplementary Quality Assurance Requirements for Installation, Inspection, and Testing of Structural Concrete and Structural Steel During the Construction Phase of Nuclear Power Plants," 1974.

ANSI N271/ANS-56.2, "Containment Isolation Provisions for Fluid Systems," 1976.

IEEE Std 317, "Electric Penetration Assemblies in Containment Structures for Nuclear Power Generating Stations," 1971, revised 1972, 1976.

Section 4.9

RG 1.6, "Independence Between Redundant Standby (Onsite) Power Sources and Between Their Distribution Systems," March 1971.

RG 1.9, "Selection, Design, and Qualification of Diesel-Generator Units Used as Onsite Electric Power Systems at Nuclear Power Plants," March 1971, revised November 1978.

RG 1.22, "Periodic Testing of Protection System Actuation Functions," February 1972.

RG 1.32, "Criteria for Safety-Related Electric Power Systems for Nuclear Power Plants," August 1972, revised June 1973, May 1977, April 1978.

RG 1.41, "Preoperational Testing of Redundant On-Site Electric Power Systems to Verify Proper Load Group Assignments," March 1973.

RG 1.47, "Bypassed and Inoperable Status Indication for Nuclear Power Plant Safety Systems," May 1973.

RG 1.75, "Physical Independence of Electric Systems," February 1974, revised January 1975, September 1978.

RG 1.78, "Assumptions for Evaluating the Habitability of a Nuclear Power Plant Control Room During a Postulated Hazardous Chemical Release," June 1974.

RG 1.81, "Shared Emergency and Shutdown Electric Systems for Multi-Unit Nuclear Power Plants," June 1974, revised January 1975.

RG 1.95, "Protection of Nuclear Power Plant Control Room Operators Against an Accidental Chlorine Release," February 1975, revised January 1977.

RG 1.97, "Instrumentation for Light-Water-Cooled Nuclear Power Plants to Assess Plant Conditions During and Following an Accident," December 1975, revised August 1977.

RG 1.118, "Periodic Testing of Electric Power and Protection Systems," June 1976, revised November 1977 and June 1978.

RG 1.128, "Installation-Design and Installation of Large Lead Storage Batteries for Nuclear Power Plants," April 1977, revised February 1978.

IEEE Std 279, "Protection Systems for Nuclear Power Generating Stations," 1968, revised 1971.

IEEE Std 308, "Class 1E Power Systems for Nuclear Power Generating Stations," 1970, revised 1971, 1974, 1978.

IEEE Std 338, "Periodic Testing of Nuclear Power Generating Station Class 1E Power and Protection Systems," 1971, revised 1975, 1977.

IEEE Std 384, "Independence of Class 1E Equipment and Criteria," 1974, revised 1977.

IEEE Std 387, "Diesel-Generator Units Applied as Standby Power Supplies for Nuclear Power Generating Stations," 1972, revised 1977.

IEEE Std 484, "Installation-Design and Installation of Large Lead Storage Batteries for Generating Stations and Substations," 1975.

Section 4.10

RG 1.21, "Measuring, Evaluating, and Reporting Radioactivity in Solid Wastes and Releases of Radioactive Materials in Liquid and Gaseous Effluents from Light-Water-Cooled Nuclear Power Plants," December 1971, revised June 1974.

RG 1.112, "Calculation of Releases of Radioactive Materials in Gaseous and Liquid Effluents from Light-Water-Cooled Power Reactors," April 1976, revised May 1977.

RG 1.140, "Design, Testing, and Maintenance Criteria for Normal Ventilation Exhaust System Air Filtration and Adsorption Units of Light-Water-Cooled Nuclear Power Plants," March 1978.

RG 1.143, "Design Guidance for Radioactive Waste Management Systems, Structures, and Components Installed in Light-Water-Cooled Nuclear Power Plants," July 1978.

RG 8.8, "Information Relevant to Ensuring that Occupational Radiation Exposures at Nuclear Power Stations Will Be As Low As Is Reasonably Achievable," July 1973, revised September 1975, March 1977, June 1978.

5
Future Nuclear Regulatory Requirements

Regulatory standards for nuclear power plants have grown increasingly stringent as a result of efforts to reduce the accident hazards per plant and to address new safety problems surfacing in operating experience and licensing reviews. These efforts have received considerable impetus from the expansion of the nuclear generating sector, as Chapter 3 demonstrates.

New nuclear plants are also subject to these efforts. They will be affected by more stringent design and construction requirements arising from new operating experience and from further pressure to reduce accident probabilities as the population of reactors increases. They will also be affected by requirements already promulgated or now being formulated that grew out of previous sector expansion but were not applied to completed plants due to "regulatory lag." In addition, design and construction of new plants will be greatly affected by the thorough re-appraisal of nuclear regulation arising from the Three Mile Island (TMI) accident. These increased requirements will cause the costs of plants now being licensed or built to significantly exceed the costs, in real terms, of reactors completed in the 1970s.

This chapter describes some of the new regulatory issues that will give rise to higher capital costs. The requirements that will add to future plant costs will be generated through the following processes:

1. implementation of requirements previously promulgated but not applied to completed plants;
2. new regulatory guides and revisions to current guides under development;
3. resolution of the "unresolved safety issues" identified by the Nuclear Regulatory Commission (NRC) and the Advisory Committee on Reactor Safeguards (ACRS);
4. resolution of other safety issues already identified in operating experience and licensing reviews;

5. new safety problems that will be identified through future operating experience and licensing reviews of new reactors;
6. correction of deficiencies revealed in the TMI accident.

Cost estimates are available for only a few of the anticipated new standards, and these estimates should not be viewed as firm since in the past it has proven extremely difficult to gauge the impacts of changes in nuclear design requirements. The standards appear to be vast in scope, and they are likely to have a major effect on costs, however. Even without the requirements stemming from TMI, implementation of recent standards and prospective new ones could easily cause the 55% real (inflation-adjusted) cost increase for 1978-88 projected from past cost relationships in Chapter 10.[a]

The effects of the TMI accident on nuclear standards are discussed in the next chapter. This chapter addresses the first five items listed above.

Section 5.1: Existing Requirements Subject To Regulatory Lag

New regulatory standards do not necessarily apply to operating reactors or those under construction. Although Chairman John Kemeny of the President's Commission on the Accident at TMI (the Kemeny Commission) considered it "shocking" that the plant was exempted from a 1975 rule requiring that reactors have redundant systems to seal off the containment building,[1] reactor regulations and standards are frequently "grandfathered." Only some new standards are imposed upon plants under construction (which industry refers to as *ratcheting,* as distinguished from *backfitting,* which pertains to changes in *operating* plants); many exempt plants being built and apply only to plants awaiting construction permits. In other cases, standards are applied on a case-by-case basis and pertain to some, but not all, plants under construction or in operation.

Because of the long time required to construct nuclear plants, many existing regulatory standards did not affect recently-completed reactors. Accordingly, the cost of a typical 1978 plant—$887/kilowatt (kW) in 1979 "steam-plant" dollars, as calculated in Chapter 10—does not reflect some requirements that have already been promulgated and will affect many plants

---

a. The statistical analysis in Chapters 8 and 10 of the costs of reactors completed during 1971-78 indicates that a 142% real increase in average reactor capital costs occurred together with the expansion of the nuclear sector from 10 GW to 55 GW during 1971-78. If past cost relationships continue and if installed nuclear capacity reaches 150 GW—the sum total of reactors currently operating and those licensed for construction—then the last gigawatt of capacity would cost 55% more to construct in real terms than a typical 1978 gigawatt. This projection does not reflect special cost impacts from Three Mile Island.

under construction. This section summarizes the major existing requirements that were not applied to recent plants. They are grouped according to the categories employed in reviewing 1970s regulatory standards in Chapter 4.[b]

Quality Assurance: Quality assurance (QA) requirements pertaining to the design, fabrication, installation, and testing of "safety-related" equipment in nuclear plants increased significantly in the 1970s and had a major effect on equipment costs and construction logistics, as Section 4.2 demonstrates. Their effect is virtually certain to be still greater for new plants, for several reasons.

First, QA standards, which for the most part are formulated by industrial engineering societies and "endorsed" in NRC regulatory guides, generally have not been applied to previously purchased equipment or to completed construction work. Some standards have been applied only to new construction starts. Regardless of whether this practice continues, new plants will be affected increasingly by existing QA standards which, due to regulatory lag, had little or no impact on even the most recently-completed plants in the data sample.

Second, NRC is likely to sharpen its efforts to gain compliance with QA requirements by licensees, their contractors, and their suppliers. Recent discoveries of major deficiencies in reactor design and construction practices, together with the TMI accident, have provoked widespread criticism of NRC's enforcement program and appear to have stimulated some changes already.

Many of these deficiencies concern seismic design and construction practices in operating plants. In March 1979, errors were discovered in an architect-engineering firm's calculations of potential earthquake stresses on coolant piping in five reactors. Recently, a dozen other plants were also found to contain piping systems and structures whose actual, "as-built" conditions differed from seismic design specifications.[2] The fact that these design and installation deficiencies occurred at many reactors and were not identified during construction indicates serious gaps in both constructors' QA programs and NRC's construction monitoring.

Other instances of inadequate implementation of QA programs have, by NRC's own admission, "appear[ed] in every facet of project activity"[3] at some sites. These have included: incorrect material specifications in procurement documents; erroneous predictions of the response of instrument systems to postulated accidents; acceptance of defective structural steel construction; failure to check vendor designs; and certification of inspection work without observation by inspectors.[4] At a number of nuclear projects, moreover, QA requirements have apparently been bypassed in major areas of construction—

---

b. The discussion is subject to the caveat that actual dates on which some requirements took effect could not be determined; many NRC regulatory guides, for example, affected design and construction practice through NRC licensing reviews before they were issued as guides.

leading, for example, to large voids in concrete walls at one site—and designs have been modified by unqualified field personnel (see Section 3.7). These deficiencies appear to corroborate the findings by the General Accounting Office that NRC inspectors "do little independent testing of construction work, . . .rely heavily upon the utility company self-evaluation, spend little time observing ongoing construction work, and do not communicate routinely with people who do the actual construction work."[5]

These revelations have begun to affect NRC's QA enforcement program. Since 1979, NRC has taken the unprecedented step of suspending large sections of safety-related construction work at several plants, including Wolf Creek, Midland, and South Texas, for up to a year until the contractor developed an adequate QA program. The most stringent NRC enforcement action, a ban on all safety-related work at Marble Hill, imposed in August 1979, was still partially in effect in early 1981. NRC has also levied fines of up to $100,000 for noncompliance with construction QA requirements—actions whose public impact on utilities far exceeds their direct costs. And NRC has recently notified all licensees that it "is becoming increasingly concerned by continuing evidence that many [licensees] are not properly implementing their [contractor] QA programs" and has called for greater compliance.[6]

NRC is also making several institutional changes aimed at reducing its reliance on the industry's self-policing of construction QA. It has begun assigning resident inspectors to all construction sites (previously they examined only facilities nearing completion) and is directing them to observe more work activities and to verify that as-built construction satisfies design requirements.[7] It is developing an information system, based on reported operational and construction defects, to enable it to more efficiently monitor the estimated 3,000 vendors of nuclear equipment and services,[8] and it is reportedly considering stationing resident inspectors at the offices of reactor suppliers and architect-engineers.[9] These reforms may improve operational safety and reliability, but they will also certainly increase the fabrication, engineering, and installation costs associated with QA.

Other prospective changes in QA programs and the expansion of the classification of equipment deemed safety-related, and thus subject to QA, are discussed in Section 6.3.

Equipment Qualification: Equipment qualification refers to testing and documentation to demonstrate that a manufactured component has been designed, fabricated, and installed so that it can function despite extremes of heat, humidity, radiation, and other stresses that could result from postulated accidents or are present in normal operation. Like quality assurance requirements, procedures for equipment qualification became more stringent in the 1970s and added greatly to reactor costs. And similarly, both regulatory lag and discoveries of widespread deficiencies in qualification programs will ensure that new reactors will bear increased costs for qualification testing and documentation.

Regulatory lag arises from the exemption of most purchased components from subsequent qualification requirements. For example, the 1974 edition of IEEE Standard (Std) 323,[10] which governs qualification of safety-related electrical equipment, surpassed the 1971 edition by adding component *aging* as a deterioration factor to be tested, and by expanding test documentation requirements and equipment tolerance margins. Because the NRC permitted grandfathering for most equipment, no plant operating as of mid-1980 will be fully qualified to the level of the 1974 standards, however.[11] Indeed, the first reactor to receive full power authorization following the TMI accident was qualified for operation under accident conditions in only 35 of 92 categories of vital electrical equipment.[c] Similarly, the "daughter standards" which apply IEEE Std 323 to specific electrical component classes such as motors and cables also were issued or revised during the mid-1970s and do not affect all equipment in recent plants.

Second, NRC investigations initiated in 1977 and continuing today have turned up serious lapses in many licensees' equipment qualification procedures. In examining the licensees' qualification *programs*, NRC found "serious problems" such as test sequences omitting important service conditions, no consideration of aging degradation, test models differing from installed components, and incomplete test documentation.[13] In examining actual *equipment*, NRC found that few components met applicable guidelines.[14] In concluding that "the nuclear industry is not devoting the resources necessary to solving the [equipment] qualification problem,"[15] the Commission further implied that existing plants have generally escaped some costs that would be incurred in complying with current requirements.

NRC's program for upgrading equipment qualification[16] essentially involves requiring most plants commencing commercial operation after 1982 to meet all of the provisions of the 1974 edition of IEEE Std 323. Exemptions from some requirements are permitted for plants currently operating or close to completion, but in the important area of aging, susceptible materials will be subject to periodic replacement and aging effects will be considered in NRC decisions on safety adequacy.[17] Earlier plants will also be required to conform with many parts of the 1974 edition of Std 323 by mid-1982. In assessing the effects of caustic sprays, for example, safety-related electrical equipment within containment at all plants must be proven capable of withstanding not only the chemicals used in the containment spray system, but also the most severe caustic spray environment that could result from a single failure in the spray system.[18]

---

c. The unit, North Anna 2, is deficient in equipment in only nine categories, with inadequate documentation of conforming equipment in 48 categories, according to the licensee, Virginia Electric & Power Co.[12]

The application of current IEEE standards will clearly add to capital costs of plants under construction (as well as requiring considerable expenditures for operating plants). Moreover, NRC's upgrading programs "do not address in detail all areas of qualification, since certain areas are not yet well understood."[19] Issues such as aging effects, synergistic effects, and effects of combustible gases on equipment with organic materials are presently being researched and may lead to changes in NRC qualification requirements.[20]

Internally-Induced Accidents: Recently-completed plants employ considerable pipe restraints, shielding, and physical separation to protect sensitive equipment from the effects of pipe ruptures and high-speed fragments ejected from rotating machinery (''missiles''). These additions were a major source of reactor cost increases during the 1970s. Although the failure of the Yankee-Rowe plant's turbine in 1980 accentuates the turbine missile concern, it is unlikely that either pipe ruptures or missiles will occasion significant cost increases. The other system included in this category, single-failure-proof reactor cranes, was only partially upgraded at recent plants, but the system is not especially expensive and its further improvement should not impose a large economic burden on new plants.

Seismic and Other Natural Phenomena: Most seismic design requirements have been applied, as practical, to plants under construction, contributing significantly to nuclear cost increases in the 1970s. There are exceptions, however. Regulatory Guides (RG) 1.92 and 1.122, which make more severe estimates of earthquakes' effects on structures, apply primarily to plants whose construction started after December 1974 and September 1976, respectively. (See Section 4.5 for names and descriptions of seismic regulatory guides.) Two guides which require more sturdy component supports for safety-related equipment, RG 1.124 and 1.130, similarly apply only to construction starts after November 1976 and July 1977. RG 1.117, which itemizes structural design features to withstand tornadoes, applies fully only to plants which received construction permits after mid-1976. Application of these requirements will cause most future plants to employ stronger structural members and pipe supports than did recently-completed plants.

Similarly, the 1975 revision of IEEE Std 344, which upgraded the qualification requirements in the 1971 edition for ensuring adequacy of seismic design of safety-related electrical equipment, will have a greater effect on new plants. Plants now being built will also be less likely to avoid some seismic-related costs through construction deviations or erroneous pipe stress calculations, as did a number of plants in the data sample (discussed above). In addition, future reactors might require ''seismic scram'' systems that would automatically shut down the reactor when triggered by earth tremors; these systems are in place at Diablo Canyon and DOE's plutonium production reactors at Hanford and Savannah River and are planned for new plants in Japan.[21]

Other new seismic considerations may offset some of these pressures. Some test data indicate that seismic design practices may be excessively conservative in calculating resistance of plant structures to earthquake motion, for example.[22] Such conclusions are for the most part preliminary and subject to conflicting interpretations,[23] however, so that any easing of seismic requirements is speculative. Nevertheless, slowing the *rate of change* of seismic standards could reduce the number of costly design and equipment changes required during construction.

Fire Protection: All generations of reactors have had to upgrade equipment and designs to improve fire protection after the 1975 Browns Ferry fire. As discussed in Section 4.6, all plants have added fire detection and suppression systems, including portable emergency lighting and communications. There are major generational differences, however, in methods used to prevent fires from propagating, especially among safety-related electric cables (which burned severely at Browns Ferry). Plants currently in early construction must provide two cable-spreading rooms to separate cables in redundant equipment ''trains.'' Cable trays (conduits) must be metallic and equipped with heat detectors, cable routings must have frequent ''fire stops'' and fireproof penetrations, and the cable materials and insulation must meet IEEE flame-resistance criteria.[24]

Operating plants and those in middle or late construction have not been required to make these provisions if they employ fire-retardant coatings and materials. Some of the plants, however, are conforming with these cable design criteria, perhaps to avoid possible backfitting later on.

Newer plants are also required to use greater distances and additional fire barriers to separate safety-related equipment items from their redundant counterparts and from fire hazards in non-safety-related (and thus less well-protected) areas. Their ventilation and drain systems must be capable of removing and sampling combustion-related heat, smoke, and water without permitting fires to spread. Noncombustible materials must be employed for many walls and structures. Again, some of these measures are being taken at plants under construction that are exempt from compliance.

Fire protection issues, methods, and regulations are complex and somewhat plant-specific, making generalization difficult. Nevertheless, there is a broad continuum away from reliance on fire-retardant coatings and toward greater physical separation and other design features, and improved cable insulation, as plants enter construction. Although recently-completed plants did have to make considerable and costly adjustments during construction, the more comprehensive fire protection measures being built into new plants appear likely to impose even higher costs.

Additional fire protection requirements for operating plants recently promulgated by NRC are discussed in Section 4.6.

Reactor Core and Core Cooling Systems: Most of the regulatory standards in this area described in Section 4.7 have been applied to recently-completed plants as well as to plants still in construction. Exceptions are several 1973 regulatory guides requiring improved welding and fabrication for steel used in critical "Class 1 and 2" fluid systems. These affected most but not all pressure-retaining and Emergency Core Cooling System (ECCS) piping, vessels, and pumps at recently completed plants. Similarly, RG 1.139, requiring redundancy, quality assurance, and improved controls for residual heat removal equipment, and RG 1.133, mandating an automatic system to detect loose primary coolant system parts, were only partially applied to recent plants.

Containment Structures and Systems: Many but not all of the improvements in containment structures and systems developed during the 1970s were employed by the later plants in the capital cost data base. Important exceptions are design changes in new BWRs to accommodate vibratory pressures on containment structures in the event of a loss-of-coolant accident (LOCA). These will require thicker walls, greater steel placement, and stronger equipment supports.

Similarly, some recent plants were not required to apply the improved concrete placement practices in RG 1.55 (issued in 1973), or the more stringent quality assurance procedures for installing concrete and structural steel in foundations, formwork, and steel reinforcement in RG 1.94 (1975, endorsing a 1974 American National Standards Institute [ANSI] standard). RG 1.141, concerning isolation valves for piping systems penetrating containment (1978, endorsing a 1976 ANSI standard)—the standard grandfathered at TMI and referred to by Professor Kemeny—did not apply to most recent plants. And most electrical penetration assemblies through containment walls in recent plants were constructed according to the 1972 edition of IEEE Std 317, rather than to the design, material, and qualification criteria in the 1976 edition.

Changes in containment systems and design features to control hydrogen generation and to cope with degraded-core and core-melt accidents are discussed in Section 6.2.

Electric Power, Instrumentation, and Control: Expanded requirements for these systems contributed greatly to nuclear capital cost increases in the 1970s. Not only were qualification procedures made more stringent for safety-related electrical and control equipment (Section 4.3), but redundancy and independence criteria were upgraded and equipment performance requirements were expanded (Section 4.9).

Many improvements dictated by NRC regulatory guides and IEEE standards applied only to plants beginning construction after the time of promulgation, however, and therefore are not reflected in the costs of the data base plants. Increasing qualification requirements were discussed above. In addition, new plants will have to provide more redundant back-up for electrical equipment and instrumentation. Prime examples are the back-up circuits for connecting the electrical system to the transmission grid specified in the 1978 edition of IEEE Std 308 and the prohibition in RG 1.81 against employing a "swing diesel" generator as a single back-up power supply for new multi-unit stations. Moreover, IEEE Std 384, issued in 1974 and revised in 1977, will require physical separation via distances or enclosed cabinets for much safety-related equipment already subject to redundancy criteria, such as instruments, electric cables, electrical penetrations through containment, and auxiliary systems essential to the operation of safety-related instruments and power systems.

Furthermore, some recent standards which expand performance criteria for instrumentation and control equipment will affect future plants more than recently-completed ones. These include the provision of a separate emergency shutdown room in the auxiliary building, and equipment to isolate the control room and maintain it in a habitable state during fires, chemical spills, or other accidents. Most importantly, RG 1.97, which specifies instrumentation to monitor plant conditions during and after accidents, is being significantly upgraded from its initial 1975 edition and 1977 revision. Those editions for the most part exempted plants in operation or under construction, and permitted the licensee to determine the instrumentation to be installed. In contrast, a second revision being drafted itemizes hundreds of required instruments and their measurement ranges. (This revision is discussed in the next chapter with other post-TMI regulatory changes.)

Conversely, technological advances should permit some cost reduction in power, instrumentation, and control equipment at future plants. Instrumentation systems are increasingly employing solid-state electronics, and digital computers are supplanting analog-based logic systems. These innovations are cheap, compact, and low in power usage compared to current equipment. Moreover, remote multiplexing and prefabricated cabling may reduce electrical circuitry equipment and installation costs.

Cost reductions from new technology are likely to be far offset, however, by the increased requirements for power supply, electrical circuitry, and instrumentation described above. In addition, operating experience continues to provoke consideration of further improvements. One example, the recent serious malfunctions in power supply systems for non-nuclear instruments at several Babcock & Wilcox (B&W) reactors, is discussed in detail in Section 5.4.

Radiation Control Systems: Equipment to reduce offsite radiation releases during normal plant operation and following accidents was significantly upgraded during the 1970s at all plants under construction (and at most operating plants). Hence, their effects have been fully felt in the later plants in the data base.

In contrast, few design measures to reduce worker exposure to radiation were made at completed plants. Some steps are being taken at plants under construction: remote systems for handling radwaste equipment, shielded cubicles and portable-shielding access in high-radiation areas, ventilation upgradings, and overall equipment design and material selection to reduce radiation fields and repair times. These may reduce downtime and maintenance burdens but will add to capital costs. Exposure reductions may be necessary, moreover, simply to keep up with prospective decreases in permissible exposure limits (see discussion in next section).

Improvements in waste solidification systems also may be required. Most systems in use leave small amounts of free-standing, corrosive liquids, but states with low-level waste repositories are pressuring NRC to require that wastes shipped to repositories be completely solidified. Cement systems that might be necessary could reportedly add up to $20/kW to plant costs.[25]

Section 5.2:   New And Revised Regulatory Guides

The previous section described regulatory standards that are already in place but which did not affect the design and cost of recently-completed plants, because of regulatory lag. New nuclear plants will bear increased costs not only from those, but from revisions of existing regulatory guides and formulation of new ones as well.

Through the end of 1978, NRC had issued 143 "Division 1" (Power Reactor) regulatory guides. These guides spanned virtually every facet of reactor design and construction, and encompassed many, if not most, of the increased regulatory standards that contributed to more than doubling real nuclear capital costs from 1971 to 1978.

Twenty-six of these guides were under revision as of March 1979, just prior to the TMI accident.[26] Among them are guides pertaining to seismic instrumentation, reactor coolant pump flywheels, protection against pipe whip, QA requirements, low-alloy steel welding, qualification of safety-related electrical equipment, loose-part detection systems, residual heat removal systems, radionuclide filtration and adsorption, containment isolation for piping, and safety-related concrete structures. Not all pending guide revisions portend sweeping changes, but the large number being revised—almost one-fifth of the total—is further indication of instability in nuclear regulatory standards.

Another particularly important regulatory guide in revision is RG 8.8, Occupational Radiation Exposures. Previous revisions expressed NRC's intent to make exposures "as low as reasonably achievable" in both operation and decommissioning activities. Current consideration of exposure reduction centers around workers who may require special protection from radiation, such as women of childbearing age.[27] Any reduction in permissible doses would disproportionately affect maintenance work efficiency, since a larger portion of workers' allowable exposure would be consumed unproductively in setting up and transit. Further design steps to control exposures, such as greater equipment separation, larger work spaces, and remote servicing, would increase capital costs. Design features to reduce decommissioning costs and exposures, ranging from extra space around components for easier dismantling to expanded record-keeping with as-built drawings and maintenance histories, could similarly add to costs.

In addition to regulatory guide revisions, NRC is developing 25 new Division 1 guides.[28] While it is not possible here to analyze the potential import of each guide, many appear capable of significantly altering design and construction practices and adding to costs. Among the subject areas they address are:

- containment spray system design;
- foundation and earthwork construction criteria;
- soil liquefaction potential at construction sites;
- ultrasonic testing of reactor vessel welds and Class 1 and 2 austenitic piping systems;
- analysis, design, construction, and testing of concrete containment structures;
- electric instrumentation and control system portions of protection systems, protective action systems, and auxiliary supporting features;
- single-failure criteria for fluid systems;
- qualification and production tests for piping and equipment snubbers;
- qualification tests for cable penetration fire stops;
- qualification of electric modules;
- safety-related permanent dewatering systems;
- earthquake instrumentation data recording and processing;
- tornado-driven missiles; and
- extreme wind speeds for coastal sites.

Gross numerical comparisons are somewhat speculative, but if the new regulatory guides have the same proportional effect on costs as the 143 existing guides, their impact would be substantial. Indeed, past cost increases do not fully reflect all existing guides, due to regulatory lag, so that the effect on costs "per guide" was probably greater than their numbers suggest.

Section 5.3: Unresolved Safety Issues

Reactor operating experience, licensing reviews, and safety research have brought to light a number of potential design deficiencies common to all reactors or to a particular reactor type. Discovery of such deficiencies has prompted many of the new regulatory standards discussed above and in Chapter 4. Many potential deficiencies have not yet been addressed by regulatory action, however. Although they particularly affect existing plants, efforts to resolve them will amost certainly impose large new costs upon plants under construction and will cause considerable, persistent uncertainty in plant design requirements.

As of September 1980, NRC listed 14 *Unresolved Safety Issues* (USIs) pursuant to the following definition:

> An Unresolved Safety Issue is a matter affecting a number of nuclear power plants that poses important questions concerning the adequacy of existing safety requirements for which a final resolution has not yet been developed and that involves conditions not likely to be acceptable over the lifetime of the plants affected.[29]

NRC staff has nominated seven additional candidate issues for Commission designation as unresolved safety issues. Moreover, the Advisory Committee on Reactor Safeguards (ACRS), a body of senior nuclear safety specialists attached to NRC, has compiled 17 other "generic items relating to light-water reactors" which are not in the NRC list of USIs but nevertheless could significantly affect reactor design criteria. Finally, although some issues in previous NRC and ACRS lists have been declared resolved, past experience indicates that genuine resolution and implementation will prove elusive and that some "resolved" issues will require further regulatory consideration.

NRC List of Unresolved Safety Issues: In December 1977, Congress amended the 1974 Energy Reorganization Act to require NRC to submit a plan to identify unresolved reactor safety issues. The first list, in early 1978, had 133 issues, of which 41 were "Category A," with the remainder designated lower priority.[30] The list was reduced a year later, and 17 issues were selected as having the greatest safety significance (see box). Some issues apply primarily to existing plants; for example, *Reactor Vessel Materials Toughness* concerns possible weakening of pressure vessel weld seams in older reactors by long-term irradiation, due to the welds' high copper content (since reduced). Many USIs pertain to both new and existing plants, however. A case in point is *Systems Interactions*.

Systems interactions are events in which different plant systems interact with each other in such a way that the performance of safety systems may be degraded. They are hard to foresee "because the design and analysis of plant

NRC Unresolved Safety Issues
(January 1979 list[31])
(Italicized issues are discussed in text. See
References 29 or 31 for complete description of each issue.)

1. Water Hammer

2. *Asymmetric Blowdown Loads on the Reactor Coolant System*

3. PWR Steam Generator Tube Integrity

4. *BWR Mark I and II Pressure Suppression Containments*

5. *Anticipated Transients Without Scram (ATWS)*

6. BWR Nozzle Cracking

7. *Reactor Vessel Materials Toughness*

8. *Fracture Toughness of Steam Generator and Reactor Coolant Pump Supports**

9. *Systems Interactions*

10. Environmental Qualification of Safety-Related Electrical Equipment

11. *Reactor Vessel Pressure Transient Protection**

12. *Residual Heat Removal Requirements***

13. Control of Heavy Loads Near Spent Fuel

14. *Seismic Design Criteria*

15. BWR Pipe Cracks

16. Containment Emergency Sump Reliability

17. *Station Blackout*

---

*Denotes issue declared resolved by NRC.

**Denotes issue declared resolved by NRC but still with pending questions, as discussed in text.

systems is frequently assigned to teams with functional engineering specialties [that are inadequately co-ordinated] to enable them to identify adverse interactions between and among systems.''[32] In one such instance, at Zion Unit 2 in 1977, test signals applied to plant sensors caused plant controls to reduce water inventory in the reactor coolant system while simultaneously disabling all automatic protection systems capable of detecting the loss of water.[33] The interaction in this case occurred between the plant's control system and protection system. Other systems interactions have involved failures (e.g., loss of power to instruments) that caused automatic reactor shutdown and also generated spurious signals to instruments and controls that temporarily defeated shutdown heat removal capability.[34]

Reactors are subject to a great number of potential systems interactions. Interconnected electrical or mechanical complexes are particularly prone since components can ''run away'' or ''hang up'' between their normal and failed state, leading to excessive voltage, frequency, flow, etc. or whatever service the system provides.[35] Nonconnected systems may also interact because failure of their pneumatic, electrical, or hydraulic lines may degrade the others, through rupture of steam lines or compressed air systems, for example. As a result, systems interactions are difficult to anticipate, and thus hard to prevent. According to the ACRS,

> [Systems] interactions are likely to be unique to each plant and are unlikely to be revealed by LERs [Licensee Event Reports filed by utilities] since the probability for such interaction to occur may be modest . . . [N]either LERs nor a study of plant diagrams and other drawings will consistently reveal the potential for such interactions between non-connected systems, because such drawings generally show single features or systems . . . *Thus, uncovering the potential for [systems] interaction . . . will usually require careful, in-situ examination of the physical plant . . .* consider[ing] all features having the potential to damage safety systems, including the safety systems themselves.[36]

NRC intends to approach the issue by using ''fault-tree'' analysis to determine whether its current review procedures adequately account for potentially harmful interactions. The Three Mile Island accident, however, has forced NRC to undertake new studies of operator actions, design errors, and maintenance procedures[37]—issues that are central to systems interaction (the Zion episode was initiated by operator error and compounded by design inadequacy). Notwithstanding the May 1981 target date[38] (postponed eight months from the year-earlier target), resolution appears far off. If NRC ultimately concludes that some harmful interactions cannot be anticipated, it may require segregating all components and circuits of certain safety systems in costly ''bunkered'' housings.[39] To prevent identified interactions, NRC might require less drastic but still costly remedies, such as improved equipment qualification or greater physical separation.

Of the other unresolved safety issues, only *Seismic Design Criteria* might have a similarly pervasive design impact, and it applies largely to existing reactors built to less stringent seismic criteria. But many USIs could significantly affect future plants. *Asymmetric Blowdown Loads* might require modifying PWRs to prevent postulated ruptures in reactor coolant pipes from overstressing reactor vessel supports, a condition that could damage both normal and emergency cooling systems while preventing insertion of control rods to shut down the fission reaction. *BWR Pressure Suppression Containments* could lead to major design changes if current designs are found to expose equipment in the "suppression pool" to unacceptably large dynamic effects from air and steam forced out of the drywell during a loss-of-coolant accident.[d] In *Station Blackout*, NRC is weighing further provision of diverse and redundant feedwater power sources to ensure that a complete loss of a/c power from both the offsite grid and the onsite emergency diesels could be accommodated safely.

Progress toward resolving the unresolved safety issues has been "disappointing," according to the report of the NRC's Special Inquiry Group chaired by Mitchell Rogovin (the Rogovin Report).[40] The Kemeny Commission surmised that "labeling . . . a problem as 'generic' may provide a convenient way of postponing decision on a difficult question" by removing the issue from contention in licensing nearings.[41] By late 1980, NRC had resolved only three of the 17 USIs it first designated in early 1978, and two of the three can hardly be considered settled.

One of these, *Residual Heat Removal Shutdown Requirements,* was declared completed in 1979 because of equipment improvements mandated in the Standard Review Plan.[42] Yet in mid-1980, NRC staff identified a new proposed USI, *Shutdown Decay Heat Removal Requirements,* which is similar to the old in both name and content.[43] The new issue is related to three post-TMI studies of heat removal systems which will assess the systems' reliability under degraded conditions (*e.g.*, complete loss of feedwater), weigh the need for a diverse heat-removal path if secondary-side cooling (*e.g.*, through steam generators) is lost, and consider requiring a complete additional heat removal system.[44] One study was to be completed by the end of 1980, another by mid-1982, and no schedule has been set for the third, indicating the uncertainty attached to this "resolved" issue.

Similarly, the research findings which "resolved" *Fracture Toughness of Steam Generator and Reactor Coolant Pump (RCP) Supports* (concerning the supports' structural integrity during accidents) pertain only to *operating PWRs.*[45] NRC is just beginning its review of new PWRs by asking licensees for

---

d. Although this issue's title refers only to Mark I and II containments, the Mark III design now in construction is also addressed through former "Task A-39," concerning safety relief valve pool loads, which has been absorbed into the issue.

information on the geometry, weld characteristics, design loads, etc. of steam generator and RCP supports. Later it will assess all vessel supports, BWR coolant pump supports, and PWR pressurizer supports.[46] Recent cracking of all but one of the 48 support bolts for the steam generators at Prairie Island 1 gives added impetus to this issue.[47]

The third "resolved" issue, *Reactor Vessel Pressure Transient Protection*, does appear to have been completed. Relief valves are to be designed for greater pressure removal, and operating procedures are to be modified to prevent subjecting a cooled-down reactor vessel to high pressure—a combination conducive to metal fatigue and fracture.[e] Indeed, true resolution of most of the unresolved issues will require increments of equipment adding to direct plant costs, aside from the indirect but high costs of design uncertainty while resolution is pending.

Advisory Committee on Reactor Safeguards List: ACRS has long maintained its own lists of unresolved safety issues under the title, "Generic Items Relating to Light-Water Reactors." The first list, in 1972, showed 25 resolved and 22 unresolved issues. Resolved issues were deleted and new ones added roughly annually until publication of the final March 1979 list of 25 issues, including ten from the 1972 compilation. NRC drew eight issues from this list in formulating its list of USIs. The other 17 issues are shown here in the box.

The 17 were excluded from the NRC list because they were deemed not urgent to safety or were considered to be moving toward resolution through other processes, such as revision of regulatory guides. Nevertheless, all concern either possible inadequacies in systems in plants operating or being built, or potential improvements in safety over current designs. Resolution of many issues could have a considerable impact on the design and cost of new plants.

Among issues involving uncertainties in present systems are questions pertaining to engineered safeguards, such as Containment Sprays (Item A-2), ECCS Valves (D-1), and Hermetic Seals (E-1); to plant designs for preventing accidents, such as Turbine Missiles (A-1) and Seismic Soil-Structure Interaction (F-1); to plant behavior during accidents, such as Fuel Behavior Under Abnormal Conditions (A-5); and to vital instrumentation and controls, such as Computer Protection Systems (C-1). These issues concern plant safety and will require regulatory attention—a situation complicated in some cases by gaps in knowledge, such as in fuel behavior (A-5).

---

e. Another issue, *BWR Nozzle Cracking*, appears close to resolution through development of a triple-sleeve sparger to stop feedwater nozzle cracking and elimination of control rod drive return lines that have suffered nozzle cracking.[48] But the BWR history of chronic pipe and component cracking, flow induced vibration, and other primary system problems argues for caution in concluding that these design changes will end cracking without introducing other defects.

Advisory Committee On Reactor Safeguards
Generic Safety Issues
(March 1979 List,[49] excluding issues in NRC list)

A. Issues From December 1972 List

  1. Turbine Missiles

  2. Effective Operation Of Containment Sprays In A LOCA

  3. Instruments To Detect Severe Fuel Failures

  4. Monitoring For Excessive Vibration Inside The Reactor Pressure Vessel

  5. Behavior Of Reactor Fuel Under Abnormal Conditions

  6. BWR Recirculation Pump Overspeed During A LOCA

  7. Advisability Of Seismic Scram

  8. ECCS Capability For Future Plants

B. Issues Added In February 1974 List

  1. Ice Condenser Containments

  2. PWR Primary Coolant Pump Overspeed During A LOCA

  3. Periodic 10-Year Review Of All Power Reactors

C. Issue Added In March 1975 List

  1. Computer Reactor Protection System

D. Issues Added In April 1976 List

  1. Locking Out Of ECCS Power Operated Valves

  2. Design Features To Control Sabotage

  3. Decontamination And Decommissioning Of Reactors

E. Issue Added In February 1977 List

  1. Long-Term Capability Of Hermetic Seals On Instrumentation And Electrical Equipment

F. Issue Added In November 1977 List

  1. Soil-Structure Interaction In Seismic Events

The potential design improvements in the ACRS list could also impose higher equipment costs. Instrumentation to detect severe fuel failure (A-3) and to monitor pressure vessel vibration (A-4) would require sensors, relays, circuitry, and other electrical equipment. Seismic scram capability (A-7) would entail an extra automatic reactor shutdown system triggered by seismic disturbances, as discussed in Section 5.1. Improved ECCS capability (A-8), which NRC staff is studying for possible designation as an unresolved safety issue,[50] could involve not only equipment modifications to enhance the system's reliability, but also design changes to facilitate reflooding of the reactor core, with attendant engineering analysis. Sabotage prevention features (D-2) could require, in addition to security and access-control systems, such design features as completely separated access controls for redundant safety systems, as well as operational rules prohibiting the same maintenance personnel from working on redundant safety "trains." Similarly, decontamination and decommissioning (D-3) involve, *inter alia*, possible changes in materials and configurations for primary system equipment to reduce long-term radioactivity buildup, thereby reducing the need for periodic decontamination and simplifying reactor dismantlement at end of life.

Furthermore, the criteria by which ACRS deems a generic issue resolved appear to be as loose as those applied by NRC in ranking its issues. ACRS defines "resolved" either "in an *administrative* sense, recognizing that technical evaluation and satisfactory implementation are yet to be completed, or in a *narrow* or *specific* sense, recognizing that *further steps are desirable . . . or that different aspects of the problem require further investigation.*"[51] The Rogovin Report confirmed that ACRS's "definition of the 'resolution' of a generic issue does not consider its implementation, and the committee does not follow up on 'resolved' generic issues to determine whether or how they are being implemented by the NRC staff."[52] Thus, among the 51 previously "resolved" ACRS issues are some that, in the words of one ACRS member, were "placed in the 'resolved' category prematurely because what was thought to be a regulatory position became 'unstuck' and was not implemented."[53]

A prime example of such resolved but unsettled issues is *Anticipated Transients Without Scram (ATWS)*. This issue, first raised in 1969, concerns the possible failure to insert the control rods immediately ("scram") to shut down the nuclear fission process following a sudden deviation from normal operating conditions (a "transient") that causes reactor pressure and temperature to rise. ACRS declared the ATWS issue resolved in 1973 by an AEC staff report that evaluated the likelihood and consequences of ATWS events and proposed remedial design changes.[54] Their implementation has been stalled ever since, however, by a concerted industry lobbying effort motivated by the conviction that only minimal improvements are necessary and that the AEC/NRC fixes would be very costly.

Thus, ATWS has not actually been resolved. In fact, NRC designated it as an unresolved safety issue in 1979, although ACRS has never restored it to its generic safety issues list. NRC currently plans to hold rulemaking hearings to consider improvements to scram systems[f] and design changes to reduce the rate of fission in the event of scram failure. These could include, for BWRs, modification of circuitry governing control-rod injection, a recirculation-pump trip, an improved standby liquid control system, and a high-capacity automated boron system, and analogous changes for PWRs. Aside from the direct costs, estimated by NRC at up to $7/kW,[55] but considered higher by industry, the delayed resolution of ATWS lends credence to the possibility that a significant fraction of the other 50 ACRS items were resolved in an ad-ministrative sense only, and that hardware changes lie ahead.

Several of these formerly resolved items stand out as candidates for substantial new cost impacts: *Hydrogen Control After a LOCA* and *Contain-ment Pressure Following a LOCA*, both receiving post-TMI consideration (see next chapter); *Performance of Critical Components in post-LOCA Environ-ment*, a pressing issue due to industry's lapses in adhering to environmental qualification requirements; *Instrumentation to Follow the Course of an Acci-dent*, under revision prior to TMI; and *Inservice Inspection of Reactor Coolant Pressure Boundary* and *Detection and Location of Primary System Leaks*, identified by the ACRS as issues resolved on a narrow basis meriting possible action.[56]

Finally, apart from the status of the NRC and ACRS issues, it needs to be recognized that even genuinely resolved issues may not have required changes in recently completed plants. Although some regulatory standards have been applied to many or all plants under construction, others have largely exempted plants in progress and have been directed primarily at new construction starts. Accordingly, truly resolved issues are not necessarily reflected in the cost of a typical 1978 plant. Implementation of some will thus show up as cost *increases* for many later plants. Of course, currently unresolved issues will carry further cost impacts from resolution during construction, with the considerable engi-neering effort, planning uncertainty, and logistical disruption which regula-tory and design changes can entail.

New Unresolved Safety Issues: NRC's current list of 14 unresolved safety issues (a figure that excludes two issues that may become "unstuck," as argued above) may soon be at least doubled. NRC staff asked the Commission-ers in late 1980 to designate another seven new issues as USIs. The ACRS and the new Office for the Analysis and Evaluation of Operational Data (AEOD), created after Three Mile Island, have together proposed four more, and NRC staff is studying six other issues for possible USI designation (see box).

Two issues, *Seismic Qualification* and *DC Power Reliability*, concern

f. A partial failure to scram accident at Browns Ferry 3 in June 1980, the most serious such episode at any U.S. commercial reactor to date, is discussed in Section 5.4

Possible New Unresolved Safety Issues
(as of late 1980)

Issues Recommended by NRC Staff*

1. Shutdown Decay Heat Removal Requirements

2. Control Room Design

3. Consideration of Melted or Degraded Cores in Safety Reviews

4. Long-Term Upgrading of Training and Qualifications of Operating Personnel

5. Operating Procedures

6. Seismic Qualification of Equipment in Operating Plants

7. Control System Reliability

Issues Being Studied by NRC Staff*

1. Reliance on ECCS

2. In-Situ Testing of Valves

3. Protective Device Reliability

4. PWR Pipe Cracks

5. BWR Main Steam Isolation Valve Leakage Control Systems

6. Radiation Effects on Reactor Vessel Supports

Other Issues Recommended by ACRS**

1. DC Power System Reliability

2. Single Failure Criterion

Other Issues Recommended by AEOD***

1. Safety Implications of Steam Generator Transients and Accidents

2. Piping and Use of Highly Combustible Gases in Vital Areas

---

*Reference 43, and Reference 57 for Control System Reliability.
**See Reference 58.
***See Reference 59.

only existing plants, while another six—the *Single Failure Criterion* and the first five issues in the list of NRC recommendations—arise directly from Three Mile Island and are discussed in Chapter 6. The other nine issues, however, have been prompted primarily by problems at other reactors (several of these are discussed in the next section). Moreover, NRC anticipates that in addition to its seven recommended new USIs, five to six more will be added the following year, with a steady rate of three anually thereafter.[60]

Section 5.4: New Safety Issues From Reactor Operating Experience

Reactor operating experience has been identified as a major source of information leading to more stringent regulatory standards (Chapter 3). Reactor operation may reveal heretofore unanticipated safety problems or demonstrate that previously identified problems are not being adequately resolved by current designs. (Conversely, favorable operating experience can lead to a *relaxation* of safety margins, but this appears to occur relatively infrequently.)

The rate at which operating experience reveals new safety risks is critically important. The detection rate of such risks *per year* strongly affects the pace of imposition of new safety requirements, and thus of higher costs, for nuclear plants. In turn, this rate will be determined by changes in the number of safety defects discovered per *reactor-year* of operation. If plant designs were static, this latter rate should constantly be falling, since the ''pool'' of undetected defects would shrink as safety problems surfaced in operation. In practice, however, the increased size, complexity, and perhaps scrutiny of reactors appear to have added many new safety defects to the pool, so that the rate of detection of defects per plant has risen, not fallen.

This key trend is indicated by records of issuance of generic (applicable to many reactors) safety-problem communications to licensees by the NRC's Office of Inspection and Enforcement (see box). After dipping briefly in 1975, the combined rate of issuance of NRC Bulletins and Circulars *per reactor* rose in each of the next three years, far surpassing the levels in the early 1970s (see Figure 5.1). The rate shot up further in 1979, in part due to the NRC's increased post-TMI emphasis on monitoring operating experience. Since the accident, NRC's generic communications to licensees, mostly drawn from operating experience, have been running at almost three times the per-reactor rate in the early-1970s.

Thus, not only the nuclear sector as a whole but the typical operating reactor as well is contributing to detection of new safety problems at an all-time high rate. And even if, as is likely, the per-reactor rate were to decline in the future, it would probably fall only gradually. Moreover, because the number of operating plants is increasing, the detection rate of safety problems per year

NRC Safety-Communications Media

NRC's Office of Inspection and Enforcement uses three media to apprise licensees of new developments, mostly arising from operating experience, pertaining to reactor safety and design.

Bulletins, issued since 1971, describe specific actions that licensees must take—analyses, tests, equipment replacements, or design changes—in response to "events in which the safety significance is of such a magnitude as to result in an immediate impact on all of a certain type of licensee."[61] Recent bulletins have concerned seismic deficiencies, feedwater system pipe cracks, inoperability of heat removal systems, and loss of BWR scram capability, among other issues.

Circulars, issued since 1976, notify licensees of actions *recommended* by NRC in safety matters that are "generally of lesser significance" than those addressed by bulletins.[62] Unlike bulletins, circulars do not require licensee response. Many have considerable safety import, however. Recent circulars have addressed, for example, lubrication problems in emergency diesel generators and ECCS turbine drives, inadequate environmental qualification of safety-related electrical equipment, and reactor vessel steam voids caused by loss of cooling water flow to reactor coolant pumps.

would still continue to rise for some time. (This occurred during 1971-74, as the number of Bulletins increased each year despite the absence of an increase per reactor, as Figure 5.2 shows.) Accordingly, a constant stream of new reactor malfunctions should be expected, raising new safety issues and reactivating old ones, requiring remedial actions that will raise reactor costs.

Although the safety issues that will emerge from (or be confirmed by) future operating experience are not now discernible, the types of problems that will arise may be illustrated by recent operating experience. One malfunction of note was the failure of 40 percent of the control rods to "scram" (insert rapidly and fully to shut down the reactor) automatically upon command at Browns Ferry 3 in June 1980.[64]

Scram failure could have extremely serious consequences. Operators required 14 minutes to achieve manual scram at Browns Ferry, but scram time was not critical since the shutdown had been planned. Failure to scram rapidly in the event of a "transient," such as a turbine "trip" (shut-off), could leave

Information Notices were begun in 1979 to provide early warning of issues that may subsequently prove important to safety. They are generally less significant than, and sometimes overlap with, bulletins and circulars. Nevertheless, NRC sometimes employs information notices to transmit original safety concerns, such as cracking in ECCS piping (Notice 80-15), violation of separation criteria for ECCS cables (Notice 79-32), and unavailability of safety systems due to design deficiencies in power supplies (Notice 79-04).

Figures 5.1 and 5.2 employ the combined number of bulletins and circulars as the best measure of NRC safety-related communications. Although circulars are usually less vital to safety and do not have as great an impact, they do convey important imformation and as a category are somewhat interchangeable with bulletins. NRC says that is has "shifted some of what were formerly bulletins into circulars and information notices."[63] The latter, less central to safety and sometimes redundant, are not included in the figures for conservatism.

A fourth communications medium is letters to licensees by NRC's Office of Nuclear Reactor Regulation. Like bulletins, NRC letters concern potential safety matters and require a written response. Unfortunately, they are filed by docket (reactor) number, and are not compiled in any available list.

the reactor fissioning at full or partial power with no heat-removal path, a condition that could cause core overheating, leading to fuel melting and, potentially, to significant core damage. (This is the subject concern of ATWS, discussed in the preceding section.)

The Browns Ferry partial scram failure was caused by undetected water accumulation in the "scram discharge volume system" (SDV) which prevented the hydraulic action necessary to force the control rods into the core. The water had accumulated after a previous scram because of blockage in the SDV vent or drain. In turn, the accumulation was not detected because water-level instrumentation was located in a separate tank that was incorrectly calibrated to the SDV.[65]

Remedial action has already begun at Browns Ferry and other BWRs, with installation of vents which open to the atmosphere to prevent creation of a vacuum that could impede water drainage. NRC is also likely to require that the two halves of the SDV be built to safety-grade standards and made indepen-

Figure 5.1
NRC Bulletins And Circulars Per Operating Reactor

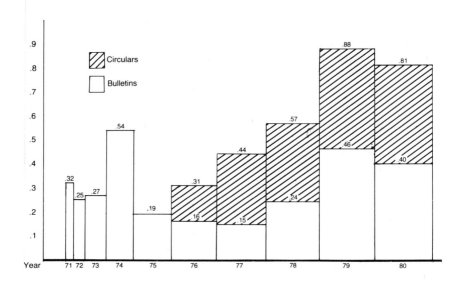

Bar widths represent numbers of licensed operating reactors (including partial years from commercial start, excluding units under 400 MW).

Supplements to Bulletins and Circulars are not included.

Circulars were begun in 1976. Upper figures denote total of Bulletins and Circulars.

dent. This would require directly connecting the scram discharge volume and the instrument volume, doubling the number of vent and drain lines, and adding about a half-dozen new valves.[66] The AEOD has proposed other corrective measures, including a diverse instrument to monitor SDV water level, advanced monitoring techniques such as differential pressure cells, and installing redundant, automatic valves on all vent and drain paths from the SDV and associated instrument lines.[67]

The implications of the Browns Ferry scram failure do not end with these equipment additions, however. The accident undercut the nuclear industry's contention that scram systems were sufficiently reliable to obviate the need for design features to mitigate an ATWS event. It will undoubtedly lead NRC to adopt more stringent ATWS requirements than would otherwise have been considered.[68] In the same fashion, the scram failure will make at least some regulators more skeptical of probabilistic analyses which purport to prove that particular safety systems are highly reliable[69]—an important issue discussed in Chapter 6.

Another problem area highlighted by recent experience is control system

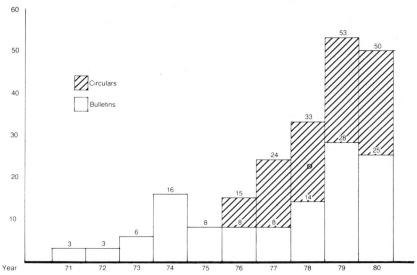

Figure 5.2
NRC Bulletins And Circulars Per Year

Figure 5.1 explanations of Bulletins and Circulars apply here.

reliability. On three occasions in as many years, a failure in power supply for *non-nuclear instrumentation* (NNI) at a Babcock & Wilcox reactor caused the *integrated control system* (ICS) to initiate an abrupt change in reactor status while leaving the operator without the information needed to understand and recover from the abnormal plant condition.[g]

The accident at Rancho Seco in 1978 started when a light bulb was dropped inside a control console, causing faults in two-thirds of the NNI signals to the ICS and the control room. Feedwater flow to the steam generators was cut off for nine minutes, causing both generators to boil dry, and correct information was not restored to the control room for 70 minutes.[70] At Oconee 3 in 1979, operation of the ICS and most instrumentation was interrupted for several minutes after a fuse in the NNI system blew, leading to excessively

g. Non-nuclear instrumentation reports critical parameters such as reactor coolant system temperature, pressure and flow; steam generator pressure and level; and pressurizer level; it excludes only core-related variables such as neutron flux. The integrated control system at B&W plants co-ordinates the reactor, feedwater supply to the steam generators, and the steam turbine.

rapid cooling of the reactor coolant system.[71] In 1980, 70% of the NNI indicators were either lost or sent false signals to the ICS after a short circuit at Crystal River 3, causing one steam generator to go dry and leading to a large spill of reactor coolant into the containment.[72] In each case, decay heat removal mechanisms which much function following reactor trip were either lost or partially unavailable.

These accidents point up serious inadequacies in industry's and NRC's approach to reactor control systems. First, they show that control system design can affect the number and types of transients which safety systems must arrest. Second, they demonstrate that failure of control systems may impede or even negate operation of the necessary safety systems.[73] Third, they raise concerns regarding the availability and accuracy of information normally supplied to the operator.[74] Previously, it was assumed that failure of control systems would neither interfere with safety systems nor affect information flow.

The accidents have led NRC staff, with considerable prodding from ACRS, to nominate Control System Reliability as an unresolved safety issue. In its pending review, NRC is likely to designate much non-nuclear instrumentation at all reactors as safety-grade, and thus subject to QA requirements. In addition, a third instrument train for some plant variables could be required to help operators resolve contradictory information which could arise if one of the present two trains fails.[75] Other design changes, including greater separation of power supplies for instrumentation and control equipment, could also result from the review.[76]

The failures of control and scram systems are only a sample of the problems emerging in reactor operating experience. During 1979 and 1980, NRC issued Bulletins and Circulars at the rate of almost one per reactor-year, or about 50 annually. Other recent Bulletins have concerned deficiencies in seismic design and construction, inadequate environmental qualification of electrical components, poor design of jet pump ''hold-down'' beams, and operability of residual heat removal systems, among many other deficiencies.

In addition, a 1979 ACRS review of utility ''Licensee Event Reports'' identified a number of operating ''events having potentially serious safety implications''[77] not discussed here. They include *flow-induced vibration* in equipment and piping carrying water and/or steam which ''frequently cause[s] failure of equipment, electrical wiring or components, pumps, valves and piping systems;''[78] *leakage between interconnected fluid systems* (through isolation valves) that can defeat differences in pressures or chemical concentrations needed by separated systems; *degradation of containment isolation* through failure of containment monitoring systems needed to actuate isolation, for example; and *possible loss of engineered safety features* through human error or equipment failure such as inadvertent isolation of safety systems when malfunctioning ventilation simulates leaks in the systems themselves.

The presence of problem areas such as these, the frequency with which they issue from reactor operating experience, and the difficulty of satisfactorily resolving previously identified issues appear to ensure a large body of pressing safety concerns to affect reactor regulation, design, and construction for many years to come.

References

1. *New York Times*, 2 June 1979, "Three Mile Island Nuclear Plant Was Exempt From Radiation Rules."

2. NRC, Office of Inspection and Enforcement, Bulletins 79-02 and 79-14.

3. NRC, Office of Inspection and Enforcement, Information Notice 80-26.

4. Reference 3.

5. GAO, *The Nuclear Regulatory Commission Needs to Aggressively Monitor and Independently Evaluate Nuclear Powerplant Construction*, EMD-78-80 (1978), p. ii.

6. Reference 3.

7. NRC, *NRC Action Plan Developed as a Result of the TMI-2 Accident*, NUREG-0660 (1980), p. II.J.2-1.

8. *Inside N.R.C.*, 16 June 1980, p. 4.

9. Reference 8.

10. IEEE, IEEE Std 323-1974, "Qualifying Class 1E Equipment for Nuclear Power Generating Stations."

11. NRC Commissioner Peter Bradford, Speech of 24 September 1980, NRC Release No. S-13-80.

12. NRC Commissioner Victor Gilinsky, Statement on North Anna 2 Full Power Authorization, 21 August 1980, NRC Release No. 80-152.

13. NRC, Memorandum and Order CLI-80-21, 23 May 1980, p. 10. See also Office of Inspection and Enforcement communications pertaining to equipment qualification: Bulletins 77-05, 77-05A, 78-02, 79-01, 79-01A, 79-01B, and three Supplements to 79-01B; and Circulars 78-08 and 80-10.

14. Reference 13 (NRC Memorandum and Order), p. 10.

15. Reference 13 (NRC Memorandum and Order), p. 11.

16. Reference 13; and NRC, *Interim Staff Position on Environmental Qualification of Safety-Related Electrical Equipment*, NUREG-0588 (1979).

17. Reference 16 (NUREG-0588), p. 15.

18. Reference 16 (NUREG-0588), p. 7.

19. Reference 16 (NUREG-0588), p. 1.

20. Reference 16 (NUREG-0588), p. 10.

21. G.E. Cummings, J.E. Wells, and H.E. Lambert, *Nuclear Safety*, *19* (No. 5), 590 (1978).

22. See, for example, Lawrence Livermore Laboratory and ANCO Engineers, Inc., *Methods and Benefits of Experimental Seismic Evaluation of Nuclear Power Plants*, NUREG/CR-1443 (1980).

23. See, for example, *Science*, *209*, 1004-1007 (1980); and Y. Jih-Ting & Y.P. Aggarwal, "Seismotechtonics of Northeastern United States and Eastern Canada," submitted to *J. Geophysical Research*.

24. IEEE, IEEE Std 383, "Class 1E Electric Cables, Field Splices, and Connections for Nuclear Power Generating Stations" (1974); endorsed by NRC Regulatory Guide 1.131, "Qualifications Tests of Electric Cables, Field Splices, and Connections for Light-Water-Cooled Nuclear Power Plants" (1977).

25. *Inside N.R.C.*, 22 October 1979, pp. 8-9.

26. NRC Office of Standards Development, "Regulatory Guides Under Development," 24 April 1979, first drafted in March 1979.

27. *Inside N.R.C.*, 28 July 1980, p. 3.

28. Reference 26.

29. NRC, *Annual Report to Congress, 1979*, NUREG-0690 (1980), p. 66.

30. NRC, *NRC Program for the Resolution of Generic Issues Related to Nuclear Power Plants*, NUREG-0410 (1978).

31. NRC, *Identification of Unresolved Safety Issues Relating to Nuclear Power Plants*, NUREG-0510 (1979), p. 16.

32. Reference 29, pp. 77-78.

33. NRC, Office of Inspection and Enforcement, Circular 77-13; and letter from Robert Pollard, Union of Concerned Scientists, to Attorney General Griffin Bell, 13 October 1977.

34. See ACRS, *Review of Licensee Event Reports, 1976-1978*, NUREG-0572 (1979), pp. D-8 through D-12 and D-19 through D-22, for detailed accounts of several systems interaction events.

35. ACRS, Letter to NRC Chairman, "Systems Interaction Study for Indian Point Nuclear Generating Unit No. 3," 12 October 1979.

36. Reference 35. Emphasis added.

37. Reference 29, pp. 79-80.

38. Reference 29, p. 67.

39. *Nucleonics Week*, 11 May 1978, p. 8.

40. NRC Special Inquiry Group, *Three Mile Island: A Report to the Commissioners and to the Public*, NUREG/CR-1250, Vol. II (1980), p. 21.

41. J.G. Kemeny *et al.*, *Report of the President's Commission on the Accident at Three Mile Island*, p. 20.

42. Reference 29, p. 80.

43. NRC Staff Paper, "Special Report to Congress Identifying New Unresolved Safety Issues," SECY-80-325, 9 July 1980.

44. Reference 7, Task II.E.3.

45. NRC, *Potential for Low Fracture Toughness and Lamellar Tearing on PWR Steam*

Generators and Reactor Coolant Pump Supports, NUREG-0577 (1979), p. 10.

46. Reference 45, p. 3.

47. NRC, Office of Inspection and Enforcement, Information Notice No. 80-36.

48. NRC, *BWR Feedwater Nozzle and Control Rod Drive Return Line Nozzle Cracking*, NUREG-0619 (1980).

49. ACRS, *Generic Items Reports*, 15 November 1977 and 21 March 1979.

50. Reference 43.

51. ACRS, *Generic Items Report*, 15 November 1977, p. 2. Emphasis added.

52. Reference 40, Vol. II, p. 50.

53. David Okrent, *On the History of Light Water Reactor Safety in the United States* (University of California at Los Angeles, 1979, unpublished manuscript), p. 6-18.

54. AEC, *Technical Report on Anticipated Transients Without Scram for Water-Cooled Power Reactors*, WASH-1270 (1973).

55. NRC Staff Paper, ''Proposed Rulemaking to Amend 10 CFR Part 50 Concerning Anticipated Transients Without Scram (ATWS) Events,'' SECY-80-409, 4 September 1980.

56. See Reference 51.

57. NRC, Memorandum from Office of Nuclear Reactor Regulation to Chairman, 10 September 1980.

58. ACRS, Letter to NRC Chairman, 12 August 1980.

59. NRC, Memorandum from AEOD to Chairman, 4 August 1980.

60. Reference 43, p. 4.

61. Reference 29, p. 159.

62. Reference 29, p. 159.

63. Telephone communication with Ed Jordan, Assistant Director for Technical Programs, NRC Office of Inspection and Enforcement, 4 September 1980.

64. This event, and subsequently discovered problems with BWR scram systems, were reported in NRC, Office of Inspection and Enforcement, Bulletin 80-17, three supplements thereto, and Information Notice 80-30. The Bulletin and supplements were counted as only one Bulletin in Figures 5.1 and 5.2.

65. NRC, AEOD, ''Report on the Browns Ferry 3 Partial Failure to Scram Event on June 28, 1980,'' 30 July 1980, Section 8.

66. Reference 63.

67. NRC, Memorandum from AEOD to Office of Nuclear Reactor Regulation, 18 August 1980; and *Inside N.R.C.*, 11 August 1980, p. 10.

68. *Inside N.R.C.*, 11 August 1980, pp. 17-18.

69. David Okrent, ''New Trends in Safety Design and Analysis,'' paper presented at International Atomic Energy Agency Conference on Current Nuclear Power Plant Safety Issues, October 1980, p. 11.

70. NRC, ''Current Events, Power Reactors,'' June 1978.

71. NRC, *Power Reactor Events*, 2 (No. 2), March 1980; and Office of Inspection and Enforcement, Bulletin No. 79-27.

72. Reference 71 (No. 3), May 1980; and Office of Inspection and Enforcement, Information Notice No. 80-10.

73. NRC, Memorandum from Robert Bernero and F.H. Rowsome to Office of Nuclear Reactor Regulation and AEOD, 14 March 1980.

74. This description of control system inadequacies was adapted from Reference 69, p. 8.

75. ACRS, Letter to NRC Chairman, 10 June 1980.

76. Reference 75.

77. Reference 34, p. 3-1 ff.

78. Reference 34, p. D-13.

6

The Impact Of Three Mile Island On Nuclear Regulatory Requirements

The March 1979 accident at the Three Mile Island (TMI) reactor in Pennsylvania will affect design and construction requirements for nuclear plants in several fundamental ways. Because the accident extensively damaged the reactor core and was "evidently a significant precursor of a core-melt accident,"[1] the Nuclear Regulatory Commission (NRC) will now make "explicit consideration of accidents involving severely damaged or molten cores"[2] in its licensing reviews. This will almost certainly cause substantial upgrading of systems for emergency core cooling, decay heat removal, radiation protection, instrumentation and control, and emergency management. It may also lead to significant new design features such as "core ladles" or "filtered, vented containments" to accommodate damaged cores.

Perhaps more importantly, both the occurrence of the accident and the errors and malfunctions that characterized it have severely challenged basic premises in NRC's approach to safety. Previously, plants were licensed according to their ability to mitigate a set of *design basis accidents*. *Safety-related equipment* was designed and built to meet the *single failure criterion* to ensure that the vital flow of cooling water to the reactor core was maintained, regardless of the status of *non-safety equipment*.[a] The accident, however, "exceeded many of the present design bases by a wide margin"[3] because of multiple failures involving both safety-related and non-safety equipment.

Finally, the accident underscored important analytical and administrative gaps in NRC's regulatory policy that will need to be remedied, at probable great cost to the nuclear industry. NRC was pre-occupied with "large-break" loss-of-coolant accidents and thus virtually ignored accidents involving limited coolant loss as occurred at TMI, even though such accidents were con-

a. These terms are explained later in the text.

sidered much more likely.[4] Its policy of "grandfathering" some regulations contributed to the release of radioactivity from within the containment to the auxiliary building, and thence to the atmosphere, even though a regulation requiring redundant containment isolation predated plant start-up by several years. Most damningly of all, NRC took no preventive steps following TMI "dress rehearsal" incidents at other reactors and failed to heed an analysis of design weaknesses in Babcock & Wilcox (B&W) reactors that anticipated the accident.

As a result, TMI has occasioned "an across-the-board re-examination of the present regulatory system [in which] the entire complex of design, manufacturing, operation, and...regulation [is] on trial," according to *Nuclear Engineering International*. This will lead to "upgraded requirements in virtually all areas and new approaches to safety in many of them,"[5] even in systems that were not directly at fault in the accident. As the director of the Nuclear Power Division of the Electric Power Research Institute has acknowledged,

> A significant number of the lessons [from TMI], perhaps the majority of lessons, will have nothing to do with the accident, but will be learned because the system is being subjected to a scrutiny considerably more intense than has been the case in the recent past.[6]

Accordingly, "licensing stability," the holy grail of fixed design requirements perenially pursued by the nuclear industry, is now "a total loss," in the words of one senior industry executive said to have long been involved in licensing issues.[12] The Three Mile Island accident "has permanently altered the regulatory process for nuclear power," states a recent Department of Energy (DOE) analysis.[13] New regulatory requirements will be added and current ones modified at accelerated rates, and costs will grow apace.

Section 6.1: Three Mile Island And NRC's Approach To Safety

Prior to TMI, NRC based its licensing of reactors upon provisions made in plant designs to prevent or mitigate a group of specific, postulated accidents. This group either included or was supposed to "envelop," by virtue of the severity of the assumed accident sequences, all conceivable mishaps except for so-called "Class 9" accidents. These are calamities considered so unlikely — and in some cases believed so difficult to mitigate — that reactor designs were not required to withstand their hypothetical occurrence.[14] Among the Class 9 accidents excluded from the "design basis" were all accidents involving significant core damage, including a core meltdown, wherein the buildup of heat from a loss of reactor coolant causes the fuel rods to melt, creating a

Bechtel On Three Mile Island And Its Aftermath

One view of TMI within the nuclear industry is that its apparently limited consequences demonstrate that current reactor designs provide sufficient safety margins. The major changes needed are operational and managerial reforms that can be enacted quickly and cheaply. For example, in June 1980 W. Kenneth Davis, vice president of Bechtel Power Corporation, told a parliamentary committee weighing Great Britain's nuclear policy:

> Probably the most important lesson learned was that the avoidance of accidents depends more on human and institutional factors than on design features.

> The most important lessons learned were implemented for operating plants within a short time after the accident.

> The main conclusion to the Kemeny review was that nuclear power should proceed, but proceed with caution.[7]

These statements are open to serious question, however. The many TMI-related design changes and equipment requirements under NRC consideration belie the notion that most of TMI's legacy is already in place in the form of low-cost "software" improvements. Similarly, the Kemeny Commission "gave no such yellow light" to the nuclear industry[8] as Davis alleges in the last excerpt above. Indeed, the Commission expressly "did not attempt to reach a conclusion as to whether . . . the development of commercial nuclear power should be continued."[9] Its report did state as its fundamental conclusion that:

> To prevent nuclear accidents as serious as Three Mile Island, fundamental changes will be necessary in the organization, procedures, and practices—and above all—in the attitudes of the Nuclear Regulatory Commission and, to the extent that the institutions we investigated are typical, of the nuclear industry.[10]

Finally, if the Kemeny Commission members "chose not to stress the need for specific technological improvements, [this was] because they thought more specialized studies now being drafted would do just that," according to *Science*.[11] And indeed, NRC's comprehensive, 500-page TMI "Action Plan," published in May 1980, promises extensive new software *and* hardware, as discussed in Sections 6.2 and 6.3.

molten core that could potentially breach the containment and release a significant fraction of the reactor's inventory of radioactivity.

To judge the reliability of systems intended to prevent design basis accidents, NRC employed a concept called the *single failure criterion*. As explained by the Rogovin Report, the detailed TMI inquiry commissioned by NRC,

> This criterion is a requirement that a system designed to carry out a specific safety function must be able to fulfill its mission in spite of the failure of any single component within the system, or failure in an associated system that supports its operation. (In reality, the single failure criterion is a *double* failure criterion: it requires that the design must be able to bring the plant to a safe shutdown despite occurrence of an accident *plus* the failure of any one additional safety component or system.)[15]

Although the single failure criterion promoted reliability by requiring that reactors contain redundant and diverse systems for preventing accidents, it also was used to set an upper bound on safety: NRC did not require that designs mitigate accidents in which two or more "safety-related" components or systems fail independently.

Moreover, the only systems and components required to meet the single failure criterion — indeed, the only ones considered in NRC's reviews — were those that the applicant, subject to NRC approval, deemed safety-related. This comprised all equipment which, according to the applicant's or NRC's analysis, would be needed to shut the reactor, remove decay heat, and contain radioactivity during design basis accidents. Non-safety items are "not reviewed by the NRC to see whether they will perform as intended or meet various dependability criteria [nor do they] receive continuing regulatory supervision or surveillance to see that they are properly maintained or that their design is not changed in some way that might interact negatively with other systems."[16]

The TMI accident revealed serious shortcomings in each of these constructs upon which NRC has based its approach to safety. First, the "sequence of events [and the degree of] core damage [were] more severe than those considered in current design basis events."[17] Core temperatures exceeded 3500°F,[18] a level at least 1300°F above the design basis for emergency cooling systems. The chemical reaction between water and the zirconium fuel cladding generated five to ten times as much potentially explosive hydrogen as is specified in designing hydrogen control systems.[19] Fuel pellets and stainless steel parts of fuel assemblies cracked, crumbled, and may have melted and fused together.[20] Although such damage was considered possible before the accident, it was excluded from the design basis since plant safety features were provided to prevent it[21] and its probability of occurrence was judged so low as to make it incredible.

Second, several multiple failures compounded the accident, demonstrating the inadequacy of the single failure criterion. Auxiliary feedwater, a safety system counted on to maintain the flow of reactor coolant when the main feedwater supply was interrupted, was lost in the initial minutes because *both* redundant discharge valves were closed. Both independent trains of another automatic safety system, the Emergency Core Cooling System (ECCS) high-pressure injection system, were turned off for several hours by the operator.

Third, equipment categorized as non-safety-related not only initiated the accident (through failure of the entire feedwater purification system due to one malfunctioning component), as NRC and industry analyses assume, but also unexpectedly aggravated it. The pressurizer relief valve that stuck open, allowing water to flow out of the reactor vessel in the equivalent of a "small-break" loss-of-coolant accident (LOCA), was not safety-related and so was not subject to strict requirements for fabrication, qualification, and redundancy. Nor was the instrumentation which indicated merely that the valve had been *signaled* to close but did not show that it had stuck open. Similarly, the thermocouples that correctly indicated dangerously high core temperatures, but which the operators apparently disregarded due to past failures, were not safety-related. In fact, control room design, operator procedures, and most instrumentation, all of which were heavily implicated in the accident, were not labelled safety-related and so were not reviewed by NRC.[b] In particular, it was assumed that operators might not *activate* safety systems, but not that they might *defeat* them.

These "failure modes" carry profound implications for nuclear safety regulation. First, because "some accidents are outside or are not adequately assessed within the [current] 'design envelope,'"[22] NRC will expand the spectrum of design basis accidents used for safety assessments. "Degraded core" and core-melt accidents will receive particular attention in both analysis and design.[23] NRC's pending rulemaking on degraded cores will consider for the first time design changes necessary in major plant systems—containment, emergency cooling, decay heat removal, and radiation control—to cope with major core damage.

Second, NRC is beginning to upgrade the single failure criterion to require that some safety systems be designed to perform despite *multiple* independent failures in addition to an assumed initiating failure. NRC has already abandoned the criterion for auxiliary feedwater systems by requiring licensees to adopt operating procedures that can compensate for loss of that system, even though it is intended to withstand single failures. The Commis-

---

b. The TMI operators did employ some non-safety-related equipment to recover from the accident. Although some equipment items did thus function beyond their design requirements, others such as the reactor coolant pumps (which augmented coolant circulation) were severely strained in doing so and cannot be relied upon to mitigate other accidents. In fact, the pumps were shut off approximately 1½ hours into the accident due to severe vibration, terminating coolant flow and leading to core damage.

sion is also considering requiring additional power supplies in the event that a plant is "blacked out" from the transmission grid *and* its onsite power systems fail (two independent failures) during a "transient" that shuts the plant generator.[24]

More fundamentally, the Advisory Committee on Reactor Safeguards (ACRS) has concluded that the single failure criterion "is more likely to be applicable only to very simple systems" such as design of electric circuitry, for which it was originally devised.[25] "For complex systems [such as nuclear plant safety systems] multiple failures may be experienced subsequent to the initial failure and some other standard of acceptability is needed."[26] Says the ACRS,

> [A]ttention must be devoted to the sequences and consequences that could result from many different combinations of multiple mistakes or failures, *two or three or five, or even six,* as necessary to determine the possible interactions and their consequences . . .[27]

This approach would significantly expand NRC failure assumptions, primarily by requiring consideration of the consequences of losing redundant safety systems for a wide range of accidents.

NRC's proposed successor to the single failure criterion is a program for developing complete taxonomies of accident sequences which allow for operator error, multiple failures of active safety-related components, and single failures of passive components.[c] The "Interim Reliability Evaluation Program" (IREP) is expected to provide the means "to identify particularly high-risk accident sequences at individual plants and determine regulatory initiatives to reduce these high-risk sequences."[28] NRC does not expect to complete the risk analyses for each plant until at least 1983, however,[29] after which the analyses and debates over proposed design and equipment changes would begin. Accordingly, although the impact of upgrading the single failure criterion cannot be charted today, the effort to do so will involve considerable uncertainty that will aggravate the many significant ongoing problems in reactor design and construction.

Finally, "[t]he current classification of systems and equipment into 'safety-related' and 'nonsafety-related' is especially unsatisfactory," said the Rogovin Report,[30] and "NRC's past emphasis on ill-defined, safety-related systems and components has caused it to miss important safety issues," according to the General Accounting Office (GAO).[31] In the future, more equipment will be subject to NRC requirements for redundancy, diversity,

---

c. Active components require mechanical motion to fulfill their safety functions, *e.g.*, opening of a valve, as distinguished from passive components such as pipes that do not require actuation. To date, NRC has not required licensees to postulate failure of safety-related passive components in mechanical systems.

Chapter 6

qualification, and strict quality assurance (QA). NRC has begun to expand the safety-related category, (*e.g.*, to include pressurizer relief valves), and it is also considering establishing intermediate classifications between systems judged most and least important to safety, based on the findings of the IREP program. Special attention will be given to potentially harmful effects of nonsafety equipment on safety equipment—a long-time concern of ACRS[32]— in part through a regulatory guide that NRC hopes to issue in late 1983 instructing licensees to apply QA criteria to nonsafety systems whose failure may pose risks.[33] (See Section 6.3 for further discussion of QA after Three Mile Island.)

Section 6.2:   Effect Of TMI On Equipment And Design

"The accident at TMI," reports NRC's TMI Action Plan, "demonstrated the reality of the risk, *previously only theoretically assessed*, of accidents that result in substantial degradation and melting of the core." Accordingly, "[t]he Action Plan calls for the development and implementation of a number of phased actions dealing with *explicit consideration* of accidents involving severely damaged or molten cores.[34]

These actions and other prospective design and equipment changes stemming from TMI are discussed in the following sequence: rulemaking on degraded cores, improvements in instrumentation, control room, and emergency preparedness, ECCS and decay heat removal systems, and radiation protection.

Rulemaking on Degraded Cores: NRC's major action to develop post-TMI design and equipment requirements is a rulemaking to establish policy concerning possible accidents that fall outside the current design basis. The rulemaking will consider:[35]

(1) *Filtered-vented systems* to relieve containment pressure by venting steam and gases through clean-up, heat-absorbing devices. These systems, still in the conceptual stage, might consist of pools of water with submerged gravel-sand filters.

(2) *Core ladles* designed to temporarily contain a molten fuel core that has burned through the reactor vessel, perhaps similar to the magnesium oxide "core catcher" once considered for floating nuclear plants.

(3) *Hydrogen control measures* to prevent hydrogen-oxygen explosions inside containment. Possible measures include maintaining an oxygen-free containment ("inerting"), controlled-ignition devices, and controlled-burning "recombiners" outside containment fed by dedicated penetrations.

(4) Expanded design criteria to ensure that decay heat removal, radwaste, and makeup and purification systems can function under degraded-core conditions.

(5) Radwaste system design features to aid in post-accident recovery and decontamination.

There can be no mistaking the rulemaking's sweeping scope which, according to an anonymous senior NRC staffer, "will make the ECCS hearings [the bitterly contested AEC rulemaking that lasted from mid-1971 to early 1973] look like a Sunday school picnic."[36] The Action Plan estimates that "as many as 40 regulatory guides . . . may have to be revised to achieve a consistent regulatory approach."[37] Insofar as the entire NRC regulatory structure assembled from the late 1960s to the present comprises less than 150 regulatory guides, the rulemaking appears capable of massively restructuring current regulatory standards.

Only a few very preliminary cost estimates are available for the equipment improvements and new systems that the rulemaking will consider. Filtered-vented containments might cost between $10 and $50 million per reactor, according to a tentative Action Plan estimate,[38] but NRC cost projections for design changes have usually proven low. Similarly, the nuclear industry concedes that its old $3 million estimate for floating-reactor core ladles is not applicable today because area and thickness requirements have increased.[39] Perhaps most tellingly, the Atomic Industrial Forum (AIF) did not estimate hardware costs to mitigate degraded cores in its cost assessment of over 50 other TMI-related measures,[40] probably reflecting industry's aversion to giving explicit consideration to particularly costly new measures.

Even speculative cost estimates are scarce because the requirements that will emerge from the degraded core rulemaking are unknown. Aside from the prospective combative nature of the rulemaking, there are *"critical phenomenological unknowns* or uncertainties that impact containment integrity assessments and judgments regarding the desirability of certain mitigating features."[41] "Research on degraded cores stopped in 1963,"[42] according to the senior NRC staffer quoted earlier, because such accidents were outside the design basis. As a result, core geometry and motion during and after a meltdown are "a major area of uncertainty," and such pivotal issues as thermal hydraulics, material motion, heat transfer, and the effects of reactor vessel failure "are not well understood."[43]

NRC's research to provide a rulemaking basis will investigate *damaged fuel behavior*, including hydrogen formation, fuel and clad melting, debris formation, and flow blockage; *molten core behavior*, including its possible interactions with reactor coolant, containment fluids, plant structures, and soil; and *effect of hydrogen explosions on containment*.[44] The program is large, with approximately 40 research projects costing $15 million in FY-82 alone.[45] NRC does not anticipate completing these studies until late 1983,[46] a schedule that augurs poorly for expeditiously completing the rulemaking and resolving the serious design uncertainties that the degraded core issue poses.[d]

The prospects for settling this issue without extensive design and equipment changes appear dim. TMI, the Browns Ferry partial scram failure and other recent reactor mishaps (Chapter 5), and re-evaluations of earlier accident analyses have convinced one ACRS member that "There are many potential paths to severe core damage or core melt so that it will be difficult to make the frequency of such an accident very much smaller [than about one in a thousand per reactor per year] with a high degree of confidence."[48] The solution, then, is not only to take steps to reduce the chances of core-damage accidents but also to "provide containment capability, as practical, for a wide spectrum of severe accidents as a separate line of defense."[49]

Improvements in Instrumentation, Control Room, and Emergency Preparedness: The demonstration at TMI of the "reality" of core-melt accidents is also motivating improvements in reactor instrumentation, control room design, and equipment to manage emergency response.

NRC has acknowledged a need for vastly improved equipment to monitor plant systems and variables during and after accidents, including those involving degraded cores.[50] A revision of Regulatory Guide (RG) 1.97 awaiting Commission approval will require capability for measuring temperatures, pressures, flow rates, water levels, gas concentrations, radioactivity levels, and valve positions at hundreds of locations in the reactor core, the reactor coolant system, containment systems, steam generators, auxiliary systems, and electrical power systems.[51] Most of the instruments will need to be qualified for accident conditions, and many will require recording capability to aid diagnosis during and after accidents. A minority of the instruments have been ordered for 1981 installation, but the schedule for the remainder is still pending. NRC's estimate of the total implementation cost is up to $6 million per reactor,[52] while industry's is "$6 million with a high probability of exceeding $14 million" (slightly less for plants in early construction).[53] Moreover, RG 1.97 is said to include only minimum requirements for instrumentation, and more may be added in the future.[54]

Control room instruments and designs will also be improved under the Action Plan. Equipment for recording vital data, testing control panel lights, and regulating control room access, for example, will be upgraded at new reactors and at most existing ones. In addition, a *Safety Parameter Display System* will report on important data, such as reactivity control and core cooling, to give operators an overview of the plant's safety status. The system and improvements are estimated to cost $2-4 million per plant, exclusive of implementation of a regulatory guide being prepared to settle the "unresolved

---

d. Similarly, NRC did not issue the rulemaking notice until October 1980, far behind its original, October 1979 objective of publishing it "within a few months."[47]

safety issue'' status pending for control room design.[55]

Simulators will probably be installed at each reactor site, in part to help operators increase their proficiency to conform with stiffer NRC requirements.[56] These extremely complex machines must mimic the plant's entire piping, wiring, and control systems, even including the plant computer, and must train operators to cope with multiple failure, incorrect instrumentation, and failure of safety systems. The Action Plan estimates their cost at up to $6 million each,[57] but one industry estimate is $9 million,[58] and a recent contract indicates a much higher cost.[e]

The Action Plan also calls for three new facilities at each plant to support emergency operations: managerial and technical staff would diagnose an accident and conduct emergency operations from a *Technical Support Center* built to control-room habitability standards and receiving most of the accident-parameter information required in RG 1.97; to avert control room crowding, support staff would report to an *Onsite Operational Support Center* for task assignments; and emergency functions and radiation measurements would be co-ordinated with federal, state and local authorities at an offsite *Emergency Operations Facility* provided with a duplicate Safety Parameter Display System. Industry cost estimates for the combined facilities are $5-11 million per site, not including communication systems linking the plant to local emergency authorities.[60]

Finally, NRC is considering requiring *data links* to convey critical information, including all instrumentation status stipulated in RG 1.97, to NRC's operations center outside Washington.[61] Although industry is bitterly contesting this proposal on cost grounds, the widespread mistrust of utility capability to handle reactor crises makes it plausible that NRC will require that the data links provide comprehensive, on-line information-exchange capability, at considerable cost.

Decay Heat Removal and Emergency Core Cooling Systems: The residual heat removal (RHR) function removes fission product "decay heat" after reactor temperature and pressure have been brought below normal operating conditions, *e.g.*, following a reactor scram. At most plants it shares pumps and heat exchangers with parts of the ECCS—an arrangement that is said to restrict plant flexibility to arrest certain accidents.[62] The Action Plan commits NRC to study the desirability of a dedicated, "bunkered" RHR system independent of other plant systems.[63] Such a system would require its own services and coolant supply, would need to be fully available on demand and immune to plant equipment failure, and would be protected against most acts of sabotage.[64] Evaluation and promulgation of RHR system requirements are expected to take two to three years.[65] (NRC staff has recommended that RHR

---

e. Consumers Power is constructing a $38 million nuclear training center with simulators and duplicate control rooms to serve two reactor stations, Palisades and Midland.[59]

function be designated an unresolved safety issue.)

Concern over the RHR system's dependence upon non-safety-qualified equipment such as the ECCS precedes the TMI accident. Because RHR systems depend on instruments, valves, and logic systems that might be lost due to power failure, the system might be unable to function following the reactor scram which such a power failure would initiate. Thus, separate from the Action Plan, the ACRS has recommended improving the reliability of the electrical services supporting the RHR system or providing the self-contained system to be studied by NRC.[66] In addition, the desirability of RHR upgradings has been highlighted by several losses of RHR capability at PWRs, most seriously at Davis-Besse in 1980.[f]

Similarly, TMI has aggravated long-standing concerns over ECCS capability by demonstrating that emergency cooling systems could be required to operate under a wider range of pressures and for longer periods of time than hitherto assumed. NRC staff is studying ECCS response for possible designation as an unresolved safety issue and is undertaking new studies to reduce performance uncertainties under diverse accident conditions.[g] Whereas previous research focused on large pipe breaks, assuming, incorrectly, that their consequences would necessarily be worse than those of more modestly initiated accidents, NRC will now examine small break LOCAs and transients. "Small-break LOCA analyses performed by the LWR vendors . . . have shown that large uncertainties may exist in system thermal-hydraulic response due to modeling assumptions or inaccuracies," says the Action Plan, and changes in analysis methods may be needed that would require modifying operating procedures and/or equipment.[69]

Radiation Protection: The TMI Action Plan also commits NRC to correct deficiencies that the accident investigations found in the design, equipment, and management of radiation protection programs.[70] The degraded-core rulemaking and other Action Plan provisions will consider a host of measures to reduce the potential for worker exposure and radiation releases during accidents and to aid in accident recovery and decontamination:[71] piping and instrumentation to detect, vent, and segregate radioactive liquids and gases in

---

f. In April 1980, RHR capability at Davis-Besse was lost for 2½ hours during refueling when power to one RHR train failed while the other was down for maintenance. The containment spray system and part of the ECCS were also down for maintenance, and the reactor vessel head was detensioned and thus might have failed under high pressure. Reactor temperatures rose from 90°F to 170°F before the heat removal pump damaged by the power failure could be repaired.[67]

g. A 1980 NRC Bulletin provides an excellent example of the effect of TMI on ECCS concerns. It reports Westinghouse's discovery that failure of the pressurizer relief valve could damage the centrifugal charging pumps—a high-pressure addition to the ECCS at recent PWRs— before they could complete coolant injection. Failure of that valve allowing reactor coolant to drain out through the pressurizer was a critical malfunction at TMI, and it is fair to surmise that it was instrumental in Westinghouse's discerning the vulnerability of the charging pumps.[68]

primary coolant that would collect in containment following some accidents; additional, upgraded filters to capture iodine and other radionuclides that might be vented to areas outside containment such as the radwaste building or the control room; and design features such as hookups for portable decontamination systems to facilitate post-accident cleanup.

Monitoring of radioactivity is also likely to be expanded. The Action Plan will consider requiring effluent monitoring for a wider selection of radioisotopes and continuous monitoring of isotope concentrations in the environment (concentrations are currently measured over three-month periods rather than continuously).[72] Monitoring systems may need to be designed to accident conditions (environmentally-qualified and equipped for a broad range of concentrations) and connected to the control room for continuous data transmission. In addition, NRC will develop more stringent performance criteria for radiation surveying and monitoring instruments, such as dosimeters, to improve worker protection and reduce personnel risks from corrective efforts during accidents.[73]

Section 6.3: Effect Of TMI On Quality Assurance And Requirements For Safety Equipment

NRC is undertaking two major initiatives to upgrade the integrity of nuclear plant design, equipment, and construction following TMI. One initiative will apply quality assurance and other stringent requirements to some equipment and systems previously judged not important to safety and thus not subject to QA requirements or NRC review. The other will strengthen the conduct of QA and expand NRC inspection of equipment design, fabrication, and installation.

Expanding the Safety-Related Classification: As discussed in Section 6.1, NRC requirements for ensuring a high degree of reliability apply only to equipment that the licensee, with NRC approval, deems necessary to prevent design basis accidents. Other, ''nonsafety'' equipment is assumed to neither enhance nor interfere with the performance of ''safety'' equipment during accidents. The TMI accident, however, has drawn attention to the fact that NRC accident analyses assume performance by a substantial body of nonsafety equipment whose design and manufacture are not reviewed by NRC and which is not built to withstand accident conditions. This category includes control systems, in-core instrumentation, pressurizer relief valves and heaters, level instrumentation for the pressurizer, steam generator and refueling water tanks, turbine bypass valves, and diesel generator support systems.[74]

NRC's program for expanding the list of safety-related equipment is based on its Interim Reliability Evaluation Program (IREP, see Section 6.1). This comprehensive assessment of the potential contribution of each plant component and system to accident risks is not scheduled for completion until late 1983, however, ensuring that the applicability of requirements for safety-related equipment will be uncertain for at least several years. The target date is uncertain, too, because of the many different cooling systems, instrumentation and controls, plant power sources, etc., in use and the difficulty of scoping the many types of interactions between safety and nonsafety equipment which can add to risk (see p. 136).

Moreover, data deficiencies may severely constrain the entire approach of probabilistic risk assessment. Many components whose contribution to risk must be assessed do not have well-defined failure rates because they are employed rarely or in heterogeneous situations. "There's not enough experience and the character of the input data changes,"[h] says one ACRS member, contrasting the recent occurrence of partial scram failure at Browns Ferry Unit 3 (p. 144) with the "unrealistically low probability" previously assumed by General Electric.[75] Because "the scram system was probably the system most studied using probabilistic techniques, . . . this occurrence must give pause to one's acceptance of any claim of high reliability for a particular system, based solely on probabilistic analysis," according to another ACRS member.[76] These methodological limitations are likely to force NRC to take a conservative approach in determining which equipment is important to safety.

Changing the Conduct and Scope of Quality Assurance: NRC will also expand its oversight of nuclear plant design, equipment-manufacture, and construction because of TMI. Because deficient design and equipment figured heavily in the accident, quality assurance "needs to be strengthened in the areas of design methodology and installation conformance with design intent," according to the ACRS.[77] The accident closely coincided, moreover, with other revelations of shortcomings in nuclear industry design and construction practices and in NRC's oversight, as discussed in Section 5.1.

NRC's corrective program emphasizes keeping industry audits independent of design and construction teams, perhaps by requiring that licensees take over QA from their architect-engineers (A-Es) and constructors, or even by making QA personnel NRC agents.[78] The Commission is also assigning inspectors to all construction sites, increasing the extent of actual construction monitoring, and initiating trial programs in which it will independently examine construction materials and methods.[79] Activities of reactor vendors and A-Es may also come under closer scrutiny; NRC is considering assigning resident inspectors to vendor and A-E headquarters[80] and is being urged to require rigorous QA procedures for probabilistic analyses used in safety-related design.[81] These initiatives will, it is hoped, make for safer plants, but

h. An example of changing data given by the ACRS member is temperature fluctuations affecting valve springs.

they will also add to direct costs, schedule delays, and overall construction complexity.

Section 6.4: Effect Of TMI On Feedback From Operating Experience

All of the Three Mile Island accident investigations roundly criticized NRC and the nuclear industry for failing to incorporate information available through reactor operating experience into the design and operation of new plants. The Kemeny Commission, which studied the accident for the President, found that "prior to the accident there was no *systematic* method of evaluating [operating] experiences, and no systematic attempt to look for patterns that could serve as a warning of a basic problem."[82] The Rogovin Report concluded that "NRC and the industry have done almost nothing to evaluate systematically the operation of existing reactors, pinpoint safety problems, and eliminate them by requiring changes in design, operator procedures, or control logic."[83] The Rogovin Report called this "an unacceptable situation that compromises safety and that cannot be allowed to continue,"[84] while the Kemeny Commission urged that current plants and past experience be combed "to assess compliance with current requirements, to assess the need to make new requirements retroactive to older plants, and to identify new safety issues."[85]

These stinging criticisms and emphatic recommendations were occasioned by the fact that the initial sequences in the TMI accident had happened twice before—at a Swiss reactor in 1974, and at Davis-Besse in Ohio (a virtual twin of TMI) in 1977. In both instances, and at TMI, instruments misled operators into thinking the reactor vessel was full while reactor coolant was actually escaping through a stuck-open pressurizer-relief valve. Although the first incident involved a Westinghouse reactor, it was not reported to NRC because foreign plants do not fall under NRC reporting requirements. The Davis-Besse case was intensively studied by the utility, the vendor, and the NRC, but no recommendations were developed and no findings were transmitted to Metropolitan Edison, the operator of TMI.[i] To the Kemeny Commission, this illustrated the "lack of 'closure' in the system" that ranked high among the deficiencies which "convinced [us] that an accident like Three Mile Island was eventually inevitable."[86]

Remedial action by NRC has included several steps. One was establishment in late 1979 of an Office for Analysis and Evaluation of Operational Data (AEOD) to collect, analyze, and evaluate operating experience and recommend NRC action. Although the extent of the new Office's clout within NRC is

---

i. The Rogovin Report (Reference 14) contains the definitive account of the TMI precursors and the failure to heed them (p. 94ff).

not yet clear, it was lauded by the President's Nuclear Safety Oversight Committee for its in-depth analysis of the 1980 Browns Ferry partial scram failure.[87] Moreover, its recommended response to that accident goes beyond NRC's initial actions,[88] and the AEOD director also has insisted that industry-sponsored data groups identify the subject plants of operating reports to NRC, even if NRC uses such information punitively.[j] "Trying to operate in a semi-secret atmosphere is an old way of doing business that shouldn't persist," he was reported as saying.[89] NRC is currently reviewing its requirements for non-licensee reporting (10 CFR 21) and is revising several regulatory guides governing licensee reporting requirements.

In addition, the Commission appears to have stepped up licensee notification of new safety problems surfacing in operating experience or elsewhere (e.g., research and testing programs, plant construction). Issuance of safety-problem bulletins and circulars by the Office of Inspection and Enforcement has risen by more than 50% since Three Mile Island, from a rate of .57 bulletins and circulars per reactor-year during 1978 and the first quarter of 1979, to .88 in the 21 months following TMI (April 1979 through December 1980; see Figure 5.1 in previous chapter). This dramatic increase probably reflects both an unrelated spurt in the occurrence of safety-related problems and a heightened sensitivity to accidents and to the importance of operational feedback after TMI. In light of the importance of past "lessons learned" in adding to regulatory requirements and costs,[k] continuance of this trend would accelerate the rate of increase in regulatory standards and costs.

Section 6.5: The Safety-Cost Tradeoff After TMI

Costs are incurred whenever equipment is added or designs are altered to reduce accident risks. Neither Congress nor NRC has established a "safety goal" for nuclear power,[l] and the Commission has few guidelines for comparing equipment costs against safety benefits. Nevertheless, NRC commissioners and staff are mindful of the potential cost impacts of new regulatory requirements, and decisions to strengthen safety requirements are usually made with at least an implicit recognition of their possible costs and effects on

---

j. The industry groups referred to are the Nuclear Safety Analysis Center and the Institute of Nuclear Power Operations, both formed in the aftermath of TMI.

k. The effect of operating experience on regulatory requirements is treated in detail in Chapter 3. Despite the lack of past *systematic* evaluation of operating experience (a qualifying term employed by both the Kemeny and Rogovin reports), some evaluation, if haphazard, was performed by AEC and NRC and contributed significantly to increased regulatory standards.

l. NRC has a program to develop a safety goal by the end of 1981.[90] The provisions for public participation and the complexity of the issue make it doubtful that the schedule can be met, however.

the nuclear industry. Accordingly, NRC's willingness to abide by the cost consequences of its regulatory actions can be significant in determining whether new regulatory criteria are adopted.

There is much evidence that the Three Mile Island accident has shifted the equilibrium between lower costs and added safety toward the side of greater safety. The many new requirements, initiatives, and proposed actions in the TMI Action Plan can be taken as signs of a stronger intent to reduce accident hazards. Another measure is the increase in safety-related communications to licensees discussed just above. A third is current statements of regulatory philosophy; these too give indication that safety concerns now loom larger relative to costs than they did prior to TMI.

What might be termed the "pre-TMI" perspective on balancing costs and safety was articulated by AEC Commissioner (later NRC Chairman) Anders to Congress in 1974:

> When one speaks of costs, *it would be irresponsible not to balance the gain from the incremental improvement in safety against the incremental cost of this improvement.* AEC is particularly concerned about... the various social and environmental costs that could result from a lack of power.[91]

Contrast that statement with then-NRC Chairman Hendrie's remarks to the Commission's chief industry adversary on regulatory matters, the Atomic Industrial Forum, in late 1979:

> We can and do consider costs in our rulemaking... but *the improvement in safety.. must be the dominant element in our considerations...* [W]hen we come to... a specific matter, on a specific plant, costs and related factors do not count for much.[92]

The treatment of safety and costs as co-equals has given way to the dictum that safety is paramount and costs are secondary.

A similar change is apparent concerning the related question of which party in the reactor safety debate bears the burden of proof. NRC Commissioner Gilinsky believes in retrospect that past policy put "the burden of proof... on the *regulators* to justify negative findings on safety matters."[93] Thus, NRC mandated only "the most conservative requirements consistent with the commercial viability of nuclear power," according to one account of pre-TMI regulation.[94] As the Kemeny Commission concluded, "the NRC has sometimes erred on the side of the industry's convenience rather than carrying out its primary mission of assuring safety."[95]

The trauma of TMI, however, has "shattered [NRC's] complacency" about reactor hazards, says Gilinsky.[96] NRC proclaims itself "in complete accord with the [Kemeny Commission's] proposition that there should be a presumption in [safety-cost] tradeoffs in favor of safety."[97] ACRS similarly has urged "a fundamental change" in the approach of the architect-engineer and the licensee "to make the safety of the plant as good as reasonably achievable, rather than merely meeting existing regulatory requirements at minimum cost."[98]

Thus, the climate at NRC has changed from one that overtly sought to weigh equally the nuclear industry's needs and public safety, to one in which safety is said to come first. Although it remains to be seen whether this climate has the lasting power to translate into effective action—and to hold up against a more energy supply-minded Administration and Congress—the attitudes noted here appear deep-rooted and unlikely to fade quickly. The reports of the Kemeny Commission, the Rogovin group, and the many other TMI investigations, moreover, have stressed that "NRC can no longer disregard its critics by citing the safety record of the industry it regulates."[99] It is fair to anticipate that the post-TMI regulatory programs and initiatives discussed here will be pursued more vigorously than their pre-accident counterparts.

Section 6.6: Effect Of TMI On Construction Scheduling And Logistics

This chapter has demonstrated that the Three Mile Island accident toppled basic precepts of nuclear safety regulation, tightened surveillance of operating experience, tempered NRC's sensitivity to the costs of new regulatory standards, and stimulated development of an Action Plan that will impose significant new design and equipment requirements and substantially upgrade the oversight of reactor design and construction. These efforts are in addition to ongoing programs that are generating and revising many regulatory guides but are registering little success in resolving significant outstanding safety issues (Chapter 5).

Those ongoing programs alone were sufficient to create a regulatory logjam and impose considerable capital cost increases, without the accident. There is now the prospect that the accident's impact may overwhelm the regulatory, manufacturing, and financial machinery necessary to build reactors, causing cost and schedule increases that could strain the system further and make it impossible to bring some plants to completion.

Regulatory Pressure: NRC has been charged with many difficult mandates in the wake of TMI. It must establish a comprehensive plan for the systematic safety evaluation of all 70 operating reactors[100]—a considerable broadening of the former Systematic Evaluation Program (SEP) covering only the 11 oldest plants—while improving lessons-integration from plant operations, settling the unresolved safety issues, and carrying out the ambitious TMI Action Plan. It must manage these programs, as well as ongoing safety research, regulatory guide issuance and revision, etc., with at best only modest increases in personnel and in real expenditures.

The effect will be a compounding of regulatory uncertainty, as indicated by the 1982-84 resolution targets of most key Action Plan items, by slippage in many targets to date (only some of which is due to industry lobbying),[101] and by statements from NRC officials. As the director of NRC's Division of Systems Safety told the Commissioners in May 1980,

> The flavor you should have from the [Action Plan] is that we did defer into 1982 a lot of things which were very important to the Kemeny Commission and to the Rogovin group, as well as to our own internal review.[102]

One indication of industry's expectations is that Commonwealth Edison will wait until "about 1985" to decide whether to backfit or retire its old Dresden 1 reactor; "at that time, NRC should have decided just what post-TMI and SEP fixes will have to be made on the unit."[103]

With NRC's attention focused upon *operating* plants, new reactors are being downgraded in the allocation of regulatory resources. Reviews of construction permit applications have been suspended since TMI, pending formulation of a licensing policy to apply accident lessons to new plants—a low-priority item compared to safety improvements at existing plants. As NRC's Director of Reactor Regulation told ACRS in justifying the suspension, "I wish I had the resources to develop new positions. I have enough to do with Browns Ferry, St. Lucie, and Crystal River [sites of serious accidents in 1980]."[104] The imbalance between NRC's resources and its responsibilities alone would perpetuate considerable licensing instability, even in the absence of the new problems which future reactor operations will inevitably generate.

Vendor Pressure: Some manufacturers that supply and build nuclear plants are experiencing difficulty in staying in business because of the acute diminution of the reactor business—a phenomenon which predates but has been exacerbated by TMI. Although none of the four suppliers of nuclear steam systems seems about to fold—indeed, the two leaders, General Electric and Westinghouse, boast of the profitability of their repair and service departments—the lack of reactor sales and the dimming of nuclear power's aura as the major future energy source are reportedly causing valued technical, engineering, and executive personnel to leave the industry. Many equipment suppliers are also reported to be withdrawing from the nuclear field, victims of "the difficulties [they] face in storing deferred equipment, maintaining their nuclear [QA] ratings, and dealing with the paperwork demanded of them by regulatory agencies" in a stagnant market.[105]

Shrinkage in the nuclear supply industry could raise reactor costs in several ways: by constricting competition and thus raising equipment prices; by precipitating equipment shortages which foul construction logistics; by dampening the willingness of firms to provide prompt, reliable supply and service in order to build long-term customer goodwill; and by reducing the flow of new talent into the nuclear field. Conversely, shrinking markets usually lead, at least temporarily, to falling prices, but this short-term effect is likely to be outweighed by the others.

Financial Pressure: The staggering cost of building new reactors— several billion dollars at least in current dollars—can put severe financial pressure on utilities to stretch out construction. (This is apart from the impact of regulatory changes on construction schedules, including the merits of deferring work while awaiting new requirements.) Such pressure can be especially acute when the completion target is far in the future, load growth is declining and uncertain, bids for higher electric rates are meeting resistance, and reactors are displacing relatively inexpensive coal rather than oil— circumstances which hold for many plants under construction. Moreover, continued construction of new reactors increasingly must compete for capital with other investments that may permit less flexibility or be more productive: construction of coal plants, oil-to-coal conversions, and post-TMI backfits of operating reactors.

Thus, construction schedules for some plants could be stretched out intentionally. Some utilities may even have no alternative if they are to maintain adequate credit ratings or even to comply with legal interest coverage requirements.[106] Although the effects of stretch-outs on plant costs are usually expressed in current dollars (unadjusted for inflation) and are grossly overstated, they do increase real interest costs, disrupt construction logistics, and expose plants to additional regulatory requirements. These effects can, in turn, foster further stretch-outs, diminish the viability of nuclear suppliers, reduce the regulatory priority assigned to new plants, etc., which would raise costs again, perpetuating the cycle.

Although the discussion in this section is admittedly somewhat speculative, it is true that reactor construction increasingly requires meshing of many financial, manufacturing, and regulatory gears. If momentum continues to be sapped from the nuclear enterprise, it is not inconceivable that the effort to co-ordinate all of the necessary parts could prove too great for some plants in progress, and that construction might permanently grind to a halt.[m]

m. The nuclear industry, for one, would brush aside this scenario with the argument that nuclear power expansion is essential to U.S. energy security and economic prosperity, so that, come what may, the means will be provided to complete reactors now being built. Nevertheless, nuclear power remains a minor energy source—providing less than 11% of U.S. electricity and under 2% of "end-use" energy in 1979-80 (3½% of gross energy supply). While nuclear generation has declined since 1978, increased use of coal and improvements in energy efficiency have significantly reduced U.S. oil use—by 600,000 barrels per day in the electric utility sector alone (a one-third reduction) during 1978-80.[107]

References

1. NRC, *TMI-2 Lessons Learned Task Force Final Report*, NUREG-0585 (1979), p. 3-5.

2. NRC, *NRC Action Plan Developed as a Result of the TMI-2 Accident*, NUREG-0660, Vol. 1 (1980), p. II-1.

3. Reference 1.

4. See, for example, NRC, *Reactor Safety Study*, WASH-1400, NUREG-75/014 (1975).

5. *Nuclear Engineering International, 24* (No. 10), 42 (1979).

6. Milton Levenson, Testimony before the House Subcommittee on Energy Research and Production, 22 May 1979.

7. W. Kenneth Davis, Testimony before the House of Commons Select Committee on Energy, 4 June 1980, pp. 13-14.

8. Russell W. Peterson (Kemeny Commission member), Letter in *New York Times,* 12 November 1979.

9. J.G. Kemeny *et al., Report of the President's Commission on the Accident at Three Mile Island* (1979), p. 4.

10. Reference 9, p. 7.

11. *Science, 206,* 796 (1979).

12. *Nucleonics Week,* 17 May 1979, p. 3.

13. DOE, *Nuclear Power Regulation*, DOE/EIA-0201/10 (1980), p. xiv.

14. This definition is from M. Rogovin *et al., Three Mile Island: A Report to the Commissioners and to the Public* (1980), Vol. 1, p. 147.

15. Reference 14. Emphasis in original.

16. *Federal Register, 45* (No. 193), 65475 (2 October 1980, advance notice of proposed NRC rulemaking on degraded cores).

17. Reference 1, p. 3-1.

18. Reference 14, Vol. 2, p. 496.

19. NRC, *Draft Programmatic Environmental Impact Statement for TMI-2 Decontamination and Disposal of Radioactive Wastes,* NUREG-0683 (1980), p. S-1.

20. Reference 19.

21. NRC, *TMI-2 Lessons Learned Task Force Status Report and Short-Term Recommendations,* NUREG-0578 (1979), p. 16.

22. Reference 14, Vol. 1, p. 150.

23. David Okrent, "New Trends in Safety Design and Analysis," paper presented at International Atomic Energy Agency Conference on Current Nuclear Power Plant Safety Issues, October 1980, p. 1.

24. *Inside N.R.C.*, 8 October 1979, p. 15.

25. ACRS, *A Review of NRC Regulatory Processes and Functions*, NUREG-0642 (1980), p. 6-5.

26. Reference 25.

27. ACRS, *Comments on the NRC Safety Research Program Budget*, NUREG-0603 (1979), p. 1-2. Emphasis added.

28. Reference 2, p. II.C-1.

29. Reference 2, p. II.C-5.

30. Reference 14, Vol. 1, p. 148.

31. GAO, *Three Mile Island: The Most Studied Nuclear Accident in History*, EMD-80-109 (1980), p. 16.

32. See, *e.g.*, Reference 25, pp. 6-2 and 6-6 through 6-10.

33. Reference 2, p. I.F-2.

34. Reference 2, p. II-1. Emphasis added.

35. Reference 16, pp. 65474-65477.

36. *Inside N.R.C.*, 28 July 1980, p. 7.

37. Reference 2, p. II.B-12.

38. Reference 2, p. II.B-15.

39. *Inside N.R.C.*, 3 December 1979, p. 13

40. Atomic Industrial Forum, "Report to the AIF Policy Committee on Follow-Up To The TMI Accident by the Working Group on Action Plan Priorities and Resources" (February 1980), Appendix A, p. D-11. The report estimated only the industry-wide cost ($57.5 million) for *conceptual designs* to mitigate degraded core accidents.

41. Reference 2, p. II.B-4. Emphasis added.

42. Reference 36.

43. Sandia National Laboratories, *LWR Safety Research Program Quarterly Report, April-June 1980*, NUREG/CR-1509/2of4 (1980), p. 62.

44. Reference 2, Task II.B. Emphasis added.

45. Telephone communication with Robert T. Curtis, Chief, Analytical Advanced Technology Branch, Office of Nuclear Reactor Research, NRC, 8 September 1980. The figures given are rough estimates including re-direction and expansion of current research as well as newly commissioned work.

46. Reference 2, p. II.B-7.

47. Reference 1.

48. Reference 23, p. 21. See also pp. 19 and 20.

49. Reference 23, p. 21. See also pp. 19 and 20.

50. Reference 2, p. II.F-1.

51. Regulatory Guide 1.97, "Instrumentation to Assess Plant and Environs During and Following an Accident," 1975, 1977, and proposed revisions in 1979 and 1980.

52. Reference 2, p. II.F-6.

53. Reference 40, Appendix A, p. C-11.

54. *Inside N.R.C.*, 8 September 1980, p. 6.

55. Reference 2, pp. I.D-8 and 9.

56. *Inside N.R.C.*, 11 August 1980, p. 6.

57. Reference 2, p. I.A.2-10.

58. *Inside N.R.C.*, 14 July 1980, p. 12.

59. *Wall Street Journal*, 10 September 1980.

60. Reference 2, p. III.A.1-11.

61. NRC, *Functional Criteria for Emergency Response Facilities* (for comment), NUREG-0696 (1980); and *Inside N.R.C.*, 8 September 1980, pp. 5-6, and 3 November 1980, p. 7.

62. See, for example, NRC, "Special Report to Congress Identifying New Unresolved Safety Issues," SECY-80-325, 9 July 1980, Enclosure 1.

63. Reference 2, Section II.E.3.

64. ACRS, *Review of Licensee Event Reports, 1976-1978*, NUREG-0572 (1979), pp. 3-2, D-5 and D-6.

65. *Inside N.R.C.*, 16 June 1980, p. 16.

66. Reference 64.

67. NRC, Office of Inspection and Enforcement, Information Notice 80-20 and Bulletin 80-12, and *Inside N.R.C.*, 5 May 1980, pp. 3 and 4.

68. NRC, Office of Inspection and Enforcement, Bulletin 80-18.

69. Reference 2, p. II.E.2-4.

70. Reference 2, p. III-3.

71. Reference 2, Section III.D.1.

72. Reference 2, Section III.D.2.

73. Reference 2, Section III.D.3.

74. Reference 14, Vol. II, p. 46.

75. *Inside N.R.C.*, 28 July 1980, p. 10, quoting member Jesse Ebersole.

76. Reference 23, p. 11.

77. Reference 25, p. 5-7.

78. Reference 2, pp. I.F-2 to I.F-3.

79. Reference 2, p. II.J.2-1.

80. *Inside N.R.C.*, 16 June 1980, p. 4.

81. Reference 23, p. 11.

82. Reference 9, p. 21. Emphasis in original.

83. Reference 14, Vol. I, p. 95.

84. Reference 14, Vol. I, p. 95.

85. Reference 9, p. 66.

86. Reference 9, p. 11.

87. Nuclear Safety Oversight Committee, Letter to the President, 26 September 1980, p. 4.

88. NRC, AEOD, ''Report on the Browns Ferry 3 Partial Failure to Scram Event on June 28, 1980,'' 30 July 1980.

89. *Nucleonics Week*, 24 July 1980, p. 6.

90. NRC, *Plan For Developing A Safety Goal*, NUREG-0735 (1980).

91. *Legislative History of the Energy Reorganization Act of 1974*, Vol. III at 3572. Emphasis added.

92. Joseph Hendrie, Speech at the AIF International Conference on Financing Nuclear Power, Copenhagen, 24 September 1979. Emphasis added.

93. Victor Gilinsky, Speech at Brown University, Providence, R.I., 15 November 1979. Emphasis added.

94. E.S. Rolph, *Nuclear Power and The Public Safety* (Lexington Books, Lexington, MA, 1979) p. 77.

95. Reference 9, p. 19.

96. Reference 93.

97. NRC, *NRC Views and Analysis of the Recommendations of the President's Commission on the Accident at Three Mile Island*, NUREG-0632 (1979), p. A-3.

98. Reference 25, p. 8-4.

99. Reference 31, p. 37.

100. NRC Authorization Act, FY80, Section 110.

101. See, for example, *Inside N.R.C.*, 22 September 1980, p. 14.

102. As reported in *Nucleonics Week*, 5 June 1980, p. 3.

103. As reported in *Nucleonics Week*, 30 October 1980, p. 13.

104. *Inside N.R.C.*, 8 September 1980, p. 8.

105. *Nucleonics Week*, 28 August 1980, pp. 1-3.

106. J. Emshwiller, ''Big Financial Problems Hit Electric Utilities; Bankruptcies Feared,'' *Wall Street Journal*, 2 February 1981, p. 1.

107. Vince Taylor, ''Electric Utilities: The Transition From Oil,'' Testimony before the Subcommittee on Oversight and Investigations of the House Committee on Interstate and Foreign Commerce, 9 December 1980, available from Union of Concerned Scientists.

7

Regulatory And Design Changes At Coal-Fired Power Plants[a]

The capital costs of coal-fired power plants in the United States increased significantly during the 1970s. The statistical analysis of recent plant costs in Chapters 9 and 10 indicates that the cost to build coal plants increased by an average of 68% from the end of 1971 to the end of 1978. This increase was in addition to inflation in construction labor and materials, and it assumes that no 1971-completed plants but all 1978-completed plants have sulfur dioxide "scrubbers." Approximately 90% of this *real* cost increase was spent for improvements in pollution control. In return, emissions of *criteria pollutants* (sulfur dioxide, or SO_2; particulates; and nitrogen oxides, or NO_x) from 1978 plants average approximately 64% less than those from 1971 plants.

For plants completed in the late 1980s, advanced control systems available or under development today appear capable of further reducing emissions by an average of about 75% below 1978-plant levels. This would require an additional 36% average increase in capital costs (in constant dollars). Compared to a 1971-completed coal plant, these advanced plants would cost approximately 130% more to build (not including construction inflation), but their emissions of criteria pollutants would average 91% less, as Table 7.1 shows.

These emission improvements are putting coal-burning at least on a par environmentally with oil-fired electric generation. Compared to 1%-sulfur oil — the average oil grade burned by utilities[1] — a typical 1978-completed coal plant emits an equal amount of SO_2, two-thirds less particulates, and slightly more NO_x. Late-1980s coal plants, with potential emission rates 70% to 80% less than a 1978 plant, could be considerably cleaner than typical oil-burning and noticeably less polluting than even the cleanest utility oil. (The coal-oil comparison is developed fully in Section 7.3.)

---

a. An earlier, shorter version of this chapter was published as an article, "Pollution Control Improvements in Coal-Fired Electric Generating Plants: What They Accomplish, What They Cost," in *Journal of the Air Pollution Control Association, 30* (No. 9), 1051-1057 (September 1980).

Table 7.1
Emission Reductions
By Typical New Coal Plants

Pollutant	Actual 1971-78	Projected 1978-88	1971-88
SO₂	74%	80%	95%
Particulates	83%	80%	97%
NOₓ	35%	69%	80%
Average Reduction	64%	76%	91%

Section 7.1:   Emission Abatement In The 1970s

Emission Standards: The use of coal to generate electricity increased rapidly in the 1960s and 1970s. U.S. coal-fired generating capacity increased by 80% from 1961 to 1971, and by another 53% to 1978. The concomitant increase in coal-generated emissions from massive new pollution sources, such as the Four Corners plant in northern New Mexico, provoked a national outcry to improve coal-plant pollution controls. Starting in the mid-to-late 1960s, some state and local authorities ordered utilities to reduce emissions of SO₂ and particulate matter by switching fuels or installing improved control devices. And in 1970 Congress amended the Clean Air Act to create a framework for reducing emissions from existing plants and set national standards for new plants.

The New Source Performance Standards (NSPS) promulgated by the federal Environmental Protection Agency (EPA) pursuant to the amendments, limited emissions of SO₂, particulates, and NOₓ from fossil-fuel plants whose construction started after August 1971. The NSPS required that new coal plants emit 55% fewer pollutants per unit of fuel burned, on average, than plants installed in 1971. Some new plants have surpassed the NSPS levels, as Figure 7.2 on page 190 shows, as a result of stricter local regulations, state limits to satisfy national ambient air quality standards, or utility efforts to keep ahead of regulations. Actual emission rates for coal plants completed in 1978 thus average approximately 64% less than for their 1971 counterparts, as shown in Table 7.1.

Emission Abatement Costs: The average capital cost of coal plants increased from $346 per kilowatt (kW) of capacity for late-1971 completion to

$583/kW for late-1978 plants, as Table 10.1 demonstrates. (These and all cost figures here are in constant mid-1979 "steam-plant dollars," that is, they have been adjusted to reflect 1979 prices of construction labor, materials, and equipment. They assume, moreover, that 1978 plants include scrubbers to remove SO_2, although about half of recent coal plants lack scrubbers, employing low-sulfur coal instead to comply with the NSPS. They also include interest during construction (IDC) in constant dollars, accounting for 8% of total costs.) Approximately 90% of the increase, or $210-215/kW, was accounted for by pollution control systems.

Sulfur Dioxide: The highest cost item added to coal plants during 1971-78 was the SO_2 scrubber. Fifteen plants in the 116-plant study sample have scrubbers, designed to remove an average of 74% of the SO_2 emitted from the boiler, or 3.7 lb of SO_2 per million Btu of fuel burned, based on the 2.3% average sulfur content of the coal used (see Table 9.4). This is sufficient to reduce emissions below the 1.2 lb NSPS limit. The scrubbers are "first-generation" devices using lime or limestone slurries that produce sludge waste.

The scrubber-equipped plants had a 26% higher average cost (controlling for chronology, location, and multi-unit siting) than the 101 non-scrubber plants in the sample (Table 9.1). Based on the $583/kW average cost of a 1978 coal plant with a scrubber (in 1979 steam-plant dollars, shown in Table 10.1), the average scrubber cost was $120/kW, including sludge handling and disposal systems. This is identical to EPA's 1979 cost estimate for an equivalent scrubber, but is 35% below the estimate in a 1977 study of coal plant costs by the Bechtel Corporation for the Electric Power Research Institute (EPRI).[2]

Particulates: Although SO_2 control has dominated most discussions of coal pollution control, utilities achieved greater proportional reductions in particulate emission rates from 1971 to 1978 for new plants. These reductions averaged 83% while SO_2 emission rates fell 74%.

Particulates from coal-fired boilers have traditionally been controlled by electrostatic precipitators (ESPs). Typical 1971 plants were equipped with 97%-efficient ESPs costing roughly $20/kW. By the end of 1978, ESP efficiencies averaged 99.5%, costing $35/kW for conventional high-sulfur coal and $85/kW for low-sulfur coal.[3] The latter coal produces highly resistive particulates requiring a much larger ESP collection area and stronger electrostatic field. The average 1978 ESP cost of $60/kW is half the cost of a typical first-generation scrubber.

The components of the 200% average real cost increase for ESPs during 1971-78 were approximately as follows:

- 130% increase for efficiency improvements from 97% to 99.5% for a specific coal grade;
- 10% increase for greater collection area needed for lower-sulfur

coal (average new-plant coal sulfur content fell 25-30% from 1971 to 1978);

- 5-10% increase for greater collection area to provide higher collection reliability.

(Note: cost increases are multiplicative, not additive.)

Nitrogen Oxides: The average 1978 coal plant emits NO_x at a 35% lower rate than its 1971 counterpart — the smallest reduction among the three criteria pollutants. This has been achieved by replacing horizontal or vertical burner locations with tangential firing, and by boiler modifications to enable boilers to be fired with "low excess air" and in two combustion stages.

These modifications reduce combustion temperatures, thereby reducing formation of NO_x. But without corrective measures they tend to corrode furnace walls and increase formation of "slag" — solidified molten ash — on boiler tubes, leading to combustion control problems and boiler tube leaks. Many new coal boilers thus have more sophisticated combustion monitors and controls — metered orifices and finely tuned nozzles to enhance air-fuel mixing — and wider spacing between boiler tubes to reduce slagging. Others enlarge the combustion volume and employ more-widely spaced burners to achieve lower temperatures which inhibit NO_x formation. These design changes added an average of $10/kW to capital costs for 1978 plants.

Other Environmental Measures: The criteria air pollutants were not the only targets of increased pollution controls in the 1970s. Other areas of expenditures were noise attenuation measures, $10/kW; pollution abatement during plant construction, $5/kW; liquid waste systems to treat normal plant waste drains for plant re-use or external discharge, $10/kW; improved ash disposal, $5/kW ("fixation" and ponding of scrubber sludge are included in the scrubber cost); air pollution monitoring systems, $2/kW; and preparation of environmental reports to state and federal agencies, $3/kW.[4] Increasing usage of cooling towers added an average of $5/kW, and recent plants incurred an average $10/kW cost for boiler improvements to accommodate variations in coal grade caused by mine-safety rules — another environment-related capital cost. The combined cost of the above "miscellaneous" pollution-control improvements for a typical 1978 plant was $50/kW, vs. only $5-10/kW for the same measures in 1971.

Total Costs: As Table 7.2 shows, environmental concerns absorbed an average of $240/kW in capital costs for 1978 coal plants, an increase of $210-215/kW above the corresponding expeditures in 1971. This increase equals 90% of the actual average 1971-78 increase of $237/kW in the real capital costs of typical coal plants reported in Chapter 10 (Table 10.1). Much of the $25/kW difference was spent on equipment to improve performance reliability: larger, more durable coal pulverizers, control systems for load-

Table 7.2

Pollution Control Costs
For New Coal Plants

(in mid-1979 steam-plant $/kW)*

Pollutant	Actual 1971	Actual 1978	Projected 1988
Particulates	20	60	65-80
SO₂		120	140-180
NOₓ		10	60-90
Solid Waste	0-5	5	30-45
Other	5	45	65-75
TOTAL	25-30	240	360-470
INCREASE		210-215	120-230

*Costs include IDC accounting for 8% of total costs.

cycling operation (increasingly needed because of reduced load growth and/or expanded nuclear capacity), increased design margins, improved quality control, and larger stocks of spare parts. The remainder of the cost increase not accounted for by environmental equipment is attributable to increased IDC as real interest costs rose and construction durations lengthened (Table 10.4).

Section 7.2:   Emission Abatement In The 1980s

Emission Standards: A revised set of New Source Performance Standards, approximately twice as stringent as the original NSPS, was promulgated by EPA in 1979 pursuant to the Clean Air Act Amendments of 1977. The new NSPS, shown in Figure 7.2, pertain to plants that commenced construction after September 18, 1978. They will affect some plants coming into service as early as 1983 and most plants completed in 1985 or thereafter.

The 1977 Amendments also require that new utility and industrial plants built in or near designated pristine ("prevention of significant deterioration" or PSD) areas or polluted ("nonattainment") areas install the "best available control technology" or achieve the "lowest achievable emission rate," respectively. These guidelines are defined, ambiguously, as the maximum reduction possible for each pollutant, taking into account energy, environmental, and economic impacts. They are intended to be "technology-forcing," *i.e.*, to push the utility industry to develop improved controls surpassing the new NSPS. EPA will determine the actual reductions required through case-by-case "new source reviews" in its permitting process.

The new NSPS will thus serve as a floor, rather than a ceiling, for pollution control practice for many new coal plants. Since over half of the country either is in a PSD or nonattainment area, or will affect such areas through airborne transport of pollutants, a majority of new plants may be required to better the NSPS. Some utilities may opt for stricter controls in any event to avert drawn-out negotiations with EPA. The NSPS are also subject to further strengthening as coal-fired generating capacity continues to expand. Although growth in sales of electricity has fallen since 1973 to less than half the historical 7% annual rate,[b] the prohibition of new oil- or gas-fired generators and the worsening prospects for nuclear power virtually ensure growth in coal-burning capacity in the 1980s and 1990s.

Emission Abatement Costs: In estimating the additional costs of pollution controls beyond those employed by 1978 coal plants, emission rates for 1988 plants have been assumed to be one-third of those dictated by the NSPS. This will ensure that the additional cost of new control improvements is not underestimated. The cost is estimated to be approximately $190/kW, almost equal to the cost of the controls added between 1971 and 1978.

Sulfur Dioxide: The new NSPS replace the former 1.2 lb/10^6 Btu standard with a set of limits varying with coal sulfur content, as shown in Table 7.3. Ninety percent SO_2 removal is required except when emissions fall below .6 lb; below that mark, only 70% reduction is needed. Any SO_2 removed by pre-combustion coal cleaning or in bottom ash or fly ash (typically 5%) is credited as a reduction.

Average 2%-sulfur coal requires an 84% SO_2 reduction (coal with a heating value below 11,000 Btu/lb requires a greater reduction, and vice versa). But since sulfur content varies among coal shipments, a higher design efficiency, perhaps as high as 90%, is needed to meet the 30-day continuous averaging requirement when coal sulfur content averages 2%.

The 15 scrubbers in the study data base averaged $120/kW in cost and 74% design removal efficiency. Studies for EPA suggest that raising SO_2 removal efficiency from 74% to 90% increases scrubber costs by only 1-1½%, or $1-2/kW.[5] This figure appears questionable, however. More efficient scrubbers must circulate more liquid to ensure that the SO_2 is contacted by the scrubber chemical reagent, requiring larger pumps and more piping. Lime-

b. Utility sales of electricity in 1980 were 22½% higher than in 1973, implying a compound average annual growth rate just under 3%.

Table 7.3
SO$_2$ Reductions Required
Under New NSPS
(assumes 11,000 Btu/lb coal)

Sulfur Content, %	Sulfur Content (lb/10^6 Btu)	SO$_2$ Reduction	SO$_2$ Emissions (lb/10^6 Btu)
3.3-6.6	3.0-6.0	90%	0.6-1.2
1.1-3.3	1.0-3.0	70-90%	0.6
Below 1.1	Below 1.0	70%	Below 0.6

Note: SO$_2$ produced has twice the weight of S in coal.

stone feed and sludge handling systems also expand proportionally with the amount of SO$_2$ removed. Additional scrubber modules may be required to back up malfunctioning modules. In addition, design improvements may be needed to eliminate the corrosion, scaling, and plugging that have affected many scrubbers to date. An additional $20-30/kW beyond the $120/kW cost of a typical 1978 scrubber appears sufficient to ensure reliable 90% collection by the lime or limestone slurry scrubbers employed to date.[c]

Newer scrubber designs should achieve even higher SO$_2$ removal efficiencies and also dispense with the hard-to-handle sludge that lime and limestone scrubbers generate in large quantities. The *Chiyoda, Wellman-Lord*, and *magnesia slurry* processes are all operating successfully in either commercial or pilot applications in the United States and thus are now or will shortly be available for commercial ordering. They produce gypsum or elemental sulfur, both of which are physically stable and saleable, *e.g.*, for cement or wallboard manufacture. Their projected costs are $155-170/kW for high-sulfur coal and $120/kW for low-sulfur coal, according to EPRI.[8] Although these estimates

c. A recent Bechtel study for EPRI estimates capital costs of approximately $140/kW and $150/kW, respectively, for 85%-efficient limestone and lime slurry scrubbers operating with high-sulfur (4%) coal.[6] The study estimates much lower costs, $105/kW and $110/kW, for very low-sulfur coal. The estimates include a very generous allowance — 27% of total costs without IDC — for general facilities, engineering, and contingency.

Another EPRI study[7] estimates that raising limestone slurry removal efficiency from 84% to 93% requires an 18% capital cost increase. This suggests that improving the SO$_2$ collection rate from 85% to 90% would add only 10% to costs, an increase within the margins used here to project a $140-150/kW cost for 90%-efficient scrubbers burning average-sulfur coal.

assume compliance with only an 85%-removal standard, the processes are considered capable of 95% removal for a modestly higher cost, estimated here at $140-180/kW.[9]

Scrubbing costs may eventually fall considerably if another scrubbing process now under development can be applied on a large scale. In the *dry sorbent injection* process, a fine-mist alkaline slurry converts SO_2 into a sulfite/sulfate mixture which is dried to a powder by the heat of the flue gas. The gas then passes through a baghouse filter (see discussion below) which collects the dry product along with particulate matter.

EPRI's cost estimates for 85%-efficient sorbent injection are only $115/kW for high-sulfur coal and an astonishingly low $41/kW for low-sulfur coal (exclusive of the baghouse).[10] All of the major equipment items are commercially available,[11] but the process has been tested only with low-sulfur coal, so that the EPRI estimates for scrubbing high-sulfur coal are very preliminary. Several major scrubber installations using dry sorbent injection in conjunction with baghouses or ESPs are being built, however: a 100-megawatt (MW) retrofit in Minnesota starting up in early 1981, and two new 500-MW-class units in North Dakota scheduled for 1981 and 1983, all using low-sulfur coal.[12] The 1983 installation, at Basin Electric's Laramie River 3 unit, is anticipated to cost only $79/kW for both the scrubber and ESP and to consume less than one-half of 1% of the unit's total power output while removing 85-90% of the SO_2 and at least 99% of the particulates.[13]

Particulates: The new NSPS reduce allowable emissions of particulates from .1 to .03 lb/10^6 Btu of fuel input. The corresponding increase in the collection efficiency required for an average coal grade (14% ash, 11,000 Btu/lb) is from 99.1% to 99.7%. Electrostatic precipitators at new 1978 plants were already averaging 99.5% design efficiencies, with 99.7% at many plants.

An average emission rate of .01 lb/10^6 Btu is assumed here, requiring 99.9% collection efficiencies. Although an improvement from 99.5% to 99.9% control may appear to have limited value, it would substantially reduce emissions of fine particulates, the most difficult to capture under current practice.[14] Fine particulates are especially injurious because they more easily bypass the lung's defenses, are the principal carriers of trace metals in coal ash, including toxic compounds containing lead, cadmium, and arsenic, and act as a conduit into the lungs for other air pollutants. They also contribute to the reduction of visibility by scattering visible light — a particular concern in PSD areas.

If the higher efficiencies are attained with electrostatic precipitators, the increase from 99.5% to 99.9% collection would almost double particulate control costs, from $35/kW to $65/kW for high-sulfur coal and from $85/kW to $160/kW for low-sulfur coal.[15] However, 99.9% efficiency can almost certainly be provided far more cheaply for low-sulfur coal by baghouses (fabric filters). This veteran particulate control device in cement- and steel-making employs numerous suspended filter bags to trap particulates from flue gases, much like common vacuum-cleaners. It is now being applied to utility boilers with the advent of synthetic fibers (primarily fiberglas) that can withstand combustion gases from coal.

Although utility use of baghouses is a relatively new phenomenon,[d] the device is being scaled up effectively and is rapidly becoming the favored particulate control method for low-sulfur coal. Baghouse emission rates averaged .02 lb/10^6 Btu, corresponding to 99.8% particulate control, at a half-dozen small (10-44 MW of electricity or non-electric equivalent) coal-fired boilers tested in the mid- and late 1970s.[17] Scaling up is simple because baghouses are built in modules. Southwestern Public Service Co. has successfully operated a baghouse with 28 12½-MW modules at its new 350-MW Harrington 2 unit since 1978,[18] and TVA reports an emission level of only .009 lb/10^6 Btu, or 99.9% control, for the first five units retrofitted with baghouses at its Shawnee plant.[19] All ten 175-MW Shawnee units will have baghouses by the end of 1981, and at Four Corners 4 and 5, twin 800-MW units in New Mexico whose massive particulate emissions sparked national concern at the start of the 1970s,[e] 98%-efficient electrostatic precipitators are being replaced by high-efficiency baghouses.

Baghouses should cost approximately $54/kW for low-sulfur coal and $48/kW for high-sulfur, according to EPA.[20] (Baghouse cost and performance efficiency are primarily a function of the fabric used and vary only slightly with coal type.) These costs apply to baghouses guaranteed to meet the new NSPS .03 lb standard, but based on TVA's experience, such devices may average .01 lb in actual operation. TVA's low-sulfur baghouses are costing only $40/kW,[21] but anticipated bag replacement at approximately four-year intervals will add roughly $10/kW (1979 dollars) over a 30-year life. Moreover, because of the embryonic status of baghouses for full-size boilers, utilities may specify more conservative design and construction to guarantee .01 lb emission rates. Hence, 25 to 50% is added here to EPA's figures, giving baghouse costs of $60-72/kW and $68-80/kW for high- and low-sulfur coal, respectively.

The high-sulfur cost is comparable to the $65/kW estimate for an equivalent 99.9%-efficient ESP. But for low-sulfur coal, a baghouse will clearly cost less than the $160/kW cost of a 99.9%-efficient ESP. (Indeed, for low-sulfur coal the cost of a baghouse appears equal to that of a 99.3%-efficient ESP.)

d. The technology is not completely new, however. Southern California Edison successfully operated a baghouse at its Alamitos plant in 1967-68 but terminated operation when it converted from oil-firing to particulate-free natural gas. The author and his colleagues at the Council on Economic Priorities strongly urged consideration of baghouses in their 1972 study of power plant pollution control.[16]

e. The worst particulate offenders at Four Corners were the first three smaller units, which operated from 1963-64 with only 78%-efficient "mechanical collectors" until they were replaced in 1972 by 99%-efficient wet particulate scrubbers.

Averaging the two coal types, 99.9% particulate control for a 1988 coal plant should cost from \$65/kW to \$80/kW, or \$5/kW to \$20/kW more than the \$60/kW average for a 1978 plant. In this instance, a new control technology is significantly reducing the rate of cost increase for improved pollution control.

Nitrogen Oxides: The new NSPS reduce the former NO_x limit of .7 lb/10^6 Btu to .6 lb for bituminous coal and .5 lb for subbituminous coal. These levels can be achieved through further application of staged combustion and low excess furnace air which, in conjunction with tangentially-fired burner design, have enabled recent plants to meet the .7 lb limit.

The only cost associated with this modest NO_x reduction would be approximately \$5-10/kW for additional design modifications to prevent the changed combustion practices from corroding boiler tubes, and for monitoring and control systems to maintain combustion parameters within the requisite narrow range. Although EPA contends that boilers can be operated within the .5-.6 limit without tube damage, utilities are likely to incorporate preventive design features.

The new NO_x standard appears to be the minimum average level achievable through combustion modification with present boiler technology. This would explain why the new NSPS require only a 45% average reduction in NO_x emissions compared to 1971 plants, versus 91% for particulates and 84% for SO_2. Lenient treatment of NO_x may end, however, as coal use expands and pressure builds to reduce the conversion by sunlight of NO_x and hydrocarbons into smog in oxidant nonattainment areas. An emission rate around .2 lb/ 10^6 Btu for new plants may be necessary to keep utility NO_x emissions constant to the end of the century,[f] and EPA is considering promulgating such a standard in the 1980s.

Reducing NO_x emissions below the new .5-.6 lb standard will require further changes in furnace design and perhaps NO_x flue gas treatment. Two new furnace designs, the *distributed-mixing burner,* which EPA is funding, and Babcock & Wilcox's *primary combustion furnace,* cosponsored by EPRI, have achieved emission rates under .2 lb in pilot testing without reducing efficiency or corroding boiler surfaces.[23] Both operate by staging combustion, first in a water-cooled, low-oxygen environment designed to retard corrosion and inhibit oxidation of nitrogen present in coal, and second in an oxygen-rich environment where carbon combustion can be completed. Costs have not been estimated but should not exceed \$20-30/kW (relative to uncontrolled 1971 plants), since essentially modifications rather than new systems are involved.

Ultimately, however, flue gas treatment of NO_x will probably be required, either if the new furnace designs prove inadequate or to reduce emis-

f. EPRI has calculated that utility NO_x emissions will rise by 60% between 1979 and 2000 even if a .2 lb standard takes effect in 1985.[22] The projection assumes 6% annual growth in fossil-electric generation, however, twice the post-1973 growth rate for all U.S. electricity.

sions well below .2 lb. The latter could be required at some new plants in PSD or nonattainment areas in the 1980s, providing an inroad for applying the new technology at all future plants.

The most advanced NO_x treatment processes are "dry" systems using gaseous ammonia to convert NO_x to harmless molecular nitrogen. These *selective catalytic reduction* (SCR) processes are operating successfully at numerous oil-fired boilers in Japan, and, since early 1980, at a 175-MW coal-fired plant there. Although these installations were designed to remove only 50% of NO_x, reductions of 80%, to about .2 lb per million Btu, are said to be achievable by employing more catalyst.

Mitsubishi Heavy Industries, which developed the system and has licensed Combustion Engineering to build and market it in the United States, estimates an installed cost of only $30/kW for 80% removal.[24] The effective cost will be several times greater, however, if the expensive titanium and vanadium catalyst must be replaced annually, as at present. An EPRI-sponsored study estimates that coal-fired SCR systems achieving NO_x emission rates of .05-.1 lb will cost $40-90/kW.[25] The lower cost would apply if new furnace designs produce low emissions with partial treatment. Adding the estimated $20-30/kW cost of furnace changes to the low end of the range, the cost to reduce NO_x emissions to .2 lb or below is projected here to fall in a $60-90/kW range.

Other Environmental Measures: New coal-fired plants are subject not only to stricter air pollution standards but also to regulations governing solid and liquid waste, noise, and construction effluent. Regulations in these areas added approximately $30-35/kW (1979 dollars) to the average cost to construct coal plants from 1971 to 1978 and will cause further cost increases in the 1980s.

Utility solid waste — fly ash, bottom ash, and scrubber sludge — came under federal regulation with the 1976 Resource Conservation and Recovery Act (RCRA). Compliance will require lining holding ponds for scrubber sludge and ash, at costs estimated by Ebasco Services of approximately $30/kW and $5/kW, respectively.[26] The former cost could be reduced through the use of regenerable scrubbers which recycle waste products, as was assumed in projecting scrubber costs above. Conversely, costs could rise if ash and sludge are designated as hazardous wastes under the RCRA regulations to be promulgated in the early 1980s. Impermeable liners would be required to curb leaching of trace metals, and disposal could be limited to special geological formations which might lie far from the plant site. Balancing these considerations, a cost estimate of $30-45/kW for improved waste disposal appears reasonable, although costs could be lower or higher than this range.

Other concerns which contributed to 1971-78 cost increases will also add to the costs of late-1980s plants. More complete treatment of waste-water will be needed to reduce effluent discharged in conjunction with ash sluicing, boiler cleaning, feedwater and scrubber makeup, and general plant usage to zero or near-zero levels. Local regulators may increasingly apply EPA noise attenuation guidelines set under the Federal Noise Control Act, adding to costs of pulverizers, fans, and other noisy plant machinery. Other issues such as construction pollution, effluent monitoring, and fugitive emissions from coal piles may also precipitate increased requirements. Based on a literature review, the cost of these "miscellaneous" environmental protection measures could double from 1978 to 1988, contributing an additional $30-40/kW for a total of $65-75/kW.[g]

Another potential source of major costs is the use of dry cooling to reduce the water loss associated with wet cooling towers. Dry towers would be extremely expensive, perhaps in a range of $140-185/kW (assuming successful development of an ammonia phase tower and including $10-30/kW to replace the generating capacity consumed by the towers during hot, peak periods).[27] Nevertheless, they could eventually be required in the water-short West if steam-electric power plants proliferate there. This would exemplify a cost that is incurred because expansion of the number of facilities encounters a resource constraint.

Total Costs: The total increase in the capital costs of environmental controls estimated above for a typical 1988 coal plant, compared to 1978 practice, ranges from $120/kW to $230/kW. The range reflects substantial conservatism in both the individual cost estimates and the projected emission targets, which are three times as stringent as the new NSPS. Moreover, the long lead times of most regulatory standards for coal plants make it unlikely that regulations not anticipated here will significantly affect the costs of late-1980s plants. Nevertheless, using past experience as a guide, actual costs are more likely to be at the upper than the lower part of the range. For purposes of cost comparison, a single figure of $190/kW is used here to project the average cost of 1978-88 control improvements.

This amount would be only slightly less than the actual $210-215/kW average cost of coal pollution control improvements from 1971 to 1978. The major sources of that increase were the first-generation scrubber (a cost actually shared by the 1971-78 and 1978-88 periods but assigned here to the prior period) and a high-efficiency electrostatic precipitator. The biggest new cost anticipated for 1978 is for improved NO_x control, with lesser increases for regenerable scrubbers and solid waste management. Despite the marked, further reductions in SO_2 and particulate emission rates projected here for late-1980s plants, the necessary cost increases are likely to be limited by the advent of new control devices, such as baghouses, which are more expensive than current systems at today's control levels but appear less expensive at very

g. The 1978 base cost of $35/kW for "other measures" that was doubled to estimate 1988 costs excludes $10/kW for increased boiler flexibility to accommodate varying coal grades.

high efficiencies.

Finally, an additional $20/kW is likely to be added to the cost of typical late-1980s plants to pay for improved operating reliability and for higher "real" interest costs as construction periods lengthen. The projected pollution control improvements would then account for 90% of the estimated cost increases (aside from construction inflation) in coal plant capital costs from 1978 to 1988 — equalling the 1971-78 percentage. The 1988 coal plant would be 36% more expensive (in real terms) but 76% less polluting than a 1978 plant, and 129% costlier but 91% less polluting than its 1971 counterpart, with pollution equipment responsible for nine-tenths of the increased real costs (Figure 7.1).

Under these circumstances, the two periods, 1971-78 and 1978-88, would show the same percentage increase in coal plant capital costs relative to expansion of coal-fired generating capacity.[h] This result would be consistent with the hypothesis pursued here that sector expansion plays a key role in engendering more stringent environmental and safety standards at coal and nuclear plants.

Section 7.3: Emissions Comparison With Oil

Very few oil-fired boilers in the United States are equipped with pollution controls.[28] Emissions from oil-firing are limited, if at all, only through use of "sweet" low-sulfur oil which produces not only less SO_2 but also less particulate matter than high-sulfur oil. The lack of further controls reflects society's preoccupation with coal-generated emissions. Compared to uncontrolled coal combustion, oil-burning produces approximately one-third less SO_2 for the same sulfur content (moreover, oil's average sulfur content is one-half of coal's), one-third less NO_x, and 99% less particulates.

With modern control devices, however, emissions from new coal plants are comparable with those from oil-burning. Figure 7.2 shows average emissions for five generations of coal-fired plants — new 1971 plants, 1976-77 plants meeting the original New Source Performance Standards, 1978 plants, 1983-84 plants meeting the new NSPS, and 1988 plants — while Figure 7.3 shows average emissions for high-sulfur oil, average-sulfur oil, and low-sulfur oil. The typical 1978 coal plant is cleaner than a plant burning the average oil (1%-sulfur content) now in use, producing roughly equal amounts of SO_2 and NO_x and two-thirds less particulate matter.

The same 1978 coal plant suffers in comparison with the low-sulfur (.3%) oil used in New York City and several other populous areas, generating

h. Installed coal-fired capacity increased 53% during 1971-78 and is projected in Chapter 10 to increase approximately 60% during 1978-88. The respective average real capital cost increases are 33% (actual) and 36% (estimated), not counting scrubbers — a cost increase that could reasonably be apportioned equally to the two periods.

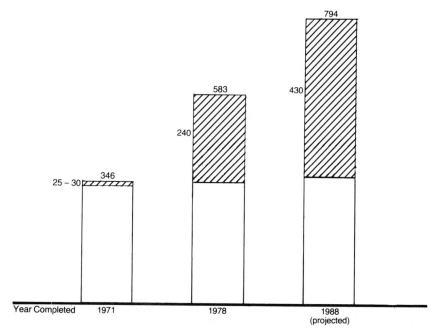

Figure 7.1
Coal Plant Capital Costs
(in mid-1979 steam-plant $/kW)

Shaded areas indicate environmental protection costs.

slightly more particulates and NO_x and three times as much SO_2. Under the new NSPS, however, coal plants entering service in the early- to mid-1980s will be considerably cleaner than typical oil-firing and will almost match low-sulfur oil, with approximately twice as much SO_2 but one-third less particulate matter and slightly less NO_x. By 1988, new coal plants could be considerably cleaner than even low-sulfur oil in all three pollutant categories, as Figures 7.2 and 7.3 show.

These data have important implications for efforts to reconcile oil-reduction goals and environmental concerns in the electric power sector. They indicate that, with current technology (for meeting the new NSPS), new coal-fired plants can replace typical oil-burning plants and improve air quality.[i]

i. Emissions of toxic trace elements from coal could offset this conclusion, however. Although advanced particulate controls collect trace-metal particulates at high efficiency (baghouses are as

Figure 7.2
Emissions Of Criteria Air Pollutants
By New Coal Plants
(lb pollutant/10^6 Btu of coal burned)

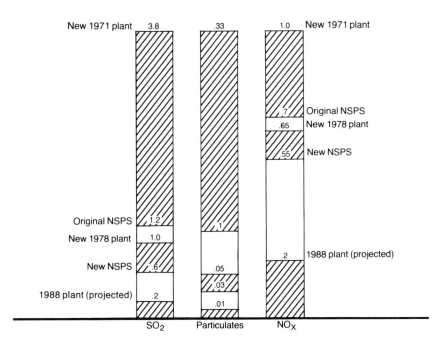

Pollutants not drawn to same scale. 1971 figures are based upon Reference 16 and assume 11,000 Btu/lb coal, 14% ash, 2.2% sulfur, dry bottom boiler, and 97% particulate collection. 1978 figures assume 99.5% particulate collection and 74% SO_2 collection. 1988 projections assume 99.9% particulate collection, 95% SO_2 collection, and 80% NO_x reduction.

Moreover, if currently available 90%-efficient scrubbers and 99.8-99.9%-efficient baghouses are employed, new coal plants could operate more cleanly than plants burning even low-sulfur oil. (Applying these controls to the typical coal in Figure 7.2 would yield emission rates per 10^6 Btu of .2-.25 lb SO_2 and .01-.02 lb particulates.)

efficient with small particles as with large), several metals in coal such as mercury may vaporize in combustion and bypass controls. Oil also contains trace metals, including vanadium, but in lesser quantities. So-called hazardous emissions from coal and oil combustion are currently being measured and studied by EPA.

Figure 7.3
Emissions Of Criteria Air Pollutants
From Current Oil Burning
(lb pollutant/10⁶ Btu of oil burned)

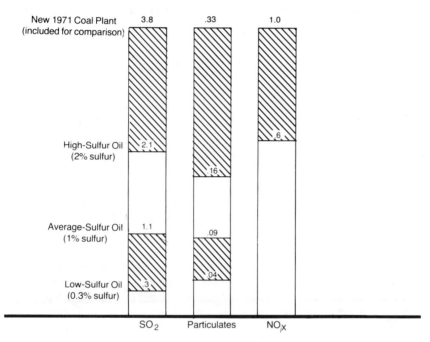

Pollutants drawn to same scales as Figure 7.2. All figures calculated from EPA, *Compilation of Air Pollutant Emission Factors*, AP-42, Part B (1977).

Heating value of 6.2 million Btu/bbl oil is assumed. NOx emissions vary with boiler design, not sulfur content, and assume 30% are tangentially-fired with remainder conventionally fired.

The data also imply that large-scale replacement of nuclear capacity with coal-fired generation need not add greatly to pollutant emissions. Assuming that new coal plants meet the 1988 emission targets discussed here, replacement of all 65,000 MW of planned nuclear capacity less than 40% complete would add only approximately 1% to 1979 nationwide emissions of SO_2, 3-4% to NO_x, and under .2% to particulates. The total SO_2 increment would be less than current emissions from a single 1,500-MW coal-fired plant in Ohio whose construction predated federal emission standards. The percentage additions would be approximately three times as great, however, if the coal plants merely met the new NSPS.[29] The additions would also vary among regions, since the nuclear plants in question are not distributed in proportion to current emissions.

Pollution-control prospects are slightly less favorable for converting oil-fired plants to coal, since NO_x control is unavailable short of major boiler overhaul, implying emission rates averaging 1.0 lb/10^6 Btu, considerably more than the average for oil. In such cases, extra effort could be invested in SO_2 and particulate abatement to increase offsetting improvements in these pollutants.

Section 7.4:   Total Pollution Control Costs

This chapter has addressed only the impact of pollution controls on coal plant *capital* costs. However, improved controls also affect fuel costs, operating and maintenance (O&M) costs, and performance reliability (capacity factor). The heat, steam, and electricity required to run pollution control equipment reduce thermal efficiency and raise fuel consumption. O&M costs are increased by the limestone and other material required by scrubbers, by disposal of ash and sludge, and by personnel to operate and maintain control devices. Control equipment breakdowns or gas and moisture ''carryover'' can restrict plant operability, although redundant scrubber modules will diminish this problem.

These ''non-capital'' costs of pollution controls are estimated in the discussions of coal fuel and O&M costs and capacity factor in Chapter 11 and are incorporated into the overall coal generating cost calculation in Table 12.1. They are summarized here in Table 7.4, which indicates that environment-related expenditures will account for about one-third (34%) of the total lifetime generating cost of typical 1988 coal plants. The percentage will probably be less if 1988 plants merely comply with the new NSPS and are not held to the more stringent emission limitations assumed in this chapter and shown in Figure 7.2. In turn, projected capital costs for control equipment will be responsible for slightly over half (54%) of all anticipated environment-related costs, and O&M costs will account for slightly under one-third (30%).[j]

Section 7.5:   Alternative Coal Combustion Technologies

This chapter has not considered the possibility that new coal combustion methods might achieve pollution control levels comparable to those specified

---

j. These figures supersede the author's statement in an earlier published version of this chapter that capital-related costs would make up two-thirds of all environmental costs for future coal plants (see Note ''a'' above). Subsequent to publishing that version, the author significantly increased his estimates of O&M pollution-control costs.

Table 7.4

Contribution Of Environmental Controls
To Coal Generating Costs

(based on cost projections for 1988 coal plants
meeting the strict control levels in Fig. 7.2)

Cost Component	Projected Cost (1979 ¢/kWh)	Share of Total Environmental Cost
1. Capital Costs	.73	54%
2. Fuel Costs	.16	12%
3. O&M Costs	.40	30%
4. Reduced Capacity Factor	.06	4%
TOTAL ENVIRONMENTAL COST	1.34*	
TOTAL GENERATING COST	3.92	

*Sum of individual costs exceeds total due to rounding.

Sources: Chapter 11 and Table 12.1.
  (1) = 430/794 (Fig. 7.1) × 1.34¢ (capital cost fixed charges, Table 12.1).
  (2) = 800 (assumed heat rate penalty) ÷ 10,000 × 1.96¢ (fuel cost).
  (3)   assumes that .5¢ O&M cost in 1979 dollars (Section 11.5) includes .02¢ for base particulate control and ash handling, .05¢ for improved ash handling, and .25¢ for scrubber, totalling 64% of O&M cost, x .62¢ (levelized lifetime O&M).
  (4)   assumes that controls reduce capacity factor by 3 percentage points, causing increase in capital fixed charges of (1-.70/.73) × 1.34.

here for 1988 plants, at lower cost. Considerable research and development effort is being devoted to new technologies, however. The most promising is *fluidized bed combustion,* in which the fuel rests on a layer of inert particles suspended by forced air. It has been used in specialized industrial applications for several decades but has only recently been examined for power generation.

Fluidized bed combustion has many prospective advantages over conventional combustion: ability to burn limestone together wth coal so that SO_2 may be converted and captured without a scrubber; combustion below the temperature of atmospheric generation of NO_x; and formation of a dry, powdery ash which is less damaging to plant equipment than ash from conventional burning of pulverized coal and also contains fewer heavy metals. Fluidized bed boilers are also considerably more compact than conventional boilers.[30]

Coal plants employing fluidized bed combustion are widely predicted to be no more costly and possibly cheaper than conventional coal-fired plants, assuming both must meet the new NSPS. Similar forecasts have been made for gas turbine cycles operating with fluidized bed combustors or integrated low-Btu coal gasifiers. Initial operation of commercial-size prototypes is unlikely until 1984 at the earliest, however, so it is doubtful that commercial plants could be operating before 1990.

Moreover, current cost estimates for these new technologies are somewhat speculative and could prove low if stringent standards necessitate design modifications or flue gas treatment. Fluidized bed combustors, for example, are considered capable of removing approximately 85% of SO_2 through contact of limestone with coal in the combustor, but scrubbers may be required to achieve higher control levels (conversely, refinements such as limestone recycle may allow 95% capture or greater).

A different technology, *coal cleaning,* holds promise for reducing flue gas treatment costs by removing pollution-generating impurities from coal at the mine. Proven physical cleaning processes using crushing and flotation-separation can remove half or more of the ash and a third of the sulfur from coal, substantially reducing the design requirements of emission control devices. Although only a small fraction of utility coal is cleaned today, increased costs for coal transportation, waste disposal, and boiler outages — all of which are mitigated by cleaning — are making cleaning more economically feasible. Coal cleaning may be especially attractive as an alternative to retrofitting controls at existing coal plants that must reduce emissions.

References

1. DOE, *Cost and Quality of Fuels for Electric Utility Plants, 1978,* DOE/EIA-0191(78) (1979), p. xiii.

2. PEDCo Environmental, Inc., *Cost Analysis of Lime-Based Flue Gas Desulfurization Systems for New 500-MW Utility Boilers* (EPA, 1979); and Bechtel Power Corp., *Coal-Fired Power Plant Capital Cost Estimates,* EPRI AF-342 (Palo Alto, CA, 1977). All cost estimates have been adjusted to mid-1979 price levels with real IDC. A subsequent Bechtel study for EPRI projects lower scrubber costs (see Reference 6).

3. EPA, *Electric Utility Steam Generating Units: Background Information for Proposed Particulate Matter Emission Standards* (1978), Table 8-2. Costs were converted to mid-1979 dollars, with $10/kW and $15/kW added for high- and low-sulfur coal, respectively, for conservatism.

4. These estimates are drawn, with modifications, from three valuable surveys of coal

pollution control costs by the power industry: R.R. Bennett & D.J. Kettler, "Dramatic Changes in the Costs of Nuclear and Fossil-Fueled Plants" (Ebasco Services, New York, NY, 1978); K.E. Yeager, C.R. McGowin, and S.B. Baruch, "Potential Impact of R&D Programs on Environmental Control Costs for Coal-Fired Power Plants" (EPRI, 1979); and C.R. McGowin and K.E. Yeager, "Advanced Programs for Utilization of Coal for Electric Power Generation" (EPRI, 1979).

5. Reference 2, PEDCo Environmental, Inc., Figures 4-1 and 4-2.

6. Bechtel National, Inc., *Economic and Design Factors for Flue Gas Desulfurization Technology,* EPRI CS-1428 (1980), Figure 1-3. Current-dollar IDC (13.4% of total cost) was removed from costs and replaced by constant-dollar IDC (8% of total cost), and 12.5% was added to convert from 1978 to 1979 dollars.

7. SRI International, *Investigation of High SO$_2$ Removal Design and Economics,* EPRI CS-1439 (1980), Vol. 2, Table 4-2.

8. Reference 6.

9. Reference 4, K.E. Yeager *et al.* estimate that 95%-efficient regenerable scrubbers will cost $120-170/kW (p. 16). Costs therein, expressed in 1976-1980 mixed current dollars with current-dollar IDC, are equivalent to mid-1979 dollars with real IDC.

10. Reference 6.

11. Reference 6, p. 6-91.

12. *Power Engineering, 84* (No. 10), 86-88 (1980).

13. R.L. Eriksen, "The Development of Dry Flue Gas Desulfurization at Basin Electric Power Cooperative," presented at Second Conference on Air Quality Management in the Electric Power Industry," Austin, TX (January 1980). Capital cost of $49.8 million in 1981 dollars (Table 8) was converted to 1979 basis @ assumed 12.5%/year inflation rate and divided by 500-MW capacity (Table 4). Power requirement of 2.45 MW (Table 8) is for both scrubber and ESP but will be slightly higher due to minor design change (telephone communication with R.L. Eriksen, Basin Electric Power Coop, 6 February 1981).

14. Reference 3, Table 7-1.

15. The "modified Deutch equation" relating ESP costs to removal efficiency predicts a 70% cost increase to improve efficiency from 99.5% to 99.9%, but it may understate cost increases at very high efficiencies.

16. C. Komanoff, H. Miller, and S. Noyes, *The Price of Power: Electric Utilities and the Environment* (Council on Economic Priorities, 1972; MIT Press, 1974), p. 16.

17. Reference 3, Table 4-5.

18. EPA, "New Stationary Source Performance Standards, Electric Utility Steam Generating Units," Part II, *Federal Register,* 33600 (11 June 1979).

19. *Power Engineering, 84* (No. 8), 95 (1980), and telephone communication with H.S. Woodson, Senior Staff Specialist, Mechanical Engineering Branch, TVA, 16 December 1980.

20. Reference 3, Table 8-1, adjusted to mid-1979 price levels.

21. J.A. Hudson *et al.*, "Design and Construction of Baghouses for Shawnee Steam Plant," presented at Second Conference on Air Quality Management in the Electric Power Industry, Austin, TX (January 1980). Cost estimate therein of $80 million was reduced to $76 million (1978-81 mixed current dollars with IDC) by H.S. Woodson (Reference 19). Figure is equivalent to approximately $40/kW in 1979 dollars with real IDC.

22. *EPRI Journal*, "Controlling Oxides of Nitrogen" (June 1979), p. 26.

23. Reference 22.

24. *Electric Light and Power*, February 1981, p. 19. Cost estimate of $35/kW was converted here to mid-1979 dollars by dividing by 1.19, representing 1½ years' inflation @ 12.5%/year.

25. Stearn-Rogers, Inc., *NOx Control for Western Coal-Fired Boilers: Feasibility of Selected Post-Combustion NOx Control Systems* (1978). EPA estimates that SCR systems will cost approximately $36/kW, or less than half EPRI's estimate for full treatment, in J. David Mobley, "Status of EPA's NOx Flue Gas Treatment Program," in EPRI FP-1109-SR, *Proceedings: Second NOx Control Technology Seminar* (1979). See also Session B Questions & Answers in same volume.

26. Reference 4. R.R. Bennett & D.J. Kettler estimate $21/kW (lined sludge pond) and $4/kW (lined ash pond), in mixed current dollars concluding in 1978. Assuming a mid-1975 midpoint for expenditures and incorporating the 37% inflation in coal plant construction from mid-1975 to mid-1979 gives the costs in the text.

27. Reference 4, K.E. Yeager *et al.*

28. The only controls in use are boiler modifications instituted by several utilities, primarily in California, to reduce NO_x formation. See Reference 16 or R.H. White, *Price of Power Update* (Council on Economic Priorities, 1977).

29. Emissions for new coal plants are calculated with emission rates for 1988 plant and new NSPS in Fig. 7.2, assuming 3.395 quads to replace 64,600 MW of nuclear @ 60% capacity factor and 10,000 Btu/kWh for coal. Megawattage is based on utility listings of nuclear construction status in DOE, *U.S. Central Station Nuclear Electric Generating Units: Significant Milestones*, DOE/NE-0030/3 (80), pp. 31-35, excluding four reactors listed as on order or under construction but subsequently cancelled (North Anna 4, Forked River, and Montague 1 and 2). Emissions for 1988-plant case, in thousand tons, are 340 SO_2, 340 NO_x, 17 particulates. NSPS-plant emissions are 1,020 SO_2, 930 NO_x, 51 particulates. 1979 utility base is calculated from 1979 coal used for electric generation (11.26 quads), using 1971-plant rates in Fig. 7.2. SO_2 base adds one-twelfth to reflect generation with oil (one-quarter as much generation as coal, half the average sulfur content, and two-thirds the emissions per sulfur concentration). NO_x base adds 25% to reflect oil and gas (half the generation and half the emission rate). Figures, in thousand tons, are 23,200 SO_2, 7,040 NO_x, and 1,860 particulates. Nationwide emissions from all sources, in turn, are approximately 30-35% greater for SO_2, 3-4 times as great for NO_x, and 5 times as great for particulates. Ohio coal plant is Ohio Power's Muskingum River, which in 1979 burned 3,817,300 tons of coal averaging 5.19% sulfur, emitting 376,000 tons of SO_2, or 11% more than the 340,000 tons calculated above from all coal generation with 1988 controls to replace nuclear capacity. Two TVA plants of approximately 2,500-MW capacity emitted even more SO_2 in 1979: 468,000 tons from Paradise (KY) and 379,000 tons from Cumberland (TN).

30. See *Power Engineering, 83* (No. 11), 46-56 (1979), for a current review of fluidized bed developmental status and design features.

8

Statistical Analysis Of Nuclear Plant Construction Costs And Durations

This chapter reports the results of Komanoff Energy Associates' statistical analysis of capital costs and construction durations of U.S. nuclear plants completed in the 1970s. A parallel analysis for coal plants is described in Chapter 9. Both analyses are employed in Chapter 10 to calculate the average increases in capital costs and construction durations experienced by nuclear and coal plants from 1971 to 1978 and to project likely further increases for future plants.

Section 8.1:   Nuclear Capital Costs

The sample for the nuclear capital cost analysis consists of 46 reactors. These are all U.S. reactors completed from December 1971 through December 1978, except for four "turnkey" plants supplied by reactor manufacturers under fixed-fee contracts. The 46 units total 39,265 megawatts (MW).

The excluded turnkey plants and their commercial operation dates are Quad Cities 1 and 2 (1972), Point Beach 2 (1972), and Indian Point 2 (1973). The data base also excludes 11 other commercial-size reactors (400 MW or larger) completed during 1968-1971. Nine were turnkey units, with reported costs that understate their actual costs. A tenth, Connecticut Yankee (1968), was subsidized under the Atomic Energy Commission's (AEC) Power Reactor Development Program. The eleventh unit, Nine Mile Point 1 (1969), predates the first units in the sample by two years and thus was excluded as a discontinuous data point.

All of the plants in the data base were ordered between mid-1965 and early 1968. All but the last received construction permits from 1967 to early

1971, prior to the 1½-year licensing suspension resulting from a U.S. Court of Appeals decision requiring the AEC to publish comprehensive environmental impact statements in reactor licensing proceedings (the "Calvert Cliffs" decision). The sample excludes several reactors that have reached commercial status since the March 1979 accident at Three Mile Island, including two that received Nuclear Regulatory Commission operating licenses shortly before the accident and subsequently achieved commercial status. Appendix 1 lists the entire data base including relevant dates.

The reactors' capital costs were tabulated from utility data reported to the federal government.[1] Because some nuclear plants incur substantial capital expenditures shortly after their designated commercial operation date, costs were tabulated from data reported for the *year following commercial operation*, where possible. Additional capital costs incurred in subsequent years were not included, however, to avoid biasing the sample in favor of later plants, which have had less time to make backfits. (The tendency for nuclear capital costs to increase after plant completion is reflected in the allowance for interim capital replacements incorporated in the nuclear fixed charge rate in Section 11.6.)

Utilities report costs in *mixed current dollars* — the sum of the dollars spent in each year of construction. They include interest during construction (IDC) paid by the utility on capital borrowed during construction. These costs were converted to mid-1979 "steam-plant" construction dollars excluding IDC through a computational process described in Appendix 3. The adjusted costs were then divided by the units' original design electrical ratings to obtain capital costs per kilowatt (kW).[a] The result was a set of 46 per-kW costs excluding IDC and adjusted for the effects of inflation in power plant construction wages and materials. Real interest during construction (expressed in constant dollars) is added back to capital costs in calculating the costs of "standard" reactors in Chapter 10.

These costs were correlated with explanatory variables in a *multiple regression* analysis — a standard procedure for explaining variance in a "dependent" variable (capital cost) as a function of variations in "independent," or "explanatory" variables. The regression equation for nuclear costs has seven "statistically significant" variables and an r^2 value (goodness of fit) of 92%. (Statistical terms are explained in Appendix 4.) This is significantly higher than the r^2 values in earlier analyses by Bupp (64%)[2] and Mooz (76%),[3] indicating that the KEA analysis statistically accounts for a higher percentage of nuclear cost variations. The variables and their correlations with capital cost are shown in Table 8.1 and described below, along with several

a. Utilities have reduced the original design electrical ratings of approximately 20 reactors in the data base. The original ratings used here and shown in Appendix 1 were reported in the NRC "Gray Books" until late 1976. The author uses these ratings to calculate capacity factors in his studies of nuclear plant performance.

Table 8.1
Nuclear Capital Cost Regression Equation

Capital Cost ($/kW, without IDC, in mid-1979 steam-plant dollars) =

	T-ratio	Significance Level (%)
6.41 ×	2.60	99.2
1.28 if Northeast ×	6.56	99.9+
A-E$^{-.105}$ ×	6.19	99.9+
MW$^{-.200}$ ×	2.54	98.5
.903 if Multiple ×	2.48	98.2
1.34 if Dangling ×	4.73	99.9+
1.20 if Cooling Tower ×	5.22	99.9+
(Cumulative Nuclear Capacity)$^{.577}$	13.55	99.9+

r^2 = .923
Adjusted r^2 = .908
F value = 64.7
Sample size = 46 units

The absolute value of all pairwise correlation coefficients is less than .34, except for .435 between cumulative nuclear capacity and multiple status.

variables of note whose correlations with capital cost were not statistically significant.

1. Northeast Location: The 46 plants are clustered in three broad geographical areas — Northeast (14 plants), Southeast (18 plants), and Midwest (12 plants). The remaining two plants are on the West Coast. There was no significant cost difference between Southeast and Midwest plants, but Northeast plants were 28% more expensive, on average, than plants in other regions. This finding is significant far beyond the 99.9% level, and there is 95% confidence that the effect lies within an 18-38% increase range.

The high costs of Northeast units appear due in part to higher construction wage rates. These are generally about 5% higher in the Northeast than in the Midwest and 25% higher than in the Southeast. The Northeast is also more densely populated and environmentally sensitive than other regions, factors that may have led to installation of more costly equipment to reduce accident hazards. In addition, the built-up Northeast may have offered fewer suitable reactor sites and thereby required more site preparation. Northeast plants had longer construction periods (see next section), leading to slightly higher costs for interest during construction, but this factor is not included in Table 8.1.

2. *Architect-Engineer Experience*[4]: Reactor costs declined as the number of reactors built by architect-engineers (A-E) increased. This reflects the growing ability of increasingly experienced A-Es to manage the complex process of nuclear design and construction. The effect is modest, with a doubling in an A-E's nuclear experience leading to only a 7% decline in costs. Nevertheless, it is statistically significant well beyond the 99.9% level. The 95% confidence interval ranges between a 5% and 9% cost reduction for each doubling of experience.

Average A-E experience almost quadrupled during the period studied, resulting in an average 13% cost reduction. This saving was swamped, however, by the trend toward increased costs as the nuclear sector expanded, as discussed below.

Six of the 46 plants were designed and built by the utility itself — Browns Ferry 1, 2 and 3, Cook 1 and 2, and Salem 1. They had considerably higher costs than other plants, suggesting that utility management of reactor design and construction is not a path to lower costs.[b]

3. *Unit Size:* The capital cost projections of the AEC, its successor agencies, and the power industry assume that nuclear per-kW costs decline by 20-30% when reactor size is doubled.[5] Virtually all government and industry analyses have applied such economy-of-scale relationships in projecting nuclear capital costs.

The KEA analysis found a far smaller economy of scale in actual experience. The decline in per-kW costs was proportional to the .20 power of unit size, so that doubling unit size led to only a 13% reduction in per-kW costs. This result is significant at the 98.5% level. The 95% confidence interval, expressed in cost reduction per size doubling, ranges from 3% to 22%.

This result was measured for direct construction costs, without interest during construction. Doubling reactor size extended construction time by an average of 28%, as shown in the next section. The resulting increase in IDC adds approximately 3% to real costs, so that the net effect of doubled size is only a 10% cost reduction — half the standard projection.

---

b. To eliminate bias in testing for the effect of utility-built plants (self-A-E), costs were correlated to self-A-E status and the six other explanatory variables described in text, without also testing for A-E number. The result was a 30% higher cost for self-A-E, with a 95% confidence interval of 13-50% and a 99.9% significance level. This effect was not included in the regression equation discussed here.

4-5. Multiple and Dangling Units: Twenty-two of the 46 units are *multiple* units — identical reactors at the same site. They are typically licensed simultaneously and constructed less than two years apart. Minor exceptions are Browns Ferry 3, licensed 14 months after units 1 and 2 and completed two years later; and Cook 2, licensed with unit 1 but completed three years later.

Utilities with multiple nuclear units were not consistent in allocating total station costs between the first and subsequent units. Some utilities assigned equal costs to the units; others weighted either the first or second unit more heavily, often for tax or income reasons unrelated to actual cost allocation. For example, the reported cost of Surry 2 was 73% *greater* than that of unit 1, while the cost reported for Calvert Cliffs 2 was 22% *less* than the unit 1 cost (in mixed current dollars). To eliminate such arbitrary variations, the constant-dollar direct costs (without IDC) of multiple units were averaged here, giving units in a multiple set, such as Oconee 1, 2 and 3, identical costs.

The statistical analysis found that multiple units averaged 9.7% lower costs than other reactors. This was the result of shared design effort, common facilities, and learning in construction. The 95% confidence interval for this effect is a reduction between 2% and 17%, and it is significant at the 98.2% level.

The data base contained four other first units whose successors had not been completed at the time of the analysis. These units, Salem 1, Hatch 1, North Anna 1 and Farley 1, had much higher reported costs than comparable plants. This appeared to result from utilities' allocating a large share of the stations' joint costs to the first unit. Pending completion of the second units, when the two units' costs can be averaged, these four units were given special designation in the statistical analysis as *dangling* units. This enabled their high costs to be explained statistically by their stations' incomplete status. Otherwise, the high costs of these recent units would have appeared as a spurious increase in the rate of cost escalation over time.

The effect of dangling status was a 34% higher cost. The 95% confidence interval is a range of 18-53%, and the significance level is beyond 99.9%. Dangling status is not assumed in any cost projections; rather, it is included in the analysis to avoid misinterpreting the effect of other variables.

6. Cooling Towers: Eighteen of the 46 nuclear units have natural-draft cooling towers. Their costs averaged 20% higher than other nuclear plants. The 95% confidence interval ranges from 12% to 28%. The effect on costs from cooling towers is significant far beyond the 99.9% level.

There is no apparent reason for the 20% cost differential between units with and without cooling towers. Most engineering studies estimate that towers added an average of 7-8% to nuclear plant costs during the period studied, and that this share declined over time since the cost of towers increased more slowly than the cost of the reactor plant. Morever, the cost of *coal* plants with towers was not different from plants without towers. Thus, the apparent effect of towers on nuclear costs may be a surrogate for some other phenomenon.

One possibility is that the presence of cooling towers resulted from public pressure for environmental protection which also led to stricter safety measures at those plants, adding to costs. This hypothesis is supported by the fact that plants with towers took 11% longer to construct than other plants. Cooling towers themselves should not add to construction time, since they are not part of the "critical path" in construction. Rather, their presence may indicate greater local environmental concern which added to safety requirements at that plant and thus to construction time and cost.

7. Cumulative Nuclear Capacity (Nuclear Sector Size): As discussed previously, all costs were converted to mid-1979 steam-plant dollars without interest during construction. Thus, costs were analyzed as if all 46 reactors had been built at mid-1979 price levels for nuclear construction labor, materials, and equipment. Despite these adjustments, costs are much higher for the later plants in the sample, as Figure 8.1 shows. This increase cannot be explained by changes in the the units' size, location, A-E experience, multiple status, etc., since these factors either did not change appreciably during the study period or did not affect costs sufficiently strongly. Indeed, changes in several variables would tend to *reduce* costs (*e.g.*, increased A-E experience), and the effect of other variables has been removed from Figure 8.1 in any event.

The cost increase appears to have been occasioned by the application of more stringent and explicit regulatory standards to nuclear plants during the late 1960s and throughout the 1970s. As described in Chapter 4, these standards added significantly to the amounts of labor, materials, and equipment required to build reactors. Moreover, they frequently were mandated during construction, causing changes in design requirements that made it difficult for utilities to control schedules and costs.

In turn, increased regulatory stringency arose from three phenomena, as described in Chapter 3: from the effort to reduce the permissible risk to public health and safety per reactor; from new information emerging from reactor licensing, design reviews, and operating experience indicating that prevailing standards were not adequate to reduce risks to desired levels; and from expansion of the regulatory effort requiring greater documentation and standardization of regulatory requirements, generally at a more stringent level.

To allow for these phenomena in the statistical analysis, two alternative regression formulations were examined. In one model, capital costs were assumed to be related to *time* in addition to the six other variables described above. In the other model, costs were assumed to be related to *the size of the nuclear sector*. Time was measured by the date each reactor received its construction permit from the AEC. Sector size was defined as the megawatts of

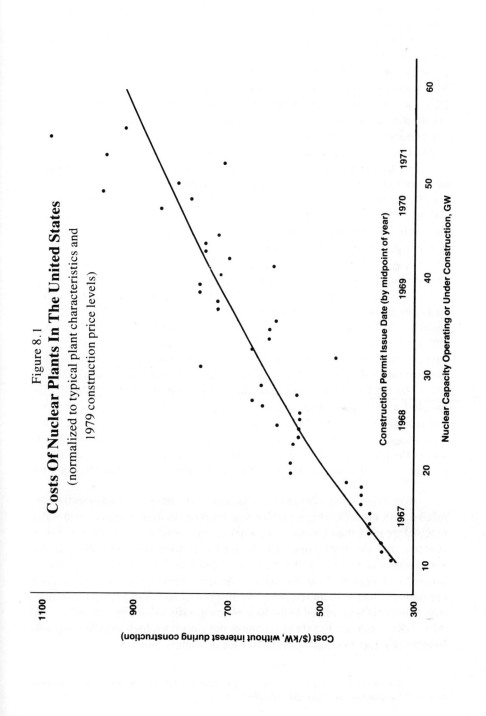

Figure 8.1

Costs Of Nuclear Plants In The United States

(normalized to typical plant characteristics and 1979 construction price levels)

nuclear capacity operating or being built on the same date, including the new plant's capacity.

Both formulations reflect the effect on costs of the upgrading of nuclear regulatory standards. The formulation with sector size, or cumulative nuclear capacity, was devised because it appears to capture more of the societal processes that give rise to new standards, as discussed in Chapter 3. It also yields a higher r^2 value in the regression model (see below). The costs calculated for standard 1971 and 1978 reactors in Chapter 10 are relatively independent of which formulation is employed, however. The choice of sector size or time is important primarily for projecting *future* nuclear capital costs.

For the sample of 1972-1978 plants studied here, cumulative nuclear capacity ranges from 10,621 MW for the first reactor receiving a construction permit, Palisades (allowing for the 9800 MW total capacity of the 15 commercial plants predating the sample, plus Palisades' 821 MW), to 55,573 MW for the last licensed plant, Farley 1. The latter figure includes the capacities of six reactors that received their construction permits prior to Farley 1 but had not been completed when the analysis was performed: Diablo Canyon 1 and 2, North Anna 2, Salem 2, and Sequoyah 1 and 2.

The correlation obtained in the regression equation shows nuclear capital costs rising at a rate proportional to the .58 power of cumulative nuclear capacity. At this rate, each doubling of the size of the nuclear sector produced a 49% cost increase.[c] The 95% confidence level is a range of 41% to 58% in the cost increase from doubling the sector — an exceedingly narrow (and thus precise) interval. The significance level is far beyond the 99.9% mark.

The nuclear sector expanded almost five-fold in the study period. Associated with this expansion was a 172% increase in costs (beyond steam-plant construction inflation) to build plants to lower levels of risk, according to the above correlation. This increase swamps the 13% reduction in costs attributable to the increase in architect-engineer experience during the sample period. (The net nuclear cost increase from 1971 to 1978 is calculated in Chapter 10.)

Non-Significant Variables: Nuclear costs were not significantly correlated with the following variables: reactor type (boiling or pressurized water reactor), reactor manufacturer, regional seismic potential, proximity to urban centers, and licensing time. The lack of a seismic-cost correlation is not surprising since none of the data base plants are in "severe" or "high" seismic risk regions,[6] so that seismic design criteria varied more with plant chronology than with geography. The absence of an urban-proximity correlation with costs was related to the lack of any heavily urban sites and also to the AEC/NRC tendency to extend standards developed for high-population plants to other reactors licensed at the same time or slightly later.

---

c. The rationale for expressing costs as an exponential function of sector size, rather than as some other function, is explained in Appendix 4.

Licensing time — the interval between construction permit application and award — increased throughout the period studied, although not by a large amount: from an average of about a year for the first plants receiving construction permits to 1½-2 years for most later plants. Because licensing time increased together with nuclear sector size (their correlation coefficient was .77), the regression equation with both variables may not authoritatively depict their effects on costs. Still, the regression result — a mere 7% decline in the size of the coefficient relating costs to sector size, while licensing time is not even remotely significant — appears to disprove the notion that funds spent during licensing added significantly to the real costs of completed plants.[d] Longer licensing periods did, of course, push back construction permit issuance and expose specific plants to greater regulatory requirements, but this phenomenon primarily affected the costs of plants whose permit sequencing was interchanged, without appreciably changing costs for the nuclear sector as a whole.

Other Treatments of Temporal Cost Increases: As mentioned above, costs were also correlated with the *date of construction permit issuance* ("CP date") rather than with nuclear sector size. In this regression equation, costs were found to increase by an average of 23.6% (in real terms) for each one-year increase in CP date. If this increase rate appears extraordinarily high, recall that all but the last of the 46 reactors received their construction permits over a four-year span, from March 1967 to March 1971. When costs are correlated with the date of plant *completion,* the average annual increase rate is 13.1% over the seven years during which the sample plants achieved commercial status (December 1971 to December 1978). An average 1971-78 increase rate of 13.5% per year of completion is calculated for "standard plants" in Chapter 10, incorporating real IDC and changes in other variables such as A-E experience and plant size.

Table 8.2 shows the regression results for the equation with construction permit date substituted for sector size. The r^2 value is 87.6%, almost five percentage points less than the 92.3% r^2 for the model with sector size. The six other variables that were significant in the regression with sector size are also significant with CP date, and their values are not significantly changed, with one exception: cooling towers add only 11.3% to cost, compared with 19.5% previously. The r^2 value for the regression with plant completion date (not shown here) is much lower, 78.4%.

---

d. When cost was regressed on sector size, licensing time, and variables 1-6 described above, it was found to be proportional to the .54 power of sector size (versus .58 previously) and to the .08 power of licensing time. T-ratios were 8.4 and 0.8, respectively, indicating only 59% statistical significance for licensing time. This result duplicates Mooz's finding as to the non-significance of licensing time (Reference 3).

	Table 8.2	
	Nuclear Capital Cost Regression Equation,	
	Alternative Model Using Date	

Capital Cost ($/kW, without IDC, in mid-1979 steam-plant dollars) =

	T-ratio	Significance Level (%)
$1.14 \times 10^{-3} \times$	4.48	99.9+
1.33 if Northeast $\times$	6.09	99.9+
A-E$^{-.125} \times$	5.61	99.9+
MW$^{-.203} \times$	2.03	95.1
.88 if Multiple $\times$	2.45	98.1
1.24 if Dangling $\times$	2.69	98.9
1.11 if Cooling Tower $\times$	2.53	98.4
$(1.236)^{CP\ Date}$	10.01	99.9+

$r^2 = .876$
Adjusted $r^2 = .853$
F value = 38.3
Sample size = 46 units

CP Date is last two digits of year, *e.g.*, July 1969 = 69.5.

Treatment of Operating Experience: Another alternative regression model correlated capital costs to both nuclear sector size and *cumulative reactor operating experience* — the number of unit-years of operation registered by all commercial reactors when the plant in question received its operating license. The intent was to measure the impact of operating experience on costs separately from that of increased sector size. The former factor leads to higher costs by revealing generic design and construction deficiencies; the latter increases public exposure to reactor risks and thus engenders a regulatory disposition to reduce the acceptable level of per-reactor risks.

The regression with both sector size and operating experience is shown in Table 8.3. The r^2 value is slightly higher than for the regression with sector size alone — 92.9% vs. 92.3%. The variable denoting cumulative operating experience ("OPEX") is significant to only the 93% confidence level, however,

Table 8.3

Nuclear Capital Cost Regression Equation, Alternative Model Using Sector Size And Operating Experience

Capital Cost ($/kW, without IDC, in mid-1979 steam-plant dollars) =

	T-ratio	Significance Level (%)
16.2 ×	3.41	99.8
1.28 if Northeast ×	6.78	99.9+
A-E$^{-.094}$ ×	5.37	99.9+
MW$^{-.266}$ ×	3.16	99.7
.897 if Multiple ×	2.70	99.0
1.31 if Dangling ×	4.37	99.9+
1.18 if Cooling Tower ×	4.89	99.9+
(Cumulative Nuclear Capacity)$^{.501}$ ×	8.62	99.9+
OPEX$^{.067}$	1.86	92.9

$r^2 = .929$
Adjusted $r^2 = .914$
F value = 60.7
Sample size = 46 units

less than the standard 95% confidence threshold. Moreover, OPEX and sector size are too closely correlated to each other (with a correlation coefficient of .66) for the measures of their effects on costs to be considered fully reliable. This regression is therefore not used here to project future costs.

The correlation coefficients in Table 8.3 are very similar to those in Table 8.1, except that a somewhat greater cost saving is associated with increased plant size. In addition, a slightly lesser cost increase is associated with sector size expansion since growth in operating experience now shares some of the responsibility for the higher costs. Based on the regression coefficients and past growth in reactor-years (from 13 when the first sample plant received its operating license to 218 for the last), a 21% cost increase was associated with increased operating experience during 1971-78 while a 139% cost increase was associated with sector size expansion. These allocations of the overall 1971-78 cost increase are not conclusive, however, for the reasons stated above.

Section 8.2: Nuclear Construction Durations

This section describes the statistical analysis of nuclear plant construction durations — the time interval between the date of construction permit award and the date the utility declared the unit to be in commercial service. These data were compiled from DOE and NRC reports, respectively.[7]

The nuclear cost analysis reported above was based on a sample of 46 plants completed by the end of 1978. The construction duration analysis is based on a nearly identical 49-plant sample: all nuclear plants that received construction permits between March 1967 and March 1971, inclusive, except Diablo Canyon 1 and 2, whose commercial operation dates were too uncertain to predict.

The sample includes all 46 plants in the cost sample except Farley 1, which received a construction permit in August 1972 following the licensing suspension that began in March 1971 pursuant to the ''Calvert Cliffs'' decision. It adds four plants which received construction permits before March 1971 but were incomplete at the time of the analysis. They were included to avoid biasing the results by excluding plants that took longer to complete than other concurrently licensed plants. Commercial operation dates for the four plants were estimated absent the post-Three Mile Island licensing suspension.[e]

The regression analysis correlates the 49 plants' construction durations to eight explanatory variables — the six that significantly correlated with costs (all except ''dangling'' status) and two others. The r^2 value for the regression is 66% — slightly higher than Mooz's 63%. (Bupp did not study construction time; see References 2 and 3.) The variables and correlations are shown in Table 8.4 and described below.

1. Unit Size: Construction time correlated with nuclear plant size at the 99.9% significance level. The effect is such that a doubling of unit size produced a 28% increase in construction time. Thus, for example, a 1000-MW unit licensed in early 1969 would require 74 months to complete vs. 58 months for a 500-MW unit licensed at the same time. The 95% confidence interval ranges from an 11% to 48% increase for doubled size.[f]

The longer time required to build large reactors results from their larger physical requirements. The quantity of labor, equipment, and materials rises

e. The four (with estimated completion dates) are Salem 2 and North Anna 2 (December 1979), and Sequoyah 1 and 2 (February and October 1979). Actually, only North Anna 2 reached commercial status before 1981 (December 1980).

f. The measured effect of size would increase if the large (approximately 1100 MW) and long-duration Diablo Canyon units were included in the sample.

Table 8.4
Nuclear Construction Duration Regression Equation

Construction Duration (in months) =

	T-ratio	Significance Level (%)
.980 ×	3.16	99.7
MW$^{.358}$ ×	3.48	99.9
A-E$^{-.111}$ ×	4.59	99.9+
(Cumulative Nuclear Capacity)$^{.185}$ ×	3.44	99.9
1.12 if Northeast ×	1.87	93.1
1.13 if Southeast ×	2.17	96.4
1.20 if Duplicate ×	3.60	99.9
1.11 if Cooling Tower ×	2.19	96.5
1.17 if Babcock & Wilcox NSSS	2.72	99.0

$r^2 = .661$
Adjusted $r^2 = .593$
F value = 9.74
Sample size = 49 units

The absolute value of all pairwise correlation coefficients is less than .3, except .552 for the correlation between Southeast and Northeast location.

All variables are significant to at least the 95% level except Northeast location (93%).

with increasing size, although the per-kW quantity declines. The longer construction time for large units adds to the cost of interest during construction. This reduces the apparent 13% cost saving from doubling plant size (obtained in the cost analysis without IDC) to approximately 10%.

2. Architect-Engineer Experience: Doubling an A-E's nuclear plant construction experience saved 7% in construction time. Although this effect is modest, its statistical significance is the highest of any variables in the regression equation — over 99.9%. The 95% confidence range is a 4-10% reduction per doubling in the number of plants built.

Shortened construction duration is a consequence of increasing architect-engineer familiarity with the physical, regulatory, and managerial problems in building nuclear plants. The 7% reduction in construction time is the same as the drop in cost per doubling of experience. A-E experience increased by 275% for a typical unit in the sample, leading to a 14% reduction in average construction time, equivalent to a nine-month reduction. This was more than offset by the lengthening of construction associated with expansion of the nuclear sector, however.

3. Cumulative Nuclear Capacity (Nuclear Sector Size): Real increases in nuclear capital costs over time were correlated above to cumulative nuclear capacity — the size of the nuclear sector. Here, regressing construction time on sector size yields a mild increase rate. Construction time was proportional to the .18 power of the amount of nuclear capacity installed or under construction. The significance level is 99.8%. A doubling of sector size thus led to a 14% lengthening of construction time; the 95% confidence interval is a 5% to 23% increase per doubling of nuclear capacity.

The nuclear sector grew almost five-fold during the study period, from 9800 MW to 55,573 MW. The statistical relation implies that an average increase of 38% in construction time was associated with this growth in capacity. The new requirements for reactor design, construction, and quality assurance brought forth by sector expansion led to extended project time. Note that this construction lengthening is not directly attributable to citizen intervention in licensing hearings; interventions did delay award of construction permits in some cases, but they had virtually no effect on the sample plants once construction began.

This 38% lengthening of construction and the 14% reduction due to A-E learning were the major changes in construction periods for the sample plants. Construction time was sensitive to reactor size, but unit sizes grew only slightly in the 1970s, measured by date of construction permit award.

4. Cooling Towers: Nuclear plants with cooling towers (19 of the 49 sample plants, or 39%) required 11% longer to build, on average, than other plants. This variable is significant at the 96.5% level. The 95% confidence interval is an increase range of 1-21%. For a typical plant in the sample, the statistically inferred impact of a tower was seven additional months.

As discussed in the cost analysis, however, cooling towers are not on the "critical path" of construction steps determining plant completion. Addition of a tower therefore should not add to construction time. Presence of a cooling tower may indicate regulatory sensitivity to environmental concerns leading to additional measures to reduce nuclear hazards, adding to construction time. This conjecture is unproven, however.

5-6. Northeast or Southeast Location: Twenty of the 49 plants are located in the Southeast, 15 are in the Northeast, 12 in the Midwest, and two on the West Coast. Southeast plants took 13% longer to construct, and Northeast plants 12% longer, than Midwest and West Coast plants.

The Southeast result is significant to the 96.4% level, with a 95% confidence interval from 1% to 18%. The Northeast result has a confidence interval ranging from 1% *shorter* duration to 26% longer, and it is significant only to the 93.1% level — below the standard 95% threshold for significance. It is included here because a separate analysis of total *project* time — the period extending from NSSS order to commercial operation — shows a 13% longer project duration for Northeast units, significant to the 98.6% level.

The longer Northeast construction duration may result from adverse weather, less favorable site conditions requiring increased site preparation, and more stringent design and construction standards in response to higher population densities and greater public concern. These factors do not appear to explain the longer construction times for southern plants. This finding is inconsistent, moreover, with the *shorter* construction period found for southern *coal* units. It also belies the widespread notion that the predominantly non-union workforces at southern plants can build plants faster than union labor.

7. *Duplicate Units:* The 15 duplicate units — second or third identical units at a single site — had an average 20% longer construction period than other units. This difference is equivalent to 13 months for a typical reactor in the sample. The result is 99.9% significant, with a 95% confidence interval of 8% to 33%.

This finding reflects the deliberate staging of construction of multiple units to complete second units one-half to two years after initial units. Such a schedule is said to optimize the sharing of unit designs, engineering staff, construction crews, and field equipment. Most multiple units receive construction permits at the same time, so that duplicate units, completed later, have longer construction periods. Construction times for initial units in a multiple set did not differ from those for units built one to a site.

8. *Babcock & Wilcox Reactors:* The nine plants using nuclear steam supply systems manufactured by Babcock & Wilcox (B&W) had 17% average longer construction durations than other units (controlling for other variables). The finding is significant to the 99.0% level, with a 95% confidence interval of 4% to 32%. It appears to have resulted from manufacturing problems that caused delays in construction of B&W reactor vessels during the sample period.[8] Since this appears to have been a one-time phenomenon, it is not built into the projections of future nuclear construction duration in Chapter 10.

Mooz also found that B&W plants took longer to construct, but neither his nor this analysis found any *cost* differences for B&W plants. Mooz hypothesized that the architect-engineers of B&W plants were able to take steps to control costs when construction times exceeded normal levels. His hypothesis is intriguing, if unproven.

References

1. Energy Information Administration (formerly Federal Power Commission), *Steam-Electric Plant Construction Costs and Annual Production Expenses* (annual).

2. I.C. Bupp *et al.*, *Trends in Light Water Reactor Capital Costs in the United States: Causes and Consequences*, CAP 74-8 (Center for Policy Alternatives, MIT, Cambridge, MA, 1974), abridged in *Technology Review, 77* (No. 2), 15-25 (1975).

3. W. Mooz, *Cost Analysis of Light Water Reactor Power Plants*, R-2304-DOE (1978), and *A Second Cost Analysis of Light Water Reactor Power Plants*, R-2504-RC (1979), (RAND Corporation, Santa Monica, CA).

4. This explanatory variable was first employed by Mooz (see Reference 3).

5. See AEC, *Power Plant Capital Costs: Current Trends and Sensitivity to Economic Parameters*, WASH-1345 (1974); any ERDA or DOE documentation of the "CONCEPT" computer program for projecting power plant costs; or Electric Power Research Institute, *Technical Assessment Guide*, EPRI PS-1201-SR (1979), p. 8-5.

6. As reported in a seismic risk map in Institute of Electrical and Electronics Engineers, *Spectrum*, November 1979, map facing p. 49, adapted from Bolt *et al.*, *Geological Hazards* (Springer-Verlag, NY, 1977).

7. DOE, *U.S. Central Station Nuclear Electric Generating Units: Significant Milestones*, DOE/NE-0030, quarterly. NRC, *Operating Units Status Report*, NUREG-0020, monthly.

8. *Nucleonics Week*, 20 September 1979, p. 9.

9

Statistical Analysis Of Coal Plant Construction Costs And Durations

This chapter reports the results of Komanoff Energy Associates' (KEA) statistical analysis of the capital costs and construction durations of U.S. coal-fired plants completed in the 1970s. The preceding chapter presented a parallel analysis for nuclear plants. Both analyses are employed in the next chapter to calculate the average increases in capital costs and construction durations experienced by nuclear and coal plants from 1971 to 1978 and to project likely further increases for future plants.

Section 9.1:   Coal Plant Capital Costs

The sample for the coal-fired plant capital cost analysis consists of 116 generating units. These are all U.S. coal-fired units of 100 megawatts (MW) or greater capacity, completed from January 1972 through December 1977. Several data sources were examined to ensure that all applicable coal-fired units were included in the sample.[1] The 116 units total 70,509 MW. They range from 114 MW to 1300 MW, and average 608 MW. Fifteen of the units have flue gas desulfurization devices, or "scrubbers."

The units' capital costs were compiled from utility data reported to the federal government[2] and are listed in Appendix 2. Because these data are tabulated by *station*, costs of "add-on" units were calculated by subtracting the cost reported for the station in the prior year from the cost reported for the year of commercial operation. The costs were then converted from reported "mixed current dollars" to mid-1979 "steam-plant" dollars, excluding interest during construction (IDC), through a computational process described in Appendix 3. These adjusted costs were divided by the units' nameplate capacity ratings, yielding capital costs per kilowatt.[a] The result was a set of 116 per-kilowatt (kW) costs excluding IDC and adjusted for the effects of inflation in power plant construction wages and materials. Real IDC is added back to the capital costs in Chapter 10, based on the inflation-adjusted cost of capital and the actual average construction durations for the coal plants in the sample.

The costs were correlated with explanatory variables in a multiple regression analysis. The regression equation has nine statistically significant variables and an r^2 value (goodness of fit) of 68%. This compares favorably with the 43% r^2 in Bupp's analysis,[4] indicating that the KEA analysis statistically explains a higher percentage of coal capital cost variations. The variables and their correlations with capital cost are shown in Table 9.1 and described below, along with several variables that did not correlate significantly with cost.

1-4. Regional Variables: Costs were correlated significantly with four regional variables, as shown in Table 9.2. The Midwest is the base region against which the regions in Table 9.2 are compared. It extends in the West from Kansas and Missouri, north to the Dakotas; and in the East from Kentucky and West Virginia, to Michigan and Wisconsin. It contains 57 (49%) of the sample plants.

The five Northeast plants were all built in Pennsylvania and averaged 14% higher cost than the base plant.[b] This is probably accounted for by higher labor costs and more stringent environmental requirements in Pennsylvania. Western plants were considerably more costly, by 26% on average, than the Midwest base plant. This resulted from many factors: environmental standards in many states exceeding federal standards; the markedly higher cost of electrostatic precipitators for controlling particulates from the low-sulfur coal used in the West (see Chapter 7); remote sites adding to transportation costs for materials and equipment; shortages of skilled labor; and high wage rates.

The lower costs of Southeast plants, 14% less than the base plants, probably resulted from weaker environmental standards, favorable construction weather, and, perhaps, predominant use of cheaper non-union labor. The considerable experience amassed by regional utilities in constructing coal-fired plants may also have contributed to lower costs.

The markedly lower cost of the Texas-Oklahoma plants — 24% below

a. Nameplate ratings have previously been employed by the author to calculate coal capacity factors[3] and are used here similarly in Section 11.2. Alternative, lower net "capability" ratings vary too greatly by recording agency to be definitive. Their use would increase per-kW capital costs by several percent but would increase coal capacity factors by an equal, offsetting amount.

b. Costs of the five Pennsylvania *nuclear* plants averaged 5% above those of other Northeast nuclear plants, which in turn averaged 28% above base nuclear costs. Thus, Pennsylvania coal plant costs should not understate coal capital costs elsewhere in the Northeast.

Table 9.1

Coal Capital Cost Regression Equation

Capital Cost ($/kW, without IDC, in mid-1979 steam-plant dollars) =

	T-ratio	Significance Level (%)
.234 ×	.89	62.3
1.14 if Northeast ×	1.81	92.7
1.26 if West ×	5.47	99.9+
.76 if South Central ×	4.51	99.9+
.86 if Southeast (not Southern Company) ×	3.09	99.7+
.73 if Southern Company ×	5.88	99.9+
1.18 if American Electric Power ×	2.26	97.4
.904 if Multiple unit ×	3.25	99.8
1.26 if Scrubber ×	4.86	99.9+
(Cumulative Coal Capacity)$^{.615}$	4.36	99.9+

$r^2 = .679$
Adjusted $r^2 = .652$
F value = 24.9
Sample Size = 116 units

All pairwise correlation coefficients have an absolute value equal to or less than .36.

the Midwest base — is noteworthy in view of the lignite and sub-bituminous coal utilized. Its average heating value, 7,436 Btu/pound, is 29% below the 10,436 Btu/pound sample average. This low-quality coal requires a significantly larger boiler, greater pulverizer capability, and a more expensive precipitator, all of which should add to capital costs. The use of non-union construction labor might help offset these factors, but it alone would not account for the 24% lower cost.

5-6. Company Variables: Table 9.3 shows that costs were significantly correlated with ownership by two large utility holding companies: the Southern Company, with ten units in the sample; and American Electric Power (AEP), with five. The cost differences between these plants and the rest of the

Table 9.2
Coal Cost Correlations To Regional Variables

Region	Effect On Cost	95% Confidence Interval	Significance Level (%)	Number of Plants/ Share of Total Sample
Northeast	+14%	− 1% to +32%	92.7	5/4%
West	+26%	+16% to +37%	99.9+	20/17%
Southeast	−16%	−22% to − 5%	99.7+	12/10%
South Central	−24%	−32% to − 14%	99.9+	8/7%

All Northeast plants are in Pennsylvania. Although the Northeast variable falls below the 95% confidence threshold, it is included in the regression because of the importance of establishing a baseline value for Northeast coal plant costs.

South Central is Texas and Oklahoma.

Southeast is Tennessee, the Carolinas, Florida, Georgia, Alabama, and Mississippi, but excludes 10 Southern Company units. These are treated separately in Table 9.3. The effect on cost of all 22 Southeast plants together is −20%, with 99.9+% significance and a confidence interval of −26% to −14%.

Table 9.3
Coal Cost Correlations To Company Variables

Company	Effect On Cost	95% Confidence Interval	Significance Level (%)	Number of Plants/ Share Of Total Sample
Southern	−27%	−34% to − 19%	99.9+	10/9%
American	+18%	+2% to +36%	97.4	5/4%

The Southern Company units do not overlap with the Southeast units represented in Table 9.2.

sample can probably be explained by the two utilities' approaches to plant design and operation. AEP builds highly reliable and efficient units, with outage rates and heat rates (fuel input per net kWh) among the lowest in the industry.[5] This performance is achieved through conservative design practices, use of high quality components and materials, and generous stocking of spare parts, all of which helped make the AEP plants 18% more expensive to construct than comparable plants.

Conversely, Southern Company's low-cost coal-fired plants — 27% cheaper than Midwest plants and 15% less than other Southeast plants — tend to have below-average reliability and fuel efficiency. It is beyond the scope of this study, however, to determine whether the more conservative design philosophy (higher first costs but better performance) gives lower costs in the long run.

7. Multiple Units: Units that share a site with another identical unit are considered *multiple* units, regardless of whether they are the first or last at the site. The constant-dollar costs of multiple units were averaged to obtain uniform costs for each unit at a site. This eliminates differences in utilities' allocation of joint costs among units.

Sixty-eight of the 116 units are multiple units (another 15 are initial units with incomplete second units at the time of the analysis). Their costs averaged 9.6% less than those of non-multiple units. This finding is significant to the 99.8% level, with a 95% confidence interval ranging from -4% to -15%. It is consistent with the power industry rule-of-thumb that a station with two identical units can be built 8-10% less expensively than a single-unit station. It also replicates the 9.7% lower cost found for multiple nuclear units. The cost savings result from common plant facilities, shared construction equipment, skill transfer in design and construction, and joint environmental review.

8. Scrubbers: Fifteen of the units are equipped with full-capacity flue gas desulfurization systems, or scrubbers. All included scrubbers in the original design — none are "retrofits." They averaged 26% higher cost than comparable units without scrubbers. The difference is equivalent to $120/kW in mid-1979 dollars (including real IDC). The significance level for the scrubber correlation is beyond 99.9%, and the 95% confidence interval runs from 15% to 38%. Units with scrubbers are volatile in costs, however, deviating by an average of 15% from the costs predicted for units with their characteristics, compared to 11% for non-scrubber units.

The scrubbers have an average design efficiency of 74% and are intended to remove an average of 3.7 pounds of sulfur dioxide (SO_2) per million Btu of coal burned. This is equivalent to a reduction of 2.1 percentage points in sulfur content for typical utility coal with an 11,000 Btu/pound heating value. Table 9.4 suggests that scrubbers designed to remove larger quantities of SO_2 may cost more, but there are considerable deviations and no clear trend is apparent.

9. Cumulative Coal Capacity (Coal Sector Size):   The increases over

Table 9.4
Cost Deviations For Coal Units With Scrubbers

Unit	Coal Sulfur %	Design Removal Efficiency (%)	Pounds SO₂ Removed Per Million Btu	Deviation From Predicted Cost
Duck Creek 1	2.75	85.0	5.1	+43%
Gardner 3	0.5	85.0	0.7	+30%
Mansfield 1	4.7	92.1	7.4	+18%
Mansfield 2	4.7	92.1	7.4	+18%
Petersburg 3	3.25	80.0	4.8	+12%
Colstrip 1	0.8	60.0	1.1	+ 4%
Martin Lake 1	1.0	70.5	2.1	+ 1%
Colstrip 2	0.8	60.0	1.1	− 2%
Southwest 1	3.5	80.0	4.9	− 4%
Young 2	0.7	75.0	1.6	− 8%
Winyah 2	1.0	69.0	1.2	− 9%
La Cygne 1	5.0	76.0	8.2	−12%
Conesville 5	4.7	89.5	8.1	−20%
Sherburne 1	0.8	50.0	0.9	−22%
Sherburne 2	0.8	50.0	0.9	−24%
Average	2.3	74	3.7	

Deviation from predicted cost is difference between actual cost and cost calculated by regression equation from plant characteristics. Although deviations expressed as percentages do not add to zero, the sum of their logarithms is zero.

time in the per-kW costs of the coal plants in the sample were considerable even after all costs were converted to mid-1979 construction dollars without interest during construction. A part of these cost increases was caused by the addition of scrubbers to some of the later plants in the data sample. But increased costs also resulted from improvements in other pollution control features, such as electrostatic precipitators, nitrogen oxide controls, and solid waste handling, as Chapter 7 shows.

The addition to coal plant costs from improvements in environmental equipment in the 1970s parallels the cost increases at nuclear plants sparked by

Chapter 9

new safety-related regulatory requirements. In both cases, expansion of the respective generating sectors raised the level of potential environmental and accident hazards associated with nuclear and coal generation of electricity and provided the impetus for remedial control measures. Three alternative regression formulations were examined to allow for this process in the statistical analysis of coal plant costs. In two models, coal capital costs were assumed to be related to *time* in addition to the preceding variables just described; in the third model, costs were assumed to be related to the *size of the coal generating sector*. Time was measured by either the date the coal boiler was ordered from the vendor (dates for actual construction starts were not available for the entire sample) or the date the plant was declared to be in commercial operation. Sector size was defined as the megawatts of coal generating capacity operating or being built on the order date, including the new plant's capacity.

All three formulations reflect the effect on costs of the upgrading of coal plant regulatory standards. The formulation wth coal sector size was devised because it appears to capture more of the societal processes that give rise to new standards, and in order to parallel the nuclear cost formulation adopted in Chapter 8. The three formulations have approximately equal statistical explanatory power, however, as reflected by their r^2 values. Moreover, the costs calculated for typical 1971 and 1978 coal plants in the next chapter do not depend heavily upon the choice of sector size or time. That choice is important only for projecting future coal plant capital costs.

For the model using sector size, cumulative coal capacity prior to the ordering of the first plant in the data base was estimated at 142 gigawatts (GW, 1 GW = 1000 MW).[c] Cumulative coal capacity of 217,348 MW was calculated for the 116th and last sample plant by adding each new plant in sequence of its boiler order (including 16 units ordered before the last plant and completed during 1978, following the December 1977 sample cut-off date), and allowing for retirements equaling 5% of new capacity additions.[d]

The correlation yielded by the model shows coal capital costs rising at a rate proportional to the .61 power of cumulative coal capacity. This rate is only .03 greater than the exponential increase rate for nuclear costs, but the latter increased much more during 1971-78 because the nuclear sector was growing proportionately much faster than the coal sector.

Based on the .61-power relationship, a 30% increase in the cost of coal plants was associated with the 53% increase in coal sector size during the study

c. U.S. fossil steam-electric capacity at the end of 1971 was 260.4 GW according to EIA (see Reference 1, 1972 edition). Coal accounted for 54.6% of 1971 fuel consumption by this capacity (1975 edition). Prorating generation to capacity — a reasonable assumption given low differences in plant running costs at that time — 1971 coal capacity is calculated to be 142.2 GW, rounded to 142 GW.

d. The ratio of retirements to new fossil capacity averaged 5.4% in 1971-75, based on the EIA reports (Reference 1). Boiler ordering dates were taken from Kidder Peabody data (Reference 1). These data are used by EIA analysts and are considered definitive.

period. This figure does not include the separately measured 26% addition to costs from scrubbers. The 95% confidence interval for the coal sector exponent ranges from .34 to .89 (implying a confidence range of 26% to 86% in the impact of sector doubling), and statistical significance is beyond the 99.9% level.

Non-Significant Variables: Coal costs were not significantly correlated with the following variables: coal heat content, supercritical (high-pressure) boilers, use of cooling towers, boiler manufacturer, designation as a "dangling" unit (the first unit in an incomplete station of several identical units), and unit size.

The last result is especially important. Larger unit sizes are widely believed to bring economies in construction. A long-standing industry-government rule of thumb holds that coal per-kW capital costs decline by 10-15% when plant size is doubled.[6] But only a weak effect, a 3% cost reduction for doubled size, was evident here, and it had only 82% statistical significance. Moreover, larger coal units take longer to build. Doubled size added an average of 13% to project duration for the sample plants, as shown in the next section. The accompanying increase in real interest during construction caused a 1% cost addition per size doubling, partly offsetting the tentative, minor scale economy observed in direct construction costs.

Other Treatments of Temporal Cost Increase: Several other regression models were tested for coal capital costs. In two, costs were correlated with time — the date of boiler order or the date of plant commercial completion. These regressions are shown in Table 9.5.

The table indicates little difference in coefficients between the two regressions employing time, and between them and the model employing sector size (Table 9.1). It also shows costs increasing at a little over 4% for each later year of plant ordering or completion. This result excludes the separately measured effect of scrubbers. That effect is added in the next chapter and contributes to a calculated average annual 1971-78 cost increase rate of 7.7% when a "standard" 1978 coal plant with a scrubber is compared to a 1971 plant without.[e]

Table 9.6 present the final coal cost regression. Like Table 9.1, it uses coal sector size to represent the accretion of environmental standards adding to costs, but unlike that model it omits the variable denoting scrubbers. The result is an increase in the exponent relating costs to cumulative coal capacity, from .61 to .86, and a corresponding increase from 53% to 82% in the apparent effect on cost of a doubling of sector size. The increases arise because the share

e. The coal sample may be said to extend to the end of 1978 because all U.S. coal plants completed in 1978 were ordered prior to the last plant included in the data base (completed in 1977).

Table 9.5
Coal Capital Cost Regressions, Alternative Models Using Date

Capital Cost ($/kW, without IDC, in mid-1979 steam-plant dollars) =

Regression Coefficient For Boiler Order Date / Regression Coefficient For Completion Date	T-ratio	Significance Level (%)
18.7/16.8 ×	4.16/4.52	99.9+
1.15/1.11 if Northeast ×	1.85/1.41	93.2/83.9
1.26/1.27 if West ×	5.51/5.80	99.9+
.76/.75 if South Central ×	4.55/4.78	99.9+
.86/.87 if Southeast (not Southern Company) ×	3.06/2.94	99.7/99.6
.73/.75 if Southern Company ×	5.77/5.50	99.9+
1.18/1.17 if American Electric Power ×	2.26/2.14	97.4/96.6
.91/.89 if Multiple unit ×	3.05/3.71	99.7/99.9+
1.25/1.26 if Scrubber ×	4.65/4.88	99.9+
(1.044/1.042) Year	4.25/4.52	99.9+

$r^2 = .677/.683$
Adjusted $r^2 = .649/.656$
F value = 24.7/25.4
Sample size = 116 units

Year is last two digits of year of boiler order or commercial completion, *e.g.*, July 1969 is 69.5.

---

of the temporal cost increase previously accounted for by scrubbers has been incorporated into the sector size variable.

This model is attractive because it implies that scrubbers are not a one-time cost but are a manifestation of an ongoing process, related to expansion of the coal sector, that continuously increases the requirements and therefore the costs of control equipment for coal plants. The frequency of scrubber units is sufficiently low, however, even near the end of the sample, that costs calculated with this model are unrealistically low — even for plants projected for the late 1980s — compared to costs calculated with models which

Table 9.6
Coal Capital Cost Regression, Alternative Model Without Scrubber Variable

Capital Cost ($/kW, without IDC, in mid-1979 steam-plant dollars) =

	T-ratio	Significance Level (%)
.0124 ×	2.54	98.7
1.22 if Northeast ×	2.45	98.4
1.25 if West ×	4.83	99.9+
.85 if Southeast (not Southern Company) ×	3.05	99.7
.75 if South Central ×	4.49	99.9+
.71 if Southern Company ×	5.96	99.9+
1.14 if American Electric Power ×	1.70	90.7
.90 if Multiple unit ×	3.09	99.7
(Cumulative Coal Capacity)$^{.861}$	5.94	99.9+

$r^2 = .608$
Adjusted $r^2 = .578$
F value = 20.7
Sample size = 116 units

measure scrubber costs separately and explicitly assign scrubbers to standard plants.

Section 9.2: Coal Plant Project Durations

This section describes the statistical analysis of coal plant project durations — the time interval between the date the boiler was ordered and the date the utility declared the unit commercially operational. These data were compiled from the Kidder Peabody data base and from EIA data, respectively (see Reference 1). *Project* duration spans a longer interval than *construction* duration, the variable measured in the reactor duration analysis in Chapter 8, since it includes any time between boiler order and start of construction (the latter data were unavailable for coal units).

In order to obtain an unbiased sample, the data base was truncated from the 116 units in the cost analysis to the 92 units ordered between January 1968 and July 1972. The first ten units in the cost sample, all ordered before 1968, are biased toward *longer* project durations because concurrently ordered units that took less time to construct were completed prior to the January 1972 data base threshold and thus are not in the sample. Similarly, the last 14 units in the cost sample, all ordered after July 1972, were constructed in a *shorter* time than concurrently ordered units that were not completed before the December 1977 data base cut-off.

The project durations of the 92 units are significantly correlated with five explanatory variables. All but one, unit size, were significant in the cost analysis. The five variables together explain 41% of the variance in project duration. Although this r^2 value is far below that of the nuclear duration analysis and the nuclear and coal cost analyses, all five explanatory variables are significant to approximately the 99% level or beyond, and all correlate logically with construction time. The variables and correlations are shown in Table 9.7 and described below.

1. Unit Size: This variable had by far the most significant correlation with project duration, well beyond the 99.9% level. A doubling of unit size extended project duration by 13%, adding, for example, seven months to the 55-month duration of a typical 500-MW unit ordered midway through the sample period. The 95% confidence interval ranges from an 8% increase to an 18% increase per size doubling.

The 13% increase in coal project time parallels the 18% construction time increase for nuclear plants from doubled unit size. Both result from the increased quantities and sizes of material and equipment required to build larger plants, although the effect was significantly greater for nuclear.

2. Cumulative Coal Capacity (Coal Sector Size): The increase in coal project durations in the 1970s is captured here by a variable representing cumulative coal capacity, as was done for coal capital costs in the previous section. The result is that the rate of increase in project time was proportional to the .53 power of coal sector size. At this rate, doubling the coal sector would increase project length by 44%. The significance level is 99.8%, and the 95% confidence range is a 15-80% lengthening of duration per doubling.

Based on this relationship, a 25% average increase in project duration was associated with the 53% expansion of the coal sector during the sample period. By the end of 1978, project time had increased by a year from the 50 months required for a typical 500-MW plant completed in early 1972. This increase may have resulted primarily from the increasing scope of environmental reviews *prior to* construction rather than from any increase in construction time. This conclusion is suggested by the lack of a correlation between project duration and use of scrubbers, discussed below, and by the expansion in pre-construction environmental reviews during the early 1970s.

Table 9.7
Coal Project Duration Regression Equation

Project Duration (in months) =

	T-ratio	Significance Level (%)
.0317 ×	2.94	99.6
MW$^{.178}$ ×	5.53	99.9+
(Cumulative Coal Capacity)$^{.526}$ ×	3.23	99.8
.88 if Southeast (not Southern Company) ×	2.65	99.0
.86 if Southern Company ×	3.27	99.8
1.08 if Duplicate	2.56	98.8

$r^2 = .415$
Adjusted $r^2 = .381$
F value = 12.2
Sample size = 92 units

All pairwise correlation coefficients have an absolute value less than .21.

3-4. Southeast Location and the Southern Company: Project duration was shorter for coal plants in the Southeast, especially those operated by the Southern Company, as Table 9.8 shows. The rapid construction of Southeast plants probably resulted from favorable weather for construction and less detailed environmental reviews (although Texas-Oklahoma location did not affect project length, despite similar characteristics). Curiously, Southeast *nuclear* plants required longer to construct than Midwest reactors.

5. Duplicate Units: Thirty-two of the 92 sample plants are duplicates — the second (or later) in a set of identical units at a single site. They required 8% more project time than other units; this finding is significant at the 98.8% level, with a 2-15% 95% confidence spread. It reflects a deliberate decision to complete duplicate units one to two years after the initial unit in order to optimize the use of common labor and construction facilities and to co-ordinate capacity expansion with system demand. Since duplicate and initial units are usually ordered within a year of each other (and sometimes simultaneously), intentional later completion causes project duration to be greater for duplicate

Table 9.8
Coal Cost Correlations To Company Variables

Factor	Effect On Duration	95% Confidence Interval	Significance Level (%)	Number of Plants/ Share Of Total Sample
Southern Co.	−14%	−22% to −6%	99.8%	10/11%
Other Southeast	−12%	−20% to −3%	99.0%	9/10%
All Southeast	−13%	−19% to −7%	99.9+	19/21%

Sample shares were computed on the basis of 92 plants in duration analysis.

units than for other units.

Non-Significant Variables: Coal project duration was not significantly correlated with other variables that affected costs: location in Texas/Oklahoma, operation by American Electric Power, or use of scrubbers. Scrubbers apparently added 4% to project time, but this result had only 52% significance and thus is not remotely conclusive. Moreover, of the last 14 units in the cost sample that were excluded from the duration analysis, the six units with scrubbers had a 5% *shorter* average project time than the eight non-scrubber units ordered in the same period (calculation includes adjustment for unit size differences). Other variables not correlated with costs — cooling towers, high-pressure boilers, and coal heat content, also were not linked with project time.

Three other variables came close to statistical significance in affecting project duration. Western plants averaged 7% shorter duration, with 90% significance. Northeast plants averaged 13% longer project time, but there were only three in the truncated sample and the correlation had only 84% significance. Publicly-owned coal units had a 7% longer project time, with 90% significance. This result may reflect different financial incentives: investor-owned utilities earn little or nothing on construction expenditures until the completed plant enters the rate base, whereas publicly-owned utilities may have less incentive to press for completion since they can raise their rates during construction. The regression equation adding these three variables had an r^2 of 47%.

References

1. DOE, *Inventory of Power Plants in the United States*, DOE/RA-0001 (December 1977); EIA, *Steam-Electric Plant Construction Costs and Annual Production Expenses*, annual; Kidder Peabody, Inc., ''Reports on Electric Utility Generating Equipment,'' annual.

2. EIA (see Reference 1).

3. C. Komanoff, *Power Plant Performance* (Council on Economic Priorities, New York, NY, 1976).

4. I.C. Bupp *et al.*, *Trends in Light Water Reactor Capital Costs in the United States: Causes and Consequences*, CAP 74-8 (Center for Policy Alternatives, MIT, Cambridge, MA, 1974). Abridged in *Technology Review*, 77 (No. 2), 15-25 (1975).

5. AEP's coal-fired units over 600 MW averaged 14% higher capacity factors (9 percentage points) than comparable units of other utilities during 1967-73. See *Power Plant Performance* (Reference 3), p. 84.

6. See Chapter 8, Reference 5.

Chapter 9

10
Changes In Nuclear And Coal Capital Costs And Construction Durations

This chapter calculates the capital costs and construction durations of standard nuclear and coal plants completed at three different times:

- the end of 1971, representing *actual* costs and durations at the beginning of the nuclear and coal samples;
- the end of 1978, representing *actual* costs and durations at the conclusion of the samples;
- the end of 1988, representing *projected* costs and durations at the likely completion date of the last nuclear plants currently holding construction permits.

These costs and construction durations are shown in Tables and Figures 10.1 and 10.2. They were calculated from the nuclear and coal capital cost and construction duration regression equations in Chapters 8 and 9, using the characteristics of the hypothetical standard plants developed below. This chapter explains the procedures used, compares the effects of the various causal factors on cost and construction time, and assesses uncertainty in the results.

Section 10.1: Hypothetical Standard Plants

Using the regression equations developed in the previous two chapters for the costs and construction times of nuclear and coal plants, one can measure actual, past changes in the costs and schedules of plants during the 1971-78 sample period and project changes in the costs and schedules of future plants.

Table 10.1
Capital Costs Of Standard Plants
(in mid-1979 steam-plant dollars, with real IDC)

	Actual		Projected
	1971	**1978**	**1988**
Nuclear	$366/kW	$887/kW	$1374/kW
Coal	$346/kW	$583/kW	$794/kW
Nuclear/Coal Ratio	1.06	1.52	1.73
Increase Over Previous Period			
Nuclear		142%	55%
Coal		68%	36%
Implied Annual Increase Rate			
Nuclear		13.5%	4.5%
Coal		7.7%	3.1%

1978 and 1988 coal costs include SO_2 scrubber.

1988 projected nuclear cost includes no allowance for special impact of Three Mile Island accident.

To do so it is necessary to posit "standard" plants whose characteristics can be applied to the regression equations.

There are two ways to select the characteristics of standard plants. The obvious method is to develop *composite* plant characteristics from the data samples. This approach would lead to misleading nuclear and coal comparisons, however, because the two samples have different characteristics. For example, Northeast plants comprise 30% of the nuclear sample but only 4% of the coal sample. Since Northeast plants cost considerably more than other plants, a composite nuclear plant would be disadvantaged in a comparison with a composite coal plant.

Considerations such as this make it necessary to base cost and duration calculations on *hypothetical* standard plants. Table 10.3 on page 232 specifies such plants for three different periods: the end of 1971, at the beginning of the sample period; the end of 1978, at the end of the sample period; and the end of 1988, the date at which a recently licensed nuclear plant could probably be

Table 10.2
Construction Times Of Standard Plants
(in months)

	Actual		Projected
	1971	**1978**	**1988**
Nuclear	65.8	77.7	97.3
Coal	52.7	60.6	73.3
Nuclear/Coal Ratio	1.25	1.28	1.33
Increase Over Previous Period			
Nuclear		18%	25%
Coal		15%	21%
Implied Annual Increase Rate			
Nuclear		2.4%	2.3%
Coal		2.0%	1.9%

Nuclear durations extend from construction permit to commercial operation; coal durations extend from boiler order to commercial operation.

1988 projected nuclear duration includes no allowance for special impact of Three Mile Island accident

completed (based on 1971-78 schedule trends exclusive of the Three Mile Island accident). Nuclear and coal plants are assumed to have the same geographical composition in each period but are not necessarily identical in other respects.

The entries in the table are explained below:

1. Sector Size: Cost escalation and schedule lengthening at nuclear and coal plants were modelled in the preceding chapters by correlating costs and construction times to the sizes of the respective nuclear and coal sectors. The rationale was that as cumulative sector generating capacity (sector size) increases, so do environmental and safety criteria which strongly affect design and construction requirements and costs.

As explained in Chapter 9, coal sector size is estimated at 142,000 megawatts (MW) for the standard plant at the end of 1971, and 217,348 MW

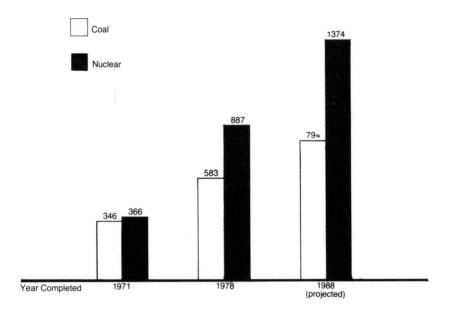

Figure 10.1
Power Plant Capital Costs
(in 1979 steam-plant $/kW, with real interest during construction)

Costs were synthesized by applying nuclear and coal cost regressions to "standard plants" as described in text. Costs of 1971 and 1978 plants are relatively independent of the regression model employed. 1978 and 1988 coal plants include scrubbers.

for the standard plant at the end of 1978.[a] For the late-1988 plant, a value of 346,516 MW was calculated by assuming annual net capacity additions of 12,917 MW, equal to 120% of average net coal additions in 1972-78. The additional 20% is intended to reflect both coal's greater share of future fossil plants and a decline in total fossil additions due to reduced load growth.

Cumulative nuclear capacity just prior to the construction permit award for the first reactor in the data base is calculated at 9800 MW. This figure includes all 15 commercial-size reactors already with construction permits but excludes small (200 MW or less) reactors. These belong to an earlier era from

a. The latter figure is actually total coal capacity ordered through 1973 and installed through 1978. Since the order for the last 1977-completed coal unit in the sample followed the orders for all units installed through 1978, the 217,348 MW figure encompasses all coal units installed through 1978.

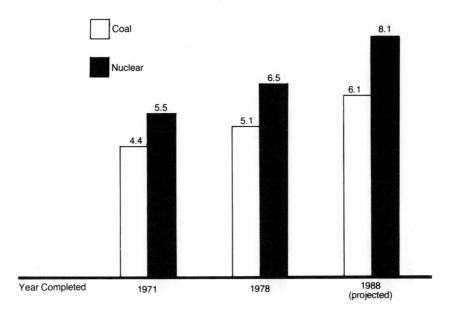

Figure 10.2
Power Plant Construction Durations
(in years)

Coal

Nuclear

Year Completed   1971   1978   1988 (projected)

Durations are measured from construction permit award (nuclear) or boiler order (coal) to commercial service date. They were synthesized by applying nuclear and coal duration regressions to "standard plants" as described in text. Durations of 1971 and 1978 plants are relatively independent of model employed.

both design and regulatory standpoints. The sector grew to 55,573 MW with construction permit awards for 52 additional reactors: the 46 completed through 1978 which comprise the data base, and six others licensed concurrently which had not reached commercial operation by the end of 1978.

Sector size for the standard nuclear plant completed at the end of 1988 was calculated by summing the capacities of all plants with NRC construction permits through 1978, the 15 pre-sample plants, and the 46 in the sample. The resulting figure of 149,648 MW represents a 169% expansion from the 55,573 MW sector size for 1978, and a 14-fold increase from the pre-sample capacity of 9800 MW. Acutal 1988 capacity could be less if the negative net ordering rate for nuclear plants persists. The figure used here has so far been virtually unaffected, however, by recent cancellations; these have befallen primarily plants *on order* rather than those with construction permits. (The sensitivity analysis in the next section shows the effect of varying this assumption.)

Table 10.3
Characteristics Of Standard Plants

	1971-78 Sample Average		Hypothetical Standard Plant					
			Nuclear			Coal		
	Nuclear	Coal	1971	1978	1988	1971	1978	1988
Sector Size, MW	23,337	175,680	9,800	55,573	149,648	142,000	217,348	346,516
Unit Size, MW	831	560	820	840	1,150	708	439	300
A-E Experience	4		2	7.5	18			
Multiple Unit	.48	.72	.50	.45	.80	.72	.72	.80
Duplicate Unit	.31	.36	.32	.27	.50	.36	.36	.50
Scrubber		.13				0	1	1
Cooling Tower	.40		.40	.40	.50			
Northeast	.30	.04	.20	.17	.14	.20	.17	.14
AEP		.04				.03	.03	.03
Southeast	.39	.19	.15	.15	.17	.15	.15	.17
South Central	0	.07				.05	.09	.09
West	.04	.17				.11	.22	.22
Midwest (Base)	.27	.49				.46	.34	.35
B&W NSSS	.17		.17	.17	0			

Figures are unit averages, except geometric means for continuous variables.
Southeast figures shown for coal include Southern Company: .09 in 1971-78 sample, and .07, .07, and .06 for standard plants.

2. Unit Size: Larger nuclear and coal units took longer to build, larger nuclear units cost less, and coal costs were unaffected by unit size in the 1971-78 data samples. The average nuclear unit in the sample is 830 MW,[b] and the typical plant size increased less than 3% during the sample period, based on the correlation between reactor size and construction permit date. Reactor sizes are increasing significantly, however, and 1150 MW is representative of plants that will come on line in the late 1980s.

Coal plant sizes actually decreased during the sample period, contrary to popular belief. A regression of unit size on boiler order date indicates that the typical size of new coal plants was 708 MW for 1971 completion and only 439 MW at the end of 1978.[c] Extrapolation of this trend would project new coal units averaging 260 MW in the late eighties. A more realistic estimate, taking into account utility attitudes, is 300 MW.

3. Architect-Engineer (A-E) Experience: This variable applies only to nuclear plants. The A-E numbers used here reflect the plant being considered, plus all previously licensed commercial-size reactors, including the 15 excluded from the data base.[d] They range from one to 23, the latter for the last plant built by Bechtel, which built 19 of the 46 plants. The other 27 plants were built by nine other A-Es (excluding utilities that designed their own units).

The average A-E number (geometric mean) increased from two at the start of the sample to 7.5 at the end of 1978. Expansion of the nuclear sector to include all plants with construction permits would raise the average figure to 18. Note that A-E experience is measured by project, not total capacity built, since each project is a learning experience.

4-5. Multiple and Duplicate Units: Identical multiple units have lower costs in both the nuclear and coal analyses. Although the percentage of multiple nuclear units fell slightly in the sample period, an increase from 45% in 1978 to 80% is assumed for future plants, reflecting the growing tendency of utilities to construct twin-unit stations. Cancellations of second units for financial reasons would reduce this figure, however. The share of multiple coal units is assumed to rise slightly from the current 72% to 80%, matching the value projected for nuclear plants. With coal units sizes falling, most coal units will be installed as multiple units. The percentage of duplicate units (the second or more in a multiple set), is relevant only to calculating construction

b. This geometric mean is used in preference to the artithmetic mean because of the logarithmic regression form employed. The arithmetic mean of nuclear unit size in the data sample is 854 MW.

c. The regression has a weak r^2 value, only 8%, but the MW coefficient is significant beyond the 99% level.

d. Also included is the Hanford plutonium production reactor, with a net electrical output of 800 MW, built by A-E Burns & Roe.

duration. It is assumed to be slightly over half that of multiple units.

6. *Scrubbers:* Only 13% of the coal units in the data sample (15 of 116) have scrubbers. All coal plants whose construction started after September 18, 1978 must have scrubbers, however, regardless of coal sulfur content, so the 1988 coal plant is assumed to include a scrubber. *A scrubber is also specified for the 1978 standard plant,* ensuring that 1978 coal plants are costed for state-of-the-art pollution controls for a conservative comparison with recent reactor costs. In reality, however, only about half of coal plants installed in the late 1970s have scrubbers (the others use low-sulfur coal to meet the New Source Performance Standard for SO_2).

7. *Cooling Towers:* This variable significantly affected nuclear costs and duration but had no apparent effect on coal costs or scheduling. The 40% penetration for cooling towers in the nuclear sample is assumed to increase to 50% for future plants, reflecting both increasing regulatory stringency and greater load growth in interior regions requiring closed-cycle cooling systems, relative to coastal areas where once-through systems are the norm.

8. *Geographical Location:* The considerable differences in geographical composition between the nuclear and coal samples make it necessary to specify compromise standard plants. The nuclear sample is dominated by Northeast plants (30%) and Southeast plants (39% of the cost sample, 41% of the construction duration sample). Only 4% and 19% of coal plants installed in 1972-77 are in these regions, respectively. Conversely, 20% of installed coal plants, but only 5% of the nuclear plants, are in the West or South Central regions. The coal plants in the study sample are also concentrated more in the Midwest than are the nuclear plants, by 53% to 26%. These figures reflect perceived regional nuclear and coal cost advantages. Although regional differences persist, they must be excised from the standard plant to prevent bias in comparing costs.

The geographical characteristics specified for 1971 and 1978 are a rough mix of the nuclear and coal samples. For the 1988 plant they are the estimated future shares of new utility construction. The Northeast contribution is set at 20% in 1971 and 17% in 1978, declining to 14% in 1988 because of low regional load growth. The West's share is assumed to double from 11% in 1971 to 22% in 1988 because of rapid growth. Other geographical percentages are shown in Table 10.3.

Section 10.2:   Changes In Capital Costs

Table 10.1 on page 228 and Figure 10.1 on page 230 give the costs of standard nuclear and coal plants installed in (late) 1971, 1978 and 1988. These costs were calculated by applying the nuclear and coal cost regression equations (Tables 8.1 and 9.1) to the characteristics of the standard plants in Table 10.3. The costs include "real" (inflation-adjusted) interest during construction (IDC) accounting for between 6% and 15% of total plant cost. The method for calculating IDC is described in Section 10.4.

Table 10.1 demonstrates that real escalation in nuclear capital costs far exceeded coal capital cost escalation during the period from 1971 to 1978. The percentage increase in nuclear capital costs, 142%, was slightly over twice that for coal, 68%, even though the entire brunt of the addition of scrubbers has been assigned to the 1971-78 period. The annual real rate of nuclear cost escalation, 13.5%, was 75% greater than that of coal escalation, 7.7%. (The latter figure includes the effect of scrubbers. Without scrubbers, the 1971-78 real increase in coal capital costs was 33%, or 4.2% per year.) As a result, the gap between nuclear and coal capital costs increased dramatically: from only 6% in 1971 without scrubbers to 52% in 1978 with scrubbers.

The table also shows that the annual rate of capital cost escalation will slow significantly in the next decade if 1971-78 cost relationships continue, falling to about one-third of the past annual rate. Nevertheless, both the nuclear and coal cost increases would be substantial, about 90% as great as the 1971-78 increases in absolute terms, because escalation will build from a larger base. The capital cost of a standard nuclear plant would then be 73% greater than the comparable coal capital cost by 1988. These projections make no allowance for the Three Mile Island accident, which, as Chapter 6 shows, appears likely to have a substantial additional impact on future nuclear costs.

Table 10.4 traces actual past and projected future nuclear and coal cost increases to changes in the factors that correlated significantly with capital costs in the regression equations. The table shows that cost changes associated with changes in plant characteristics other than scrubbers during 1971-78 were hardly noticeable next to the cost increases associated with expansion of the nuclear and coal sectors. Changes in common status, cooling tower use, and geographical location affected the costs of standard nuclear and coal plants by only 1% or less. This was due to both the modest correlation between changes in these factors and costs and to the fact that the standard plants changed little in the factor values. Projecting over the entire 1971-88 period, increases in architect-engineer experience and unit size will reduce nuclear costs by 21% and 7%, respectively. But these factors will be overwhelmed by escalation associated with nuclear sector expansion, which added 172% to nuclear costs from 1971 to 1978 and will cause a total 1971-88 cost increase of 382%, based on past trends.

Uncertainty Analysis: The 1988 cost projections in Table 10.1 are subject to several sources of uncertainty. One is the possibility that the underlying trends which brought about the 1971-78 cost escalation will perform differently in the future. Chapters 5 and 6 suggest that future nuclear

Table 10.4
Components Of Capital Cost Escalation

	Actual	Projected	
	1971-78	**1978-88**	**1971-88**
Changes in Nuclear Plants			
Multiple Unit	+ ½%	+ 4%	− 3%
Cooling Tower Use	——	+ 2%	+ 2%
Northeast Location	− 1%	− 1%	− 2%
A-E Experience	−13%	− 9%	−21%
Unit Size	− ½%	− 6%	− 7%
Nuclear Sector Size	+172%	+77%	+382%
Real IDC	+ 3%	+ 5%	+ 8%
Net Nuclear Escalation	+142%	+55%	+275%
Changes in Coal Plants			
Scrubber	+26%	——	+26%
Multiple Unit	——	− 1%	− 1%
Geographical Location	+ ½%	——	+ ½%
Coal Sector Size	+30%	+33%	+73%
Real IDC	+ 2%	+ 3%	+ 5%
Net Coal Escalation	+68%	+36%	+129%

Figures indicate changes in costs attributable to changes in standard plant characteristics shown in Table 10.3. For example, Northeast location adds to nuclear costs but it is declining in frequency and thus appears in table as a negative change in costs. Cost factors are given in Tables 8.1 for nuclear and 9.1 for coal.

Percentages are combined multiplicatively. For example, the combined effect of increased nuclear A-E experience in 1971-88 is calculated by multiplying .87 (from − 13% in 1971-78) and .91 (from −9% in 1978-88), yielding .79, or −21%.

Real IDC is the inflation-adjusted cost of interest during construction. It reflects construction duration and the real cost of capital and is calculated in Section 10.4.

regulatory standards, and thus costs, are likely to increase more rapidly in the future than is indicated by past trends. But even if trends adhere precisely to their past course, two other sources of uncertainty would remain in the cost projection methodology. One is uncertainty in the future rate of expansion of the nuclear and coal sectors; the other is the uncertainty inherent in the measurement of factors affecting costs.

Nuclear and coal sector sizes play a central part in projecting costs since they have been assigned the role of explaining the increases in costs over time resulting from increased environmental and safety standards. According to trends inferred from 1971-78 data, doubling sector sizes would raise costs by 53% for coal and by 49% for nuclear. Because the nuclear sector is considerably smaller than the coal sector, comparable capacity additions should have a far greater impact on nuclear costs. The projected 94 gigawatt (GW) expansion of the nuclear sector from 1978 to 1988 is far larger in percentage terms than the estimated 129 GW increase in coal capacity over the same period, so that nuclear costs should rise more sharply than coal plant costs.

The effect on projected costs of varying anticipated sector growth is shown in Table 10.5. The table indicates that both nuclear and coal future costs are not overly sensitive to the rate of expansion of their respective sectors. The effect on costs is approximately one-third as great as the variation from the assumed increase in sector capacity for nuclear, and one-fifth for coal. (The effect appears greater if measured against the *increase* in costs from 1978 to 1988.) Thus the rates of increase in nuclear and coal capacity could differ considerably from those assumed in defining the 1988 standard plants without drastically changing projected costs.

Uncertainty in projecting future costs from past trends also arises from statistical uncertainty in measuring the rate at which sector expansion affects costs. The regression coefficients used to calculate costs are merely estimates of the most likely values of the various factors affecting costs. The factors may actually be stronger or weaker than the values assumed, causing actual costs to differ from the calculated figures.

This uncertainty may be estimated by applying the "95% confidence intervals" of the coefficients for sector expansion reported in Chapters 8 and 9. For example, while the best estimate is that coal costs increase with the .61 power of coal sector size, there is 95% likelihood that the rate of cost increase falls between the .34 and .89 powers of sector size. This range and an analogous range for the nuclear sector coefficient are used to produce Table 10.6. The table indicates that there is only modest statistical uncertainty associated with the sector size variables used to project future nuclear or coal costs.

Results of Alternative Regression Models: Several alternative statistical models of nuclear and coal capital costs were described in Chapters 8 and 9. It is of interest to calculate the costs that these models yield for 1988 plants.

Table 10.5
Effect On Projected 1988 Costs
Of Varying 1978-88 Sector Growth

	Nuclear	Coal
1978 Sector Size	55,573 MW	217,348 MW
1988 Projected Sector Size	149,648 MW	346,516 MW
1978-88 Projected Growth	94,075 MW	129,168 MW
Effect on 1988 Cost of Varying 1978-88 Growth by:		
−50%	−17%	−12%
−25%	− 8%	− 6%
−10%	− 3%	− 2%
+10%	+ 3%	− 2%
+25%	+ 7%	+ 5½%
+50%	+14%	+11%

In calculating nuclear costs, the increase in A-E experience was prorated according to the variation in 1978-88 sector growth. Calculations do not reflect the additional minor effect of sector size on cost through its impact on construction duration. They also omit effect on cost of reduced percentage of multiple units if duplicate units are disproportionately cancelled.

A nuclear cost model in which cost increases over time are explained by growth in cumulative reactor operating experience as well as increases in sector size is presented in Table 8.3. The model was rejected because the variable denoting operating experience was not statistically significant and because the two explanatory variables were very highly correlated. The model is nevertheless attractive because it explicitly recognizes the role played by operating experience in engendering new regulatory standards and higher costs. When it is applied to a 1988 standard plant,[e] it yields a calculated cost of

e. Cumulative operating experience for the 1988 plant is assumed to be 1400 reactor-years. This is based on 218 commercial reactor-years realized through June 1977, when the last data base plant was awarded an operating license, and assumed linear growth over the ensuing 11 years from 60 plants in 1977 to the 150 currently operating or with construction permits.

Table 10.6
Sensitivity Of Projected 1988 Costs
To Uncertainty In Effect of Sector Size

95% Confidence Range Relative To Base Cost Projections	Nuclear	Coal
High Range	+ 9%	+14%
Low Range	− 8%	−12%

Figures were calculated by applying confidence intervals to the projected sector sizes for 1988 standard plants. Interval for coal sector cost effect is .335 to .894; effect for nuclear sector ranges from .491 to .664.

$1450/kilowatt (kW), including real interest during construction (IDC), 6% more than the $1374/kW cost adopted here based on the regression with sector size alone.

Table 9.6 presents a coal cost model without a variable denoting use of scrubbers, in which all temporal cost increases are represented by growth in sector size. This model yields a 1988 coal plant cost projection of $773/kW with real IDC, 3% less than the $794/kW cost calculated from the model including the scrubber variable. Although cost is more sensitive to sector size in the model without scrubbers, its steeper slope is not enough to offset the large (26%) "one-time" cost impact of scrubbers in the model used here, based on the assumed 1988 coal sector size of 347 GW.[f]

The other alternative models portray nuclear and coal costs as a function of time: reactor construction permit (CP) date, coal boiler order date, and coal plant completion date. They show nuclear costs increasing at 24% per year of CP date and coal costs rising at a little over 4% annually (exclusive of the impact of scrubbers). *These rates are the product of the particular measurement conventions employed and are not meaningful for representing past cost increases.* They differ substantially from the annual escalation rates synthesized from the regressions applied to the 1971 and 1978 standard plants — 13.5% for nuclear, 7.7% for coal — shown in Table 10.1. Future costs calculated from the alternative models would be unrealistically high for nuclear plants and low for coal plants. Moreover, there is not a sufficient basis for extending past annual percentage rates of cost increase into the future.

f. The breakeven point for the two models is approximately 385 GW of coal capacity.

Section 10.3:   Changes In Construction Durations

Table 10.2 on page 229 and Figure 10.2 on page 231 give the nuclear construction times and coal plant project times of standard plants installed in (late) 1971, 1978 and 1988. These figures were calculated by applying the nuclear and coal duration regression equations (Tables 8.4 and 9.7) to the characteristics of the standard plants in Table 10.3. Nuclear construction time extends from construction permit award to declared commercial operation; coal project time extends from boiler order date, which precedes or coincides with construction start, to commercial operation. (For convenience, coal project duration is sometimes referred to below as construction duration.)

Table 10.2 demonstrates that construction times increased moderately over the 1971-78 period: by 18% for nuclear, 15% for coal. The difference between typical nuclear and coal construction times increased slightly during this period, from 13 months to 17 months.

The table also shows that if 1971-78 construction duration relationships continue during the 1980s, construction times will increase at about the same annual rates as in the past — 2½% for nuclear (or about two months per year), and 2% for coal (about 1½ months per year). The construction time for a standard nuclear plant would then be one-third greater than for coal, or two years longer, by 1988. It would take 48% longer to complete a nuclear plant at that time than in 1971, and 39% longer for a coal plant.

These findings probably overstate *construction times* for coal plants since they are based on data for *project time*, as explained above. Actual coal construction time is generally slightly less than total project time. More importantly, some of the apparent 1971-78 growth in coal plant construction times that is extrapolated here into the eighties may have resulted from increases in licensing time. Conversely, the nuclear projections make no allowance for the Three Mile Island accident, which is likely to cause substantial stretch-outs of future nuclear construction, as Chapter 6 shows.

Table 10.7 traces the actual past and projected future increases in construction times to changes in the factors that correlated significantly with construction time in the regression equations. The table indicates that unit size, sector size, and nuclear architect-engineer experience have been the only factors causing significant changes in construction durations. Over the entire 1971-88 period, sector expansion is projected to effect approximately equal increases in nuclear and coal construction durations — 66% and 60%, respectively. Nuclear expansion, however, brings about an increase in architect-engineer experience leading to a 22% projected *reduction* in construction time from 1971 to 1988. Accordingly, nuclear expansion has a lesser overall impact than coal expansion on construction time.

Table 10.7
Components Of Construction Lengthening

	Actual	Projected	
	1971-78	1978-88	1971-88
Changes in Nuclear Plants			
Duplicate Status	− 1%	+ 4%	+ 3%
Cooling Tower	——	+ 1%	+ 1%
A-E Experience	−14%	− 9%	−22%
Unit Size	+ 1%	+12%	+13%
Nuclear Sector Size	+38%	+20%	+66%
Net Nuclear Construction Lengthening	+18%	+25%	+48%
Changes in Coal Plants			
Unit Size	− 8%	− 7%	−14%
Coal Sector Size	+25%	+28%	+60%
Net Coal Construction Lengthening	+15%	+21%	+39%

Figures indicate changes in standard plant durations attributable to changes in standard plant characteristics shown in Table 10.3. Construction length factors are given in Tables 8.4 for nuclear and 9.7 for coal.

Factors with less than 1% impact on duration are not shown. They are duplicate status for coal, and Northeast and Southeast location for both coal and nuclear. Babcock & Wilcox NSSS is reflected in 1971 and 1978 nuclear durations but not 1988.

Percentages are combined multiplicatively. See note to Table 10.4.

The major difference between 1971-88 nuclear and coal construction time trends is that the 40% *increase* in typical nuclear unit sizes will *add* 13% to nuclear construction time, while the 58% *decrease* in typical coal sizes will *reduce* construction time by 14%. Coal construction times would be 27% greater if the standard 1988 coal size matched nuclear at 1150 MW rather than the projected 300 MW. Alternatively, nuclear construction times would be 38% less at 300 MW than at 1150 MW. Since the projected 1988 nuclear construction time is 33% greater than that of coal, the difference is essentially attributable to the different plant sizes. The projected coal construction duration is probably somewhat overstated, however, as described above.

Uncertainty Analysis: Aside from the effect of variations from the assumed unit sizes, uncertainties in projecting 1988 construction durations may arise from questions pertaining to future nuclear and coal sector expansion rates. Increases in cumulative nuclear and coal capacity affect construction schedules in much the same way that they affect costs (although at a lower rate) by adding to the stringency of environmental and safety standards affecting design and construction work. Variations from the sector sizes assumed in specifying 1988 standard plants would alter the construction times projected in Table 10.2. These effects appear minor, however, particularly for nuclear plants, as Table 10.8 shows. Nuclear plant durations are especially insensitive to the assumed sector size because of the slight relationship between duration and sector size for the data sample and the offsetting reduction in construction time from increased A-E experience as the nuclear sector expands.

Uncertainty in projecting future construction schedules from past trends also arises from statistical uncertainty in the correlations of construction time to the sizes of the nuclear and coal sectors. As in the previous section, this uncertainty is estimated by applying the 95% confidence intervals of the regression coefficients for sector expansion. Table 10.9 indicates that statistical uncertainty in the effect of sector size has only a modest bearing on projected construction times. The 95% confidence range is about plus-or-minus one year for both nuclear and coal plants.

Section 10.4:   Interest During Construction

Utilities finance power plant construction by issuing bonds and selling stock. They must pay bond interest and stock dividends in order to satisfy legal obligations and retain investors' confidence. In many states, however, utilities are not permitted to earn revenue on new plants until they have been completed and enter the "rate base." As a result, utilities must raise additional capital to enable them to meet their interest and dividend requirements during construction. The value of this capital is part of the total capital cost of plant construction and is referred to as interest during construction, or IDC (or, alternatively,

Table 10.8
Effect On Projected 1988 Construction Durations
Of Varying 1978-88 Sector Growth

	Nuclear	Coal
1978 Sector Size	55,573 MW	217,348 MW
1988 Projected Sector Size	149,648 MW	346,516 MW
1978-88 Projected Growth	94,075 MW	129,168 MW
Effect on 1988 Duration of Varying 1978-88 Growth by:		
−50%	−3%	−10%
−25%	−1%	− 5%
−10%	——	− 2%
+10%	——	+ 2%
+25%	+1%	+ 5%
+50%	+2%	+ 9%

Notes to Table 10.5 apply here.

Table 10.9
Sensitivity Of Projected 1988 Construction Times
To Uncertainty In Effect Of Sector Size

95% Confidence Range Relative To Base Schedule Projections	Nuclear	Coal
High Range	+11%	+16%
Low Range	−10%	−14%

Figures were calculated by applying confidence intervals to the projected sector sizes for 1988 standard plants. Interval for coal sector duration effect is .203 to .850; effect for nuclear sector ranges from .076 to .294.

Table 10.10
Calculation Of Interest During Construction (IDC)

	Actual		Projected
	1971	**1978**	**1988**
Real Cost of Capital (per year)			
Nuclear	2.8%	3.3%	3.8%
Coal	2.8%	3.3%	3.6%
Construction Duration (months)			
Nuclear	65.8	77.7	97.3
Coal	52.7	60.6	73.3
IDC Increment to Cost			
Nuclear	8.0%	11.3%	16.8%
Coal	6.3%	8.7%	11.6%
IDC Share of Total Cost			
Nuclear	7.4%	10.2%	14.4%
Coal	5.9%	8.0%	10.4%

Construction durations are from Table 10.2. Cost of capital is from Section 11.6. Nuclear construction durations used to calculate IDC conservatively exclude licensing time, reflecting author's belief that few expenditures predate construction start.

IDC was calculated with the "Comtois formula":

$$\text{IDC Increment to Cost} = \frac{(1 + e)^N - (1 + i)^N}{N \ln (1 + e/1 + i)} - 1, \text{ where}$$

N = project time, years

e = annual escalation rate (set equal to zero here to calculate real IDC)

i = annual cost of capital

Although the Comtois formula was derived for projects with symmetric payout functions, it is applicable to typical actual nuclear and coal cash flow curves with very little error. See W.H. Comtois, "Escalation, Interest During Construction and Power Plant Schedules" (Westinghouse Power Systems Marketing, September 1975).

IDC increments to costs were added to direct costs calculated from regressions, to yield standard plant costs in Table 10.1. For example, direct nuclear cost for 1978 standard plant was $797/kW. Increment of 11.3% gives total cost of $887/kW.

AFUDC — allowance for funds used during construction).[g]

IDC reflects the cost of tying up capital during construction. It is a genuine cost and has been included, in real terms, in the costs of standard plants in Table 10.1. The percentage of total costs contributed by IDC is determined by construction duration and the real cost of capital to utilities. The greater these are, the more capital is tied up unproductively at a cost to the utilities and society.

IDC's contributions to the costs of the standard plants are shown in Table 10.10. They were calculated on the basis of the construction durations of standard plants in Table 10.2 and the "real," inflation-adjusted cost of utility capital. Both factors increased during the 1970s and are anticipated to increase in the 1980s (see Section 11.6). The table shows that the contribution of IDC to plant cost rose by about 40% from 1971 to 1978 and is projected to double overall from 1971 to 1988. Nevertheless, Tables 10.4 and 10.10 both show that increases in IDC account for only a small fraction of real past and projected future increases in nuclear and coal costs. Accordingly, increases in construction duration explain very little of real capital cost escalation (although their impact appears large when costs are calculated in *current* dollars).

g. Some states now permit utilities to add some of the capital cost to the rate base during construction, an allowance known as construction work in progress (CWIP). The resulting revenue obviates the need for IDC to the extent of the allowance.

11
Non-Capital
Cost Factors

This chapter considers factors other than capital costs which will affect the life-cycle generating costs of new nuclear and coal-fired plants: generating performance (capacity factor), fuel costs, operating and maintenance (O&M) costs, financing charges, and waste disposal and decommissioning costs. Estimated cost values for these are combined with forecasts of capital costs to yield projected total generating costs in Chapter 12.

Several of these factors will either contribute heavily to total costs, or are highly uncertain, or both. The most problematic is probably coal fuel cost, which varies considerably among different regions and whose future cost increases are hard to predict. Nuclear plant capacity factors significantly affect generating costs and have been extremely volatile among different reactors, although the industry-wide average consistently appears to be within a narrow (55-65%) range. Conversely, nuclear fuel costs, coal capacity factors, and O&M costs have only a moderate bearing on total power costs and also appear more easily predictable than other cost factors.

Although nuclear waste disposal and decommissioning costs will likely exceed utility and government estimates, they are unlikely to constitute a major part of nuclear costs (in constant dollars). Both costs are extremely uncertain at this time, however. Conversely, financing charges, which convert capital costs into annualized payment streams, are extremely important but are predictable with great precision.

The discussions of the above cost factors are necessarily brief compared to the extensive treatment of capital costs in this study.

Section 11.1:   Nuclear Capacity Factors

Capacity performance strongly affects nuclear power economics. A reactor running at a high capacity factor (electrical output as a percentage of plant design capability) accumulates considerable fuel cost savings. Under favorable circumstances, these will pay for the unit's construction *vis-a-vis* other electric-generating plants. Conversely, a utility with poorly performing reactors must build additional expensive nuclear capacity or forego much of

246   Chapter 11

the fuel cost savings.[a] For each successive ten percentage point drop in capacity factor below 80%, nuclear generating costs increase by 11%, 13%, and 16%. (These figures are sequential and assume that non-variable expenses—capital, decommissioning, O&M, and interest on fuel costs—account for 80% of total costs. See Table 12.1.)

Until the mid-1970s, utilities and the Atomic Energy Commission (AEC) generally assumed that nuclear plants would operate at 80% capacity factors. That target has not been met, of course, but it has probably been overused as a straw man by industry critics. Some existing reactors were built at low cost and have proven to be cheaper than power-generating alternatives, especially oil-firing, despite capacity factors as low as 50%. Most future nuclear plants, however, are likely to be uneconomical even at very high capacity factors because of huge construction costs. (A 78% nuclear capacity factor is needed for nuclear-coal breakeven for new plant construction—even without incorporating the Three Mile Island (TMI) accident into reactor costs—based on the cost figures in Table 12.1.)

Most utilities now estimate that new reactors will achieve 70% capacity factors, although some projections are as low as 65%. In practice, however, U.S. reactors averaged only 60% capacity factor through mid-1980[b] and have demonstrated compelling reasons to expect that future plants will fail to attain even this average. For one thing, large reactors (800 megawatts [MW] or above) have had different performance characteristics than smaller units. For another, technical and safety-related constraints on reactor operation are apparently more than offsetting design and managerial efforts to improve performance.

Past Performance Trends: The reactor performance record provides

a. Some observers describe reactor shutdown costs in terms of *both* idle high-cost plant *and* expensive replacement power, but these effects are two sides of a single coin. Large bills for replacement power reflect not only reactor unreliability but also the high cost of fossil fuels.

b. Historical plant-by-plant capacity factor data are presented in References 1-4. They employ *original* "design electrical ratings"—the same ratings used here to calculate per-kW capital costs—and include all commercial reactor experience (units over 400 MW) starting with each plant's first full calendar year. Most industry and government data employ different conventions, measuring capacity factors with reduced "revised" design ratings or "maximum dependable capability" (see Reference 2). In addition, NRC capacity factor tabulations have excluded Three Mile Island Unit 2 since late 1979. These deviations cause most industry and government performance averages to exceed those here by one to two percentage points.

Separately, performance studies by the Electric Power Research Institute show that "load-following"—deliberate power reduction due to insufficient system demand—has curtailed nuclear power production by less than 1%. Capacity factor is thus an accurate index of potential nuclear performance. "Availability factors" often cited by industry measure only the percentage of time available for operating at *any* output, ignoring partial outages that have diminished reactor capacity factors by an average of approximately 12 percentage points. See Reference 5 for a dissection of availability factors.

the most reliable basis for forecasting future capacity factors. To be sure, this record is merely the product of underlying causal factors: regulatory pressure to reduce reactor mishaps, the rate at which safety-related defects surface in operation, utility willingness to commit funds to improve performance, and the plants' intrinsic design and mechanical integrity. However, no study has managed to isolate the effects of these factors to date, much less gauge their future impacts. Performance data, moreover, indicate whether anticipated trends such as *maturation* (improvements with age) or *learning* (improvements with later reactor "vintages") have materialized.

U.S. nuclear capacity factor experience is shown in Figure 11.1 and Table 11.1. Figure 11.1 shows that reactor performance has failed to improve *over time*, in part because of a marked downturn after 1978 that coincided with, but was not solely attributable to, the Three Mile Island accident. Table 11.1 points up the striking performance difference between *large and small reactors*, especially among pressurized water reactors (PWRs) supplied by Westinghouse (W).

Past analyses by the author have determined that only a small fraction of the performance gap between large and small reactors has been statistically attributable to age differences. [6] The power industry has countered that the data bases were too small to give meaning to this finding, and it has insisted that large reactors would eventually mature to the performance levels of smaller reactors. But even though reactor operating experience has now reached the level deemed sufficient by the industry to support statistical inference, [7] the gap between large and small reactor capacity factors has remained constant at 12 percentage points.

Although the author's most recent regression analysis covered data only through 1977, there have been no appreciable changes in reactor performance since then. At that time, large reactors had a cumulative performance average of only 53%, but they had averaged a mere three years of operation; most were just entering their "maturity," as the age data in Figure 11.2 show, and were expected by the industry to register large capacity factor gains. Instead, during 1978 through mid-1980, the large reactors averaged only 55% capacity factor, raising their cumulative record by only one percentage point. Smaller reactors gained by the same minuscule amount, from 65% to 66%. Although average reactor age increased from approximately four to six years from 1977 to mid-1980, the industry-wide capacity factor average has remained at exactly 60%. [c] Anticipated performance maturation has not materialized.

These data indicate that no significant improvement in the performance of large reactors should be anticipated. The burden of proof thus lies with the industry to justify capacity factor projections well over 60% in view of the 54%

c. The average reactor capacity factor remained at 60% despite one percentage point improvements in both large and small reactor averages, because the proportion of large reactors in the mix increased.

Table 11.1
U.S. Nuclear Capacity Factors Through June 1980

	Plants Under 800 Megawatts	Plants Over 800 Megawatts	All Plants
Pressurized Water Reactors	69% *13/104*	53% *27/120*	61% *40/224*
Westinghouse	70% *11/91*	52% *13/57*	63% *24/148*
Babcock & Wilcox	(none)	54% *9/39*	54% *9/39*
Combustion Engineering	65% *2/13*	56% *5/24*	59% *7/37*
Boiling Water Reactors (General Electric)	62% *10/70*	56% *12/68*	59% *22/138*
All Reactors	66% *23/174*	54% *39/188*	60% *62/362*

Figures in italics are, respectively, number of plants and cumulative number of plant-years.

Table excludes plants under 400 Megawatts.

large-reactor average to date.

The Meaning of the 1979-80 Performance Decline: Figure 11.1 shows that industry-wide capacity factors tumbled from 65% in 1977-78 to around 60% in 1979 and 50% for the first half of 1980. (Preliminary second-half 1980 data show a rebound to about 60%.) Figure 11.2 shows the same trends for large reactors at a level about five percentage points less.

Because the downturn coincided with the March 1979 accident at Three Mile Island, some have inferred that it is transitory and that a return to the higher capacity factors of 1977-78 is in the offing. A more sober interpretation, however, may be that the decline reflects two negative performance trends: an upsurge in occurrence and detection of mechanical and design failures, and the advent of a more stringent NRC regulatory stance.

The increase in mechanical and design defects was noted in Section 5.4 (Figs. 5.1 and 5.2). It has been especially apparent for Westinghouse reactors. Many of the 24 <u>W</u> plants have experienced prolonged shutdowns since early 1979 because of problems unrelated to TMI. Approximately a dozen plants required repairs of cracked ·welds connecting feedwater piping (carrying

Figure 11.1
Average Nuclear Capacity Factor By Year, Percent
(reactors 400 megawatts or greater)

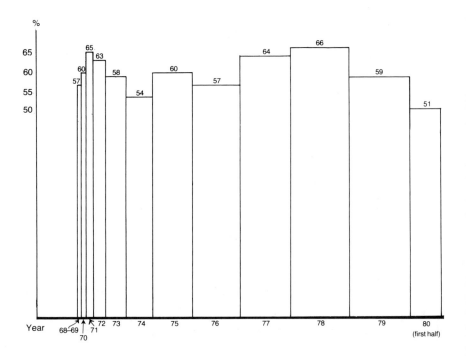

Bar width denotes number of commercially operating plants.

---

heated reactor water) to steam generators. Many have also undergone inspections and repairs of cracked low-pressure turbine disc assemblies. Two W plants were closed for eight months to repair cracks in concrete bolts used to anchor pipe supports, and another for a year to repair reactor coolant pumps and piping and to make seismic modifications. Another plant was closed for over a year to replace its corroded and cracked steam-generator tubes, a process now in progress at that reactor's twin and scheduled for at least two other plants before mid-1982.

These problems caused the average W capacity factor for 1979 and the first half of 1980 to fall to 56%, from 65% previously. The 13 large W plants (over 800 MW) ran at only 46%, reducing their cumulative average from 55% to 52%. In contrast, the eight 500-MW-class W reactors have averaged 72% both to date and in the same 18-month period, indicating that Westinghouse

Figure 11.2
Average Large Reactor Capacity Factor By Year, Percent
(reactors 800 megawatts or greater)

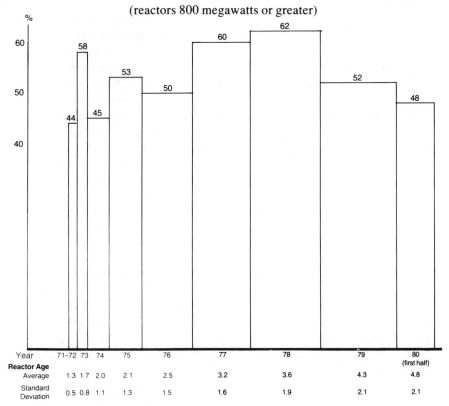

Year	71–72	73	74	75	76	77	78	79	80 (first half)
Reactor Age									
Average	1.3	1.7	2.0	2.1	2.5	3.2	3.6	4.3	4.8
Standard Deviation	0.5	0.8	1.1	1.3	1.5	1.6	1.9	2.1	2.1

Bar width denotes number of commercially operating plants.

design and mechanical deficiencies have been most pronounced at the large plants.

Similarly, many of the 22 General Electric (GE) boiling water reactors (BWRs) have been shut recently to repair cracks in primary system piping and emergency core cooling system (ECCS) sprays (a problem dating back many years) and to modify "suppression pool" containment structures to improve their stability during postulated accidents (addressing a longstanding safety problem). Outages such as these reduced the average GE reactor capacity factor to 53% for the first half of 1980 from the all-time 66% highs in 1978 and 1979. The seven Combustion Engineering plants have experienced a similar downturn. The nine reactors manufactured by Babcock & Wilcox (B&W), including the two at Three Mile Island, have suffered the worst decline, from 60% to only 41% capacity factor during January 1979 through June 1980.

TMI-related safety checks and backfits contributed greatly to B&W plant shutdowns, but so did failures in instrumentation and control systems, reactor coolant pumps, and other equipment.

The effect of mechanical problems on nuclear plant performance has apparently intensified, then, even apart from TMI, following a pronounced dip in 1977-78. In addition, regulatory constraints on reactor operation appear to be increasing. The TMI accident has legitimized regulatory-required backfits, a step which NRC had previously taken infrequently. Most reactors have sustained only brief (one- to four-week) shutdowns or outage extensions since TMI for minor equipment modifications, but NRC has committed itself in its post-TMI Action Plan to weigh major plant changes involving instrumentation, containment, and heat-removal systems (Chapter 6). TMI also takes some credit for NRC's recent establishment of compliance schedules for equipment installation (with attendant outages) to address longstanding safety issues such as environmental qualification of electrical equipment and fire protection. In addition, the accident has directed NRC's attention away from reactor *licensing* toward reactor *operations* (Section 6.6), making it less likely that licensees will be able to operate plants with equipment problems or shorten maintenance and repair outages.

TMI's impact has been reinforced by other reactor mishaps in 1979-80, as Chapter 5 shows. The discovery of errors in seismic stress calculations at five reactors and of deficiencies in ''as-built'' piping systems at a dozen others has diminished NRC's confidence in the capacity of design margins to compensate for analysis errors or equipment failure. Interactions at several Babcock & Wilcox plants between ''non-safety'' and ''safety'' instrumentation and control systems which initiated ''transients'' while cutting off critical status indicators have underscored reactor vulnerability to ''common-mode'' failures. The partial failure of automatic scram at Browns Ferry and subsequent discovery of design deficiencies in BWR scram systems have also heightened NRC's consciousness of the potential for serious accidents. As Section 5.4 notes, the rapid rise in NRC's issuance of safety-problem Bulletins and Circulars since TMI demonstrates greater attention to reactor operations as well as the presence of problem areas likely to require correction.

Outlook: The discussion above indicates that nuclear capacity factors are subject to pressures similar to those affecting capital costs. Increased operating experience provides valuable information to improve performance, but it also demonstrates that, as designed, reactors are prone to mishaps and need closer surveillance and continual backfits. Growth in the population of plants, moreover, gives greater visibility to reactor problems and adds to pressure for remedial fixes. Thus, nuclear capacity factors may be destined to fall—or at least stagnate—rather than rise as the nuclear sector expands, in much the same way that the capital cost reduction gained through learning in building more reactors has been more than offset by the costs of new safety measures needed to keep the chances of a serious accident from growing as fast as the sector itself.

It is too early to tell whether these negative forces will prove stronger than countervailing efforts to improve performance. The latter include incorporating greater design margins in new plants, stocking more spare parts, employing more overtime repair, and generally refining designs to correct emerging problems. But in light of the 54% average capacity factor to date for large reactors, it will be surprising if new plants manage to exceed 60% on a longterm average basis. The total nuclear cost calculation in Chapter 12 employs a 60% capacity factor,[d] with 55% and 65% in sensitivity analyses.

Section 11.2: Coal Capacity Factors

Fixed costs—capital charges and non-variable operating and maintenance costs—account for only 40-45% of typical coal generating costs, compared to about 80% for nuclear plants. Plant reliability is correspondingly less important to coal power economics. For each successive ten percentage point drop in capacity factor below 80%, coal generating costs rise by 5½%, 7%, and 9%. This is little more than half the sensitivity of nuclear costs. The discussion of coal performance is therefore less detailed than that of nuclear plants.

Coal plant performance, like reactor performance, has been proven sensitive to plant size (see Reference 1). Two sizes are discussed here: 300 MW, which utilities are increasingly selecting as an optimum mix of construction economies and operational reliability; and 600 MW, which is still the average choice of most large utilities.[e] Average capacity factors of 70-75% appear achievable for new 300-MW units, based on past experience, but a 70% projection is used in calculating costs in Chapter 12. For 600-MW plants, 60-70% average capacity factors should be anticipated.

These figures assume no "load-following." In practice, output cutbacks due to insufficient system demand have diminished the capacity factors of large (600-MW or larger) coal units by approximately five percentage points,[9] and by slightly more for 300-MW units. However, new coal units on oil- or

---

d. A report in preparation by Sandia Laboratories for NRC gives strong implicit support to this projection.[8] It projects capacity factors for years two through ten of only 61% for large PWRs and 65% for all BWRs, even though it considered neither first-year data nor 1980 operation. These excluded categories together account for almost half as much data, at six to seven percentage point lower capacity factors, as the data used to generate the report's predictive regression equations.

e. A suggestive but inconclusive regression on this study's coal plant data base indicates that typical unit size declined from approximately 700 MW for 1971 installation to 440 MW in 1978 (Section 10.1). Less dramatically, the average size of new fossil boiler orders reported annually in *Electrical World* fell from approximately 500 MW in 1976-78 to 450 MW in 1979. Further declines should be expected as utilities adapt capacity planning to reduced load growth.

gas-based systems will have a considerable fuel cost advantage and should undergo virtually no load-following. Load-following by new coal units on coal-based systems will probably be minimized by interconnections to oil or gas systems and load-management efforts, both of which are far more economical than in the past, and through "environmental dispatching" of new low-pollution plants in preference to older, poorly controlled sources.

The primary data sources for performance of 300-MW coal units are the author's 1976 *Power Plant Performance* (see Reference 1) and Edison Electric Institute's annual "Reports on Equipment Availability." Although these documents employ disparate data bases and capacity measures, they both indicate a 70% performance capability, or *capacity performance* (capacity factor corrected for load-following), for 300-MW coal units, and approximately 62% for 600-MW units, over the past decade (see box).

Coal Plant Performance Changes: The average capacity performance of coal-fired plants fell markedly in the late 1960s and early 1970s, as Fig. 11.3 shows, primarily because of the following four factors:

- introduction of larger, less reliable units (see size trend in Fig. 11.3);
- pollution control backfitting (primarily precipitator upgrading) beginning around 1970;
- reduced design margins (*e.g.*, smaller furnaces) to hold down capital costs;
- mismatches between boiler characteristics and coal grade because of air quality-related fuel switching and coal quality declines attributable to changed mining practices and supplier-utility relationships.

An Electric Power Research Institute (EPRI) analysis of the 1976-77 performance of large, primarily coal-fired plants suggests that the latter two factors are still taking a heavy toll. Forty percent of identifiable capacity factor losses resulted from boiler-related problems, such as slagging and fouling of boiler tubes, that are directly attributable to insufficient design margins and fuel quality problems.[f] Similar figures probably apply for other coal plant sizes as well.

Much of this chronic capacity factor loss appears retrievable through design improvements. EPRI notes that "[t]he majority of fossil boilers now

f. Reference 9, Table 5-1, shows 34.6% average capacity factor loss excluding "omitted" losses (1.5%). 13.8% (40%) is the sum of: 5.8% for boiler tubes, 1.1% for slag, ash and fouling, 1.9% for fuel handling equipment, and 1.5% for continuous deratings, all from Table 5-3; and an estimated one-third share of overhaul outages (10.8%, Table 5-1). Other boiler problems, involving fans, burners, air preheaters, or miscellany, total 4.3% (Table 5-3) but are not included here since they are not directly caused by reduced design margins or coal quality.

Chapter 11

300-MW Coal Unit Capacity Performance

Data compiled by the author (Reference 1, Chapter 6) establish that 200-400 MW coal units averaged slightly over 73% capacity performance from 1961 through 1973. Edison Electric Institute data indicate an average "equivalent availability" (availability factor corrected for partial outages—essentially the same as capacity performance) of 76% for 200-400 MW fossil units during 1968-77. (Both figures were derived by separately averaging the 200-300 and 300-400 MW size groups so as to avoid giving greater weight to the more numerous, higher-performing 200-300 MW units.)

These data require several downward adjustments. Two percentage points should be deducted from the EEI figure to eliminate higher-performing oil and gas units, and another four to convert EEI's plant capability ratings to the 5-6% higher "nameplate" capacity ratings used in the present study. The author's figure should be reduced by several percentage points to compensate for early coal plant data which preceded the recent performance decline.

These adjustments establish an approximate 70% performance average for 200-400 MW coal units over the past decade. Similar calculations on both the author's and EEI data establish a 62% performance capability for 400-800 MW coal units which is confirmed by an EPRI analysis of recent data (see Reference 9, p. 5-6).

being designed and constructed are designed for poorer fuel qualities and generally have increased design margins and so would be expected to perform better than large units now operating."[10] Boiler improvements such as spare coal crushers, more soot blowers, thicker boiler tubes, larger furnaces, and temperature controls already accounted for an increase of approximately $10/ kilowatt (kW) (1979 steam-plant dollars) in average coal plant capital costs from 1971 to 1978 (see Section 7.1).

Elimination of half of all boiler outages related to design margins and coal quality would restore as much as seven percentage points of lost capacity factor for typical 600-MW units and six points for 300-MW units. Other major EPRI and Department of Energy (DOE) research geared to performance improvements—the first such major effort ever applied to coal plants—may bring additional gains. Programs currently underway pertain to coal composition monitoring, turbine blade failure, reliability of auxiliary machinery such as feed pumps and fans, and analysis of coal combustion parameters and process-

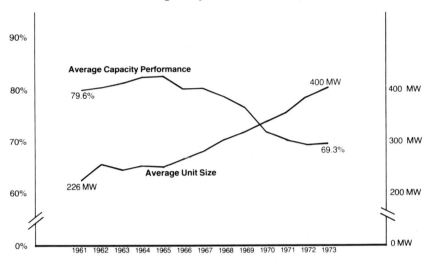

Figure 11.3
Coal Plant Capacity Performance, 1961-73

Source: Reference 1, p. 106.

es such as temperature, mixing, and kinetics.

Balanced against these putative gains, of course, are possible further operational constraints from stricter pollution control requirements. This study envisions that future (1988 and after) coal plants will have 99.9% particulate controls (probably through baghouses), 95%-efficient regenerable sulfur dioxide (SO_2) scrubbers, and nitrogen oxide (NOx) controls requiring advanced burners or flue-gas devices or both. This equipment is massive and complex and can affect other plant equipment. Although modular construction of baghouses and scrubbers will reduce plant downtime due to equipment failure, monitoring of control efficacy will almost certainly increase as use of coal expands, raising the potential for regulatory shutdowns.

Nevertheless, the negative factors on coal performance reliability appear likely to be outweighed by the positive forces, and, for that matter, by the new constraints on nuclear plant performance. Future coal plants designed for the very stringent emission levels assumed here should require few backfits, unlike inadequately controlled older coal plants or reactors whose safety has been called into question by the Three Mile Island accident. Moreover, most problems with coal pollution control systems are confined to those systems, and thus easier to fix, whereas reactor safety problems frequently "ripple through" much of the plant, as discussed in the first part of Chapter 4.

On balance, it seems reasonable to anticipate improvements of several percentage points over past levels for future coal plants. (Nuclear costs are calculated on the basis of a 60% capacity factor, despite the 54% large-reactor average to date.) This would put capacity factors (absent load-following) of 600-MW coal plants in the middle of the 60-70% range, and of 300-MW plants within the 70-75% range. Nevertheless, a 70% capacity factor, equalling past performance capability, is assumed in calculating coal generating costs for 300-MW plants in Chapter 12.

Section 11.3: Coal Fuel Costs

The cost of mining coal and delivering it to generating stations will account for half of the average life-cycle cost of future coal-generated electricity, as Table 12.1 shows (based on this study's cost assumptions, including "real" average annual coal price escalation of 2.3%). This far exceeds the 35% share of total power costs accounted for by capital costs, even assuming the advanced pollution controls just described. Accordingly, projected coal generating costs are extremely sensitive to the assumed coal cost. Indeed, varying the future annual rate of real escalation in coal prices between zero and 4% produces an extremely wide range, from 1.5 to 1.0, in the ratio of projected lifetime costs of new nuclear to coal plants (neither ratio assumes any significant cost impact from the Three Mile Island accident, as discussed in Chapter 12).

An in-depth analysis of future coal cost trends is beyond this study's scope, however. They are subject to a multitude of factors, some of which are sketched below, that cannot be predicted confidently and whose effects on costs are hard to gauge in any event. In fact, published, "expert" cost projections show considerable divergence, compounded by geographical variations in mining and transportation costs.

Instead, the 1979 U.S. average cost of coal burned by utilities, $1.20/ 10^6 Btu, is used here as a cost basis, and it is assumed to increase at an average of 2.3% per year in real terms (relative to inflation in other industrial commodities)—the average increase rate from 1974 to 1979.[11] This period is employed for extrapolation because it contains all empirical cost data subsequent to the steep run-up in prices that followed the 1973-74 oil price rise. Moreover, an examination of underlying conditions affecting coal prices suggests that price escalation should be no more severe in the future than during 1974-79.

Mining Productivity: The marked drop in mining "productivity" (tonnage per worker-shift) shown in Figure 11.4 is statistically responsible for much of the real increase in coal prices during the 1970s. Labor costs make up half of the cost of mining deep (underground) mined coal and a smaller but growing percentage of the cost of strip mined coal, which now accounts for

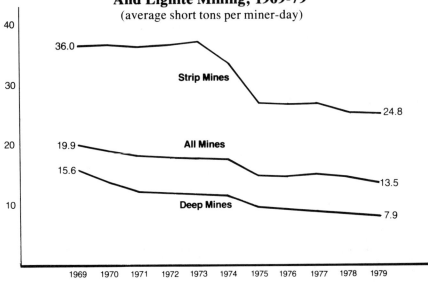

Figure 11.4
**Labor Productivity In Bituminous Coal
And Lignite Mining, 1969-79**
(average short tons per miner-day)

Source: Reference 12.

two-thirds of utility coal.

The causes of the productivity decline have been described as follows in studies for DOE and the Department of Labor (DOL), and EPRI:[13,14]

- *Health and safety and environmental regulations:* The Coal Mine Health and Safety Act of 1969 has reduced deep mine productivity by requiring added workers who do not directly extract and transport coal, *i.e.*, personnel to inspect coal faces, perform additional roofbolting and rockdusting, check for methane, etc., and also by slowing production to conduct inspections and conform with maintenance and equipment-reliability standards. State land reclamation laws and the federal Surface Mining Control and Reclamation Act of 1977 have reduced strip mine productivity by requiring operational adjustments and additional workers to reseed land and restore contours, particularly in smaller Appalachian and midwestern strip mines.

- *Labor strife and variability:* Both contract and wildcat strikes as well as absenteeism were endemic in coal-mining in the 1970s and diminished output rates. They also contributed to boom-bust fluctuations which indirectly reduced productivity by fostering marginal mining practices (*e.g.*, overtime) and creating a market for coal from marginal, high-cost mines. Poor labor-management relations also eroded work incentives.

- *High coal prices:* According to the DOE-DOL analysis, these "may not have caused inefficiency [but] have allowed it to exist,"[15] both by making marginal mines profitable and by supporting inefficient management and work practices.

Neither study found resource depletion to be a significant cause of reduced productivity—not surprisingly in view of the enormous resource base remaining in all producing areas. Both studies conclude that the productivity decline has bottomed out—a judgment strongly supported by preliminary 1980 data[16]—and anticipate future gains. The EPRI study projects average 1990 production rates of 10.8 net tons per miner per day for deep mines and 37.4 tons for strip mines,[17] 37% and 51% above respective 1979 levels.[18] The DOE-DOL analysis offers no projections but does conclude that the productivity decline in the 1970s was "transitional . . . not chronic" and that both deep and strip mines will show modestly higher productivity.[19]

These expectations are based on gradual workforce maturation, increases in mine size (particularly for strip mines), shifts to producing areas with more favorable geology (primarily the West), a judgment that new health and safety rules will not further depress productivity, and anticipated technological improvement.[20] Corroborating the last two assumptions, new mines are being built with ventilation systems and ground supports designed to permit roofbolting and mine dust control without the productivity losses absorbed by existing mines.[21] In addition, some well-financed new mines are deploying advanced technologies such as coal-bed degasification and longwall mining that may boost profits and productivity while meeting safety rules.[22]

Finally, labor-management relations in the coal fields have apparently rebounded significantly from their deteriorated status in the 1970s. New onsite grievance mechanisms and profit-sharing incentives established in the 1978 contract have reduced the number of miner-days lost to wildcat strikes by 90% from their 1977 high, and 84% from the 1974-76 average, based on 1978 data.[23] Absenteeism is down as well.[24] In addition, the mining industry has begun developing formal, systematic programs to replace the haphazard on-the-job training traditionally used to break in new miners. This "new training infrastructure"[25] should improve both miners' incentives and their ability to safely extract coal.

Demand for Coal: The rate of increase in the tonnage of coal mined will also affect coal prices. U.S. coal mining increased at an average 5.4% annual rate during 1974-80, from 610 to 836 million tons, although the increase rate would have been considerably higher had the 100 million tons/

year or more of unused mining capacity been employed in 1980. A too-rapid expansion in future mining could accelerate price increases by creating bottlenecks in the supply of mining equipment. At least as importantly, rapid expansion could create tight market conditions conducive to producer profit-taking.

Future coal demand will be determined by numerous variables, including oil and gas availability and price, government support for coal-derived synthetic fuels, nuclear power growth, the coal export market, resolution of environmental concerns, and, especially, total demand for electricity and energy. Although the rising price of oil and gas will increasingly open their markets to coal, it will also depress total energy demand below previous projections—a phenomenon consistently underestimated by most past forecasters (in assessing nuclear power growth, for example). Thus, as Table 11.2 shows, U.S. coal mining would increase at an average annual rate of only 6.1% during 1979-90, less than 1% per year above the 1974-80 rate, even if half the nuclear capacity under construction and all reactors on order are cancelled, *and* coal exports triple, *and* coal substitutes for one-quarter of present industrial consumption and one-half of present utility consumption of oil and gas, *and* coal provides half the feedstock for a one million barrel-per-day synthetic fuels industry—provided that industrial non-electric energy use remains constant and electricity generation grows at 2% annually. Most of these assumptions appear quite liberal,[g] and thus actual growth in coal demand could easily fall short of the past 5.4% annual rate that in turn lagged behind increases in capacity and permitted output to grow without precipitous price rises.

Coal Transportation: Shipping coal from mines to generating stations adds only one-fourth to the cost of typical Appalachian coal, but it exceeds mining costs for much western coal used in the South Central and Middle West regions (and, in future, on the Pacific Coast). Haulage charges for rail transport, by far the dominant mode, increased faster than general inflation in the 1970s, although this was partly attributable to the railroads' failure to anticipate the increase in demand—a condition not likely to recur—which resulted in the use of antiquated rolling stock and track.

Although federal deregulation of freight rates will lead to further price rises, average coal transportation costs will probably not increase rapidly in real (inflation-adjusted) terms in the future. Most of the future increases in coal traffic will be carried by the financially healthier railroads, and they have already begun planning for rapid growth in coal traffic, according to a recent report by the Transportation and Energy Departments.[28] Efficiencies will be gained by converting many eastern routes to dedicated "unit trains." Possible competition from coal slurry pipelines—indeed, even the mere threat of their

g. Electricity sales grew at just under 3%/year during 1973-80 and are widely projected (*e.g.*, by Exxon[26]) to increase by less than 2%/year after 1980. Industrial non-electric energy consumption fell by 3.6% from 1973 to 1979.[27]

Table 11.2
Disposition Of U.S. Coal Mining
(quads/year)

	1979	1990
Utilities (p. 23)	11.26	17.12
Industrial		
Coal (p. 21)	3.56	3.56
Share of Oil and Gas (p. 21)	0	4.02
		(¼ share)
Residential/Commercial (p. 20)	0.23	0.23
Exports (p. 54)	1.58	4.74
Synthetic Fuels	0	1.64
Addition to Stocks (p. 54)	0.76	0.30
TOTAL	17.39	31.61

Page numbers refer to data in DOE, *Monthly Energy Review*, August 1980.

1990 utility use was calculated as follows: 1979 net generation of 2,247,372 MWh (p. 60) is increased by 2%/year to 1990, yielding 2,794,325. Nuclear share is 551,880, based on 105 GW (54 GW in 1980 + half of 102 GW under construction—see Chapter 1) @ 60% capacity factor. Hydro remains constant at 279,783 (p. 60). Oil and gas is halved from 1979's 633,010 (p. 60). Remaining 1,646,157 MWh is provided by coal @ 10,400 Btu/kWh.

Exports were calculated assuming 12,000 Btu/lb. Synthetic fuels projection assumes two-thirds conversion efficiency and six million Btu/barrel. 1979 stocks addition assumes 11,200 Btu/lb. 1990 projection assumes addition of 60-day supply for 6% growth increment. 1979 total agrees with production of 17.41 quads (p. 4).

Annual mining growth measured in quads/year averages 5.6%. Assumed reduction in Btu/lb from 11,200 in 1979 to 10,600 in 1990 adds 0.5%/year to tonnage requirement, for 6.1%/year total growth.

introduction—will probably moderate increases in western long-haul rates. Higher costs for diesel fuel should not require significant rate hikes since fuel accounts for a small share of total rail costs and will increasingly be conserved through use of more efficient locomotives.

    Outlook: The preceding discussion suggests the difficulty of forecasting future coal prices and indicates that both inflationary and restraining forces are at work. Extrapolation of the 1979 average utility coal cost at the 1974-79 real increase rate appears to provide a reasonable middle ground. This approach is conservative (leading to high forecasts) in two modest respects. First, the 1979 base price, $1.20/10^6 Btu, reflects some utilities' use of premium low-sulfur coal to meet air-quality standards without scrubbers, whereas new air quality rules requiring scrubbers at all new plants will allow use of cheaper higher-sulfur coal. Second, preliminary 1980 data[29] show an inflation-adjusted decline of 1-1½% in that year's coal prices, reducing the post-1974 annual rate of real increase from 2.3% through 1979 to 1.7% through 1980.

These assumptions, combined with an assumed 10,000 Btu/kWh heat rate, lead to a projected "levelized" average coal fuel cost (see Section 12.1) of 1.96¢/kWh, in 1979 constant dollars, during 1988-2017, the years in which a coal plant commenced today is assumed to operate. This is 63% greater, in real terms, than the actual 1979 average cost of utility coal. Use of a 1.7% annual real increase rate instead of 2.3% would result in a fuel cost of 1.72¢/kWh, vs. 1.96¢/kWh forecast here.

Table 11.3 indicates that the power industry's projections of coal fuel costs fall in a narrow range and are less than this study's forecast. DOE's estimate exceeds this study's, but it assumes no mine-mouth plants except lignite-fired, and thus overstates transportation costs, and also employs "marginal prices" while admitting that "[a]ctual prices could increase at a slower rate if long-term contract arrangements keep some prices at below the marginal cost."[30] Finally, although most of the estimates in Table 11.3 predate the 1979 oil price rise and the concomitant move to substitute coal for oil, it is not clear that these factors will significantly swell coal demand, as discussed above.

Section 11.4: Nuclear Fuel Costs

Nuclear fuel costs, from uranium mining and milling through enrichment, fuel manufacture, and eventual disposal of spent fuel, account for 23% of projected nuclear generating costs. The breakdown of this share in Table 11.4 shows that nuclear power costs are only slightly sensitive to mining and enrichment costs and will barely be affected by disposal costs, *unless disposal costs are significantly greater than projected here*. Accordingly, nuclear fuel costs are treated only briefly.

Uranium Mining and Milling: The price of uranium on the spot (non-contract) market jumped from $8 to $40 per pound in 1975-76, the result of a host of factors, including price manipulation by a worldwide uranium cartel, DOE stockpiling of enriched uranium, and fears of future shortages. The spot

Table 11.3
Coal Fuel Cost Projections

Source	Estimate		Ratio To This Study's Estimate
EPRI[31]	Year 2000:	1.80¢/kWh	0.93
Sargent & Lundy[32]	1988-2017 levelized:	1.63¢/kWh	0.83
Bechtel[33]	1988-2017 levelized:	1.63¢/kWh	0.83
DOE[34]	Year 1995:	2.21¢/kWh	1.28
Gibbs & Hill[35]	1988-2017 levelized:	1.56¢/kWh	0.80

The sources used here comprise the projections available to the author and are not intended to be representative, although they appear to be. All projections except the last were expressed in 1978 dollars and were multiplied by 1.125 (1978-79 industrial commodities inflation) to convert to author's 1979 base. Reference dates or periods shown for first four projections are those employed by source and are compared against this study's estimate for same date or period. Gibbs & Hill figure in 1979 dollars was escalated to 1988 and levelized over 1988-2017 by author at 1%/year, although source's p. 4 suggests zero real escalation was assumed.

Table 11.4
Shares Of Nuclear Fuel Costs

Fuel Process	Share Of Fuel Cost	Share Of Total Nuclear Cost
Uranium Mining & Milling	50%	11%
Enrichment	28%	6%
Manufacture	8%	2%
Storage & Shipping	2%	*
Disposal	13%	3%

*Denotes less than .5%. Total does not add to sum due to independent rounding.

price has since settled back to $35 as new mines have opened, anticipated nuclear capacity has fallen precipitously, and DOE has increased the efficiency of uranium utilization in its enrichment facilities.

Uranium prices are assumed here to rise at 2% per year (in real terms) from a $35/lb base 1979 price. Costs will increase due to: (1) more stringent regulations to reduce accidents and radiological exposure in mining; (2) environmental regulations to reduce land, air, and water pollution and radiological contamination from uranium mining, milling, and mill tailings; and (3) resource depletion, which has already caused the average grade of U.S.-mined ore to decline by half from 1966-67 to 1979, from .22% to .11%.[36] The last factor reinforces the others by increasing the volume of ore processed and accompanying environmental disruption. Conversely, the rate of cost increase will be moderated by prospective spare production capacity as anticipated nuclear capacity declines further and by potential technological advances such as *in-situ* uranium leaching from phosphate rock.

Enrichment: Uranium will be enriched for future reactors at the three existing U.S. gaseous diffusion facilities and at a centrifuge plant under construction. The gaseous diffusion plants are used here as a cost basis although the centrifuge plant will be more expensive due to construction inflation.

The price of enrichment under utilities' *fixed-commitment* (long-term) contracts with AEC/DOE rose from $32 per *separative work unit* (SWU) in 1972 to $89 in 1979—a real increase rate of 4.8% per year relative to industrial commodity prices. The primary cause was the rising cost of electricity, which accounts for approximately 60% of enrichment charges. This cost will continue to rise, although not necessarily at the past increase rate, because of the large construction program and extensive pollution control retrofits being undertaken by TVA, the major supplier of power to the gaseous diffusion facilities. Real escalation at 1.5%/year is assumed here from a base of $94/SWU—the average enrichment cost in 1979.

Fuel Manufacture: This "step" involves conversion of uranium metal to a gas suitable for enriching, and fabrication of enriched uranium into oxide fuel pellets that are stacked into fuel assemblies. Both processes are technically mature. DOE cost estimates are employed here[37] with zero real cost escalation.

Spent Fuel Storage and Shipping: Irradiated fuel assemblies are assumed to be stored for ten years, at an annual cost of $6 per kilogram (kg) of uranium, and shipped to either an away-from-reactor storage site (AFR) or a final repository for $16/kg. Both costs are from DOE,[38] in 1978 dollars. Although costs could increase if siting restrictions raise transport distances and if storage and shipping regulations are made more stringent, multiple escalations would be required to materially affect nuclear costs since these steps account for only 2% of nuclear fuel costs and less than ½% of nuclear generating costs, as estimated here.

Spent Fuel Disposal: The cost to permanently bury irradiated fuel assemblies is drawn from a 1978 report by MHB Technical Associates for the Natural Resources Defense Council.[39] It assumes interim storage of 25% of the fuel at an AFR and includes government regulatory and research and development (R&D) costs and contingencies such as retrieval and relocation of some of the first buried fuel. Moreover, the "high" cost case is used here because of continuing technical and institutional uncertainties surrounding waste management,[h] although the low nuclear capacity case (105 gigawatts [GW] in 2000), with 23% higher unit costs than the reference case (200 GW), was not adopted. The cost derived here is 120% above DOE's on a per-kg basis,[42] but it nevertheless accounts for only 13% of nuclear fuel cost and 3% of total cost, in part because of accounting conventions that halve effective "back-end" costs, as discussed directly below.

Inventory Costs: These represent the cost of utility capital which pays for fuel cycle services and is thus "tied up" as working inventory prior to the utility's recovery of costs through selling the nuclear electricity to its customers. Inventory costs are important to nuclear fuel costs because the fuel cycle includes substantial lead times. These range from three years for fuel rod fabrication to four or more years for purchase of uranium, including the average of two years that a fuel element spends in the reactor before fissioning. A real annual fixed charge rate for capital of 9%—a little over one percentage point lower than that for capital costs—is applied here, resulting in the inventory costs shown in Table 11.5 below.

Inventory costs for uranium purchases, which are assumed to occur two years before fuel insertion, are 1.6 mills/kWh (one mill = .1¢). Conversely, inventory costs are *negative* 1.4 mills/kWh for spent fuel disposal—half of direct disposal costs—because utilities will be able to earn investment income with the money they receive from their customers before paying the federal government to remove the fuel (assumed to be six years after discharging the fuel). Net inventory costs are .9 mills, or 9% of direct fuel costs.

Conversion Efficiency: A "heat rate"—the number of Btus required per net kWh generated—of 10,600 is assumed. Average nuclear heat rate in

---

h. Although decontamination of Three Mile Island Unit 2 obviously poses different specific problems from spent fuel disposal, recent schedule slippage and cost increases provide another illustration of the power industry's under-estimation of the complexities inherent in dealing with irradiated nuclear materials. A year after the accident, General Public Utilities, the owner of TMI, estimated the cost of decontamination at $300 million with completion in early 1983.[40] Eight months later, the company raised its estimate to $1 billion and targeted completion for mid-1985 or later.[41] (Neither figure includes reactor "recovery" to service.)

Table 11.5
Nuclear Fuel Cycle Costs

(1) Process	(2) Unit Cost In 1979 $	(3) Number Of Units Per 1 kgU	(4) 1979-88 Escalation Factor	(5) 1988-2017 Escalation Factor	(6) Direct Cost $/kgU	(7) Years From Payment To Fission	(8) Inventory Cost Factor	(9) Total Cost $/kgU	(10) Mills Per kWh Costs Direct	(11) Mills Per kWh Costs Inventory	(12) Mills Per kWh Costs Total
1 Uranium	35	16.60	1.195	1.280	888.7	4	1.412	1254.8	3.83	1.58	5.41
2 Conversion	2.25	14.08	—	—	31.7	4	1.412	44.7	.14	.05	.19
3 Enrichment	94	4.208	1.143	1.201	543.0	3	1.295	703.2	2.34	.69	3.03
4 Fabrication	112.5	1	—	—	112.5	3	1.295	145.7	.49	.14	.63
5 Storage	11.25	6(yrs)	—	—	67.5	−5	.650	43.9	.29	−.10	.19
6 Shipment	18	1	—	—	18.0	−8	.502	9.0	.08	−.04	.04
7 Disposal	652	1	—	—	652.0	−8	.502	327.3	2.81	−1.40	1.41
TOTAL								2528.8	9.98	0.92	10.90

Column 3 assumes 0.25% enrichment tails assay. Column 4 brings 1979 costs to 1988 for uranium (2%/y) and enrichment (1.5%/y). Column 5 is levelized average cost factor with 3.8% real discount rate. Column 6 is product of previous four columns. Column 7 is referenced to middle of 4-year core residence. Column 8 is based on 9% fixed charge rate. Column 9 is product of Columns 6 and 8. Columns 6 and 9, respectively, by conversion ratio of 231.826 kWh/kgU based on heat rate and burnup in text.

Unit Costs for Rows 1 and 3 are from text. Costs for Rows 2, 4, 5 and 6 are from EIA (Ref. 30) × 1.125 (1978-79 industrial price inflation). Cost for Row 7 is from MHB (Ref. 39), $511/kgU (pp. 5-4 and 5-9), times 1.134 for 200 GW case (p. 8-3), times 1.125 to bring to 1979.

1978, the last year for which data are available, was 10,965, or 3% worse than projected here.[43] The range was 10,355-12,368 Btu/kWh, indicating that deviations from the mean are greater in the direction of inefficiency.

The other determinant of nuclear electric conversion efficiency is "burn-up"—the amount of heat produced by each unit of fuel in the reactor. PWRs employ a more highly enriched (and thus expensive) fuel than BWRs and so are designed to reach a higher burnup—32,600 thermal "megawatt-days" (MWD[t], equal to 24 MWh[t]) per metric ton of enriched uranium (MTU), versus 27,000 MWD(t)/MTU for BWRs. Actual burnup has averaged considerably less than design levels for both reactor types,[44] because some fuel assemblies have been discharged prematurely due to fuel failure and erratic capacity performance at variance with refueling schedules. The former problem has largely been solved but the latter remains. Accordingly, a burnup of 30,000 MWD(t)/MTU for a reference PWR is assumed here—8% below industry expectations.

Overall Fuel Cycle Costs: Table 11.5 shows the projected costs of each fuel cycle step and the overall total.

Section 11.5: Operating And Maintenance Costs

Operating and maintenance costs subsume all expenses incurred in running a power plant, other than fuel costs. They include operating labor, maintenance labor, maintenance materials, operating supplies, chemicals, lubricants, and water.

O&M costs have risen rapidly in recent years, in part because of increasingly stringent environmental and safety requirements. The 1976-79 average annual real escalation rates were 11% for nuclear and 7% for coal, relative to inflation in industrial commodities and adjusted for capacity factor (in the case of nuclear). Coal O&M costs are likely to increase faster than nuclear costs because of the high maintenance, material, and waste-disposal costs of SO_2 scrubbers.

Accordingly, although nuclear O&M costs averaged over 80% more than those of coal in 1979—4.11 vs. 2.26 mills/kWh—both plant types are assumed to average five mills/kWh in 1979 dollars, with 1%/year real escalation thereafter. Levelized O&M costs for the 1988-2017 plant lifetimes would then be 6.2 mills/kWh, accounting for 13% of total nuclear generating costs and 16% of total coal costs.

Nuclear Operating And Maintenance: Average nuclear O&M costs in recent years are shown in Table 11.6. Because nuclear O&M costs do not vary significantly with the amount of electricity generated, the per-kWh cost tends to be inversely related to plant capacity factor. The next-to-last column of

Table 11.6
Nuclear Operating And Maintenance Costs

Year	O&M Mills/kWh	Average Capacity Factor (CF), %	O&M Adjusted To 60% CF	Real Increase In Adjusted O&M
1976	2.39	57	2.27	—
1977	2.46	64	2.62	8%
1978	2.95	66	3.25	16%
1979	4.11	59	4.04	10%

1977-79 O&M costs are from Reference 45. 1976 is author's calculation from Reference 46. Capacity factors are from Figure 11.1. Deflator for last column is industrial commodities price index.

Table 11.6 adjusts for this by calculating hypothetical O&M costs assuming 60% capacity factor. The last column shows that adjusted O&M costs have increased steadily in real terms in each of the three past years.

This trend appears to be the result of increased efforts to improve plant performance and to comply with NRC regulatory requirements. The former have involved expanded maintenance and increases in repairs performed during overtime and/or by specialized contract personnel. The latter have required additional staffing for operations, maintenance, inspections, and security.

A cost base of five mills/kWh (1979 dollars) is assumed here for new reactors in anticipation of further efforts along these lines. This base would be reached with only two additional years of cost escalation at the 1976-79 real increase rate (10%/year) from the 1979 cost of 4.1 mills. For larger reactors—which usually have lower capacity factors and, therefore, higher per-kWh O&M costs—the 1979 average O&M cost is 4.55 mills/kWh, only one year of escalation away from the projection of five mills.[i] A nominal 1%/year real escalation rate is assumed from the five mill base, although the increase rate could easily be steeper, especially in view of more stringent NRC operational requirements and surveillance after the Three Mile Island accident.

Coal Operating And Maintenance: Average coal O&M costs in recent years are shown in Table 11.7. They have increased one-third less rapidly than reactor O&M costs, but the increase has been sizeable, nevertheless. Much of

i. Figures in this paragraph are unadjusted for capacity factor. Calculation for reactors 750 MW and above (size class used by DOE) is from DOE (Reference 45, Table 2).

Table 11.7
Coal Operating And Maintenance Costs

Year	O&M Mills/kWh	Real Increase
1976	1.42	—
1977	1.70	12%
1978	1.95	7%
1979	2.26	3%

1976-77 O&M costs are author's calculations from Reference 46, based on all coal stations 450 MW or larger completed 1970 or later. 1978 is author's rough estimate from trends in AEP and TVA costs. 1979 is from Reference 45, Table 4. Figures were not adjusted for plant generation because average coal capacity factor data were unavailable.

it apparently resulted from increased environmental requirements. The cost to dispose of recovered waste ash may have risen from roughly .1 mills/kWh to .2 mills (1979 dollars), and the frequency of sulfur dioxide scrubbers appears to have doubled to about 15%, accounting for another .1 mill/kWh increase. (Scrubbers approximately double O&M costs, based on recent data.) These two factors together would then account for half of the real 1976-79 increase.

The 1979 cost in Table 11.7 implies an O&M cost of approximately 1.8 mills/kWh for a non-scrubber plant, adjusted to the 70% capacity factor assumed for future coal plants.[j] Scrubbers will add an average of two to three mills/kWh for operating and maintenance personnel, chemical reagents, and waste disposal,[k] with costs for particular plants varying with the process employed and coal sulfur content. Further measures to segregate waste ash

j. Assuming 15% of the 1979 sample plants have scrubbers, with O&M costs double those of non-scrubber plants, the average O&M cost for the latter is 1.96 mills. If half of this cost is fixed and half variable, correction from an estimated 60% capacity factor to 70% reduces the total cost to 1.82 mills.

k. Estimate based on a report for EPRI by Bechtel.[47] O&M costs for eight scrubber processes in its Chapter 6 were adjusted to exclude steam and electricity (already incorporated in fuel cost as heat-rate penalty and confirmed as inadvertent double-counting by EPRI project manager in 3 October 1980 telephone communication with author) and converted from 1978 to 1979 by multiplying by 1.125 (1978-79 industrial commodities inflation). 70% capacity factor is assumed both in report and here. Range of costs (averaging low- and high-sulfur) is 2.2 to 3.0 mills/kWh with 2.6 mill average.

from the environment may add another one-half mill to costs, but increasing use of regenerable scrubbers should limit cost sensitivity to prospective hazardous-waste regulations. (Only three of the eight current scrubber processes produce hard-to-dispose sludge [see Note "k"], and these had the highest projected O&M costs. Three others are regenerable, one yields potentially salable gypsum, and another produces calcium salts which can be handled relatively easily.)

Combining the 1.8 mill/kWh base 1979 cost, two to three mills projected for the scrubber, and .5 mills for improved ash disposal, the estimated O&M cost for new coal plants is approximately five mills/kWh, in 1979 dollars—equal to the estimate for nuclear O&M. This too is escalated at 1% per year in real terms, giving a levelized 30-year cost in 1979 dollars of 6.2 mills/kWh. No credit is included for possible productive uses of wastes, despite improving economics. Recovered fly ash and sulfur are increasingly being used for construction and road-paving, and TVA is testing a process to extract aluminum and other metals from flyash.[48]

Section 11.6: Financing Costs

Utility plants must earn revenue to pay back investors—bondholders and stockholders—for providing capital to finance construction. The revenue requirements are proportional to the plant's capital cost, as are the corporate income taxes on net revenue, property taxes assessed by local municipalities, allowances for interim replacement of equipment, and insurance. These costs are together referred to as *fixed charges*, and the ratio of annual fixed charges to total capital cost is the *fixed charge rate*.

Fixed charge rates of 10.3% and 9.8% are assumed here for nuclear and coal plants, respectively, in real terms. (The 15-20% fixed charge rates in the literature are expressed in *current* dollars, unadjusted for inflation, and are not compatible with the inflation-adjusted methodology of this study.) The 10.3% and 9.8% rates are based on assumed real costs of capital of 3.8% for nuclear investments and 3.6% for coal, *i.e.*, on the expectation that the average rate of return on the equity (stocks) and debt (bonds) issued by utilities to finance nuclear/coal construction will exceed the inflation rate by 3.8/3.6 percentage points.

The real cost of utility capital averaged 2.8% during 1967-71 and 3.3% during 1972-77.[49] Higher values are anticipated here for new plants because growing uncertainty in the electric utility business is increasing the "risk premium" necessary to attract investment capital. A greater risk premium is anticipated for nuclear investments because of their greater capital costs (even as perceived by investors, notwithstanding limited industry candor), longer construction durations, greater exposure to regulatory delay and escalation, and vulnerability to abandonment due to public opposition or reactor accidents. These factors led to a small premium in the rates of return for some nuclear utilities in the late 1970s, and this has been widened by the Three Mile Island accident. Nuclear construction has been a major factor in recent credit-rating reductions of utilities by Standard & Poor's,[50] and there is broad agreement that nuclear investments require higher rates of return.[l]

The assumed .2 percentage point difference between nuclear and coal rates of return translates into a .25 percentage point difference in fixed charge rates. (See box for methodology used to compute fixed charge rate.) Another .25 point difference arises from the assumed allowances for interim capital replacement: .25% for coal and .5% for nuclear. The former figure is the nominal value used in power plant economic evaluations. In this case it implies that 1000-MW coal plants will require $2 million per year (1979 dollars) during plant life to replace worn-out equipment and make any necessary design changes. The nuclear figure implies interim replacements of approximately $7 million per year. Half of this is likely to be required for major equipment modifications and replacements such as chemical cleaning (decontamination) of primary system piping and new steam generators.[m] The other half would be required for other interim replacements and regulatory backfits.

Other contingencies not assumed here could further increase the difference between nuclear and coal fixed charge rates. First, the cost of regulatory-related reactor backfits could exceed the $3-4 million allowed for in the .5% interim replacement rate. Second, nuclear plant life could be shortened by radioactive wear or inability to meet advancing safety standards; reducing plant life by five or ten years from the assumed 30 years would add .6 or 1.5 percentage points, respectively, to the real nuclear fixed charge rate. Third, the nuclear risk premium reflected in the rate of return could exceed the assumed .2 percentage points; assuming a .5 point premium instead would add another

l. Most comparisons of rates of return for ''nuclear'' and ''non-nuclear'' utilities fail to differentiate between utilities with *operating* reactors and those with reactors *under construction*. The latter are both the subject of this study and the more susceptible to uncertainty. Nevertheless, a typical such analysis found that 59 utilities with planned or installed nuclear capacity had a .25 percentage point higher average common stock yield than 41 non-nuclear utilities one month before TMI, and this gap increased to .65 percentage points two months after.[51] Moreover, the effect of TMI has apparently persisted. A December 1979 study by Paine Webber Mitchell Hutchins, Inc. concluded that stock of a non-nuclear utility would sell at almost a 5% higher price-to-book-value ratio than a half-nuclear utility—equivalent to about a one-half percentage point difference in rate of return on equity.[52] The same firm ranked hydroelectric power and coal as the ''most desirable fuel sources'' and nuclear as ''least desirable,'' even below oil. Similar evidence is provided in Reference 50.

m. A DOE report[53] estimates that major equipment replacements during plant life will cost $77-117/kW for PWRs and $75-111/kW for BWRs (1979 dollars), or approximately $3 million per year. This covers two chemical cleanings, retubing of condensers, PWR steam generator replacement, and replacement of BWR control rod drives and reactor vessel piping and internals.

Calculation Of Fixed Charge Rate

Precise calculation of fixed charge rates is extremely detailed. A good approximation can be obtained through the formula:

$$FCR = \frac{C(i,N)}{1-t} \times 1 - \frac{t}{NxC\,(i,N)} + t_p + r_r + r_i$$

where: i = real cost of capital (assumed .038 nuclear, .036 coal)
N = plant life (30 years)
t = federal/state income tax (.50, minus 10% investment tax credit, or .45)
t_p = property tax rate (.02)
r_r = capital replacement rate (.005 nuclear, .0025 coal)
r_i = insurance rate (.0025)

$$C(i,N) = \frac{i}{1 - (1+i)^{-N}} = \text{capital recovery factor}$$

The formula yields a fixed charge rate of 10.28% for nuclear and 9.78% for coal.

.4% to the nuclear fixed charge rate. Any of these factors would far outweigh the minor (.1 percentage point) reduction in nuclear fixed charge rates attributable to the shorter reactor life permitted for computing utility income taxes.

Section 11.7:   Reactor Decommissioning

NRC regulations stipulate that following retirement of a reactor, it must be "decommissioned" in a manner compatible with public health and safety. This is assumed here to entail prompt, complete dismantlement. Mothballing or entombment, in which only some radioactive components are removed and the plant is sealed and monitored for radiation, appear to present unacceptable radiological hazards due to institutional uncertainties. NRC has rejected permanent entombment because, for example, "[t]he radiation dose rates from the long-lived radionuclides Nickel-59 and Niobium-94 in the activated reactor vessel internals [would] remain well above unrestricted release levels [one-half rem/yr] for a period of time far exceeding the known lifetime of any man-made structure."[54] Separately, NRC and former President Carter have advocated total dismantlement shortly after plant retirement.[55]

Because no commercial reactor has ever been dismantled, cost estimates must be based either on experience with small, experimental reactors or on conceptual studies of decommissioning conventional plants. Both approaches have drawbacks. The only power reactors completely dismantled to date, the 20-MW Elk River BWR in Minnesota and the 10-MW "Sodium Reactor Experiment" facility near Los Angeles, bear little resemblance to today's reactors. Although both may have incurred unusual costs as decommissioning pioneers, they accumulated less radioactivity than will today's plants (Elk River operated for only four years) and thus should have been easier to dismantle. Both plants cost approximately one-fourth as much to decommission as to build, adjusted for inflation.[56] Although it would be speculative to apply this ratio to current plants, the figure does suggest that dismantlement may be costly.

The most careful assessment of decommissioning costs appears to be a 1978 study prepared by General Public Utilities (GPU) for its Three Mile Island units, prior to the accident at unit 2.[57] Unlike widely reported consultant studies for NRC[58] and the Atomic Industrial Forum (AIF),[59] the GPU study includes estimates of the cost to dismantle each major structure—the reactor vessel, reactor building, auxiliary building, fuel handling building, diesel generators, etc.—made by the utility in conjunction with the original architect-engineer. A recent DOE assessment,[60] moreover, found the GPU cost estimate, $125/kW in 1979 dollars, to be "representative of the most current dismantlement assessments."[n]

Actual costs to decommission future reactors could exceed GPU's estimate. Equipment for cutting thick reactor vessel steel has not been fully developed (although improvements have been reported recently[61]); more stringent occupational exposure regulations would increase labor requirements and necessitate greater use of remote-handling machinery; and "hands-on" experience, especially important in nuclear work involving irradiated materials and environments, is almost completely lacking for decommissioning. Moreover, plants under construction will contain more equipment and structures than GPU's reference plant and thus may require more effort to dismantle.

Decommissioning costs in *current* dollars will be much greater, of course, due to monetary inflation. For example, decommissioning in the year

---

n. The NRC and AIF consultant studies estimated dismantling costs of approximately $45/kW and $50-70/kW, respectively, converted from 1978 to 1979 dollars as per Reference 60. The NRC study, however, primarily involved "scaling up" the work involved in entombing and dismantling very small plants, notwithstanding the marked differences between old 20-MW plants and today's 1100-MW reactors (*e.g.*, 3-inch vs. 9-inch thick reactor vessel steel). The AIF study omitted removal of fuel and control rods and provided for only minimum decontamination before dismantlement.

2020, 40 years after construction start for a reactor assumed to begin operating in 1988, would cost $2000/kW, based on GPU's $125/kW estimate (1979 dollars) and assuming 7% annual inflation in labor and materials. In theory, however, these funds could be raised by investing a far smaller sum today which would appreciate during the intervening 40 years.

NRC and state regulatory authorities are currently considering a number of mechanisms for financing decommissioning: issuance of bonds in interest-bearing accounts at the start of plant operation, "sinking funds" with periodic payments by customers, and accrual of reserves through depreciation accounting on the utility's books as a negative net salvage value.[62] The methods vary considerably in their treatment of ratepayer payments, tax obligations, contingencies for premature decommissioning or cost overruns, etc.

An 8% fixed charge rate is applied here to calculate the cost of decommissioning on a per-kWh basis. This is 2.3 percentage points less than the rate applied to nuclear capital costs (the difference arises from eliminating the interim replacement allowance and property tax) and appears to be at the lower end of the range of fixed charge rates implied by different financing methods.[63] It is multiplied by a 1988 decommissioning cost of $137/kW (in 1979 dollars) calculated by escalating the base 1979 cost at 1%/year (the assumed real escalation rate for construction work—see Chapter 12) only to 1988. At a 60% capacity factor, this yields a projected decommissioning cost of 2.1 mills/kWh—slightly less than 4.5% of total projected nuclear generating costs for new plants.

References

1. C. Komanoff, *Power Plant Performance: Nuclear and Coal Capacity Factors and Economics* (Council on Economic Priorities, New York, NY, 1976).

2. C. Komanoff and N.A. Boxer, *Nuclear Plant Performance Update* (Council on Economic Priorities, New York, NY, 1977).

3. C. Komanoff, *Nuclear Plant Performance Update 2* (Komanoff Energy Associates, New York, NY, 1978).

4. C. Komanoff, "U.S. Nuclear Plant Performance," *Bulletin of the Atomic Scientists*, *36* (No. 9), 51-54 (November 1980).

5. EPRI, *Nuclear Unit Productivity Analysis*, NP-559-SR (1977), Table 2.

6. Reference 1 covers data through 1975; Reference 2 adds 1976 data and responds to critiques; Reference 3 includes data through 1977. Reference 4 reports capacity factors through June 1980 but does not present any regression analysis.

7. M.E. Lapides, director of 1975-77 EPRI reactor performance studies, wrote the author on 24 May 1976: "There is some hope for your [statistical] approach when about three times as much data as now exist are available . . . (*i.e.*, about four years hence)." The mid-1980 data base of 362 reactor-years discussed here is more than triple the author's end-of-1975 data base of 110 reactor-years upon which Lapides commented.

8. R.G. Easterling, *Statistical Analysis of Power Plant Capacity Factors Through 1979*, NUREG/CR-1881, SAND 81-0018 (draft, 1981), p. 3. Capacity factors therein, calculated with "generator nameplate rating," were multiplied here by 1.049 (BWRs) and 1.062 (PWRs) to convert to original design electrical rating.

9. EPRI, *Nuclear and Large Fossil Unit Operating Experience*, NP-1191 (1979), p. 4-3. See also Reference 1, p. 72.

10. Reference 9, p. 5-3.

11. DOE, *Cost and Quality of Fuels for Electric Utility Plants*, 1978 and 1979 editions, DOE/EIA-0191. Table 11 shows that the average cost of coal delivered to utilities increased from 71¢/million Btu in 1974 to 122¢ in 1979, a 72% increase. (Average cost of coal burned reflects earlier shipments and was slightly less, 120¢.) Wholesale price inflation for the same period (producer price index) was 53%, giving a real increase of 12.3%, or 2.3% per year.

12. DOE, *Coal Data: A Reference*, DOE/EIA-0064(80), Table 12.

13. J.G. Baker, *Determinants of Coal Mine Labor Productivity Change*, Departments of Energy and Labor, DOE/IR/0056 (1979).

14. The Conference Board, *The Labor Outlook for the Bituminous Coal Mining Industry*, EPRI, EA-1477 (1980).

15. Reference 13, p. 26.

16. *Business Week*, 17 November 1980, pp. 122-130, reports preliminary Labor Department data showing first-half 1980 productivity gains of 17% at deep mines and 5% at strip mines.

17. Reference 14, Table 3-8.

18. Reference 12 shows 1979 productivity (preliminary data) of 7.9 tons per miner-day for deep mines and 24.8 for strip mines.

19. Reference 13, p. 29 and pp. 33-34.

20. Reference 14, pp. S-3 and 3-26.

21. Thomas Falkie (former director, U.S. Bureau of Mines) in *Actions to Increase the Use of Coal: Today to 1990*, Mitre Corporation (1978), p. 19.

22. See "Gas Found in Nation's Coal Beds Attracts Interest as a New Source of Heating Fuel," *Wall Street Journal*, 31 August 1977, p. 28.

23. Reference 14, Table 6-5. The early-1980 rate is said to be 86% below the 1977 peak, in "Both Sides Want to Avert Coal Walkout; Their Success May Decide Industry's Fate," *Wall Street Journal*, 8 August 1980, p. 30. See also Reference 16.

24. Reference 14, Table 6-6.

25. Reference 14, pp. 1-2 and 1-3.

26. Exxon Corp., *World Energy Outlook* (December 1980), p. 12.

27. DOE, *Monthly Energy Review*, DOE/EIA-0035(80/11) (November 1980), p. 21.

28. Departments of Transportation and Energy, *National Energy Transportation Study* (1980), pp. 37 and 65.

29. DOE, *Cost and Quality of Fuels for Electric Utility Plants, August 1980*, DOE/EIA-0075(80/08) shows a 13.9% increase in average delivered coal prices from August 1979 to August 1980. The index of industrial commodities rose 15.4% over the same interval, indicating a 1.3% inflation-adjusted decline in coal prices.

30. Energy Information Administration, *Annual Report to Congress, 1978*, Vol. 3, p. 233.

31. EPRI, *Technical Assessment Guide*, PS-1201-SR (July 1979), Exhibit 6-1, non-weighted average for eight consuming regions.

32. R.N. Bergstrom & W.W. Brandfon, *Trends in Electric Generating Costs* (February 1979), p. 8, average of low- and high-sulfur coal.

33. W.K. Davis & R.O. Sandberg, *LWRs: Economics and Prospects* (February 1979), average of levelized costs for low- and high-sulfur coal with scrubbers.

34. Reference 30, Vol. 3, Table 13.4.

35. Gibbs & Hill, Inc., *Economic Comparison of Coal and Nuclear Electric Power Generation* (January 1980), p. 14.

36. DOE, *Statistical Data of the Uranium Industry, January 1, 1980*, GJO-100(80).

37. Reference 30, Vol. 3, p. 220.

38. Reference 30, Vol. 3, p. 220.

39. MHB Technical Associates, *Spent Fuel Disposal Costs* (Palo Alto, CA, 1978).

40. *Wall Street Journal*, 18 March 1980.

41. *Wall Street Journal*, 10 November 1980.

42. Reference 30, Vol. 3, p. 220.

43. EIA, *Steam-Electric Plant Construction Cost and Annual Production Expenses, 1978*, DOE/EIA-0033/(78). Unweighted average for all 42 reporting commercial-size nuclear stations in service for all of 1978.

44. Southern Science Applications, Inc., *Historical Survey of Nuclear Fuel Utilization in U.S. LWR Power Plants*, DOE/ER/10020-TI (1979).

45. DOE, Office of Nuclear Reactor Programs, *Update*, July/August 1980, Table 1.

46. Reference 43, 1976 edition. Author's similar calculation for 1977 yielded 2.47 mills/kWh, essentially equal to DOE figure for 1977 in Table 11.6.

47. Bechtel National, Inc., *Economic and Design Factors for Flue Gas Desulfurization Technology*, EPRI CS-1428 (1980).

48. *Electrical World*, 15 August 1980, p. 75.

49. NRC, *Treatment of Inflation in the Development of Discount Rates and Levelized Costs in NEPA Analyses for the Electric Utility Industry*, NUREG-0607 (1980), Table 1, column, "Weighted Cost of Capital, based on price earnings ratio, without inflation" (implicit price deflator).

50. John R. Emshwiller, "Some Investors Shun Nuclear-Power Utilities, Jeopardizing Funds to Build New Atomic Plants," *Wall Street Journal*, 20 November 1980, p. 56.

51. New York Public Service Commission Staff, "Brief on Exceptions in Case 80003" (1979), pp. 73-76.

52. Charles A. Benore, "Status Report," 16 April 1980, pp. 1 and 6.

53. EIA, *Nuclear Power Regulation*, DOE/EIA-0021/10 (1980), Tables 16 and 17.

54. Battelle Memorial Institute, Pacific Northwest Laboratory, *Technology, Safety and*

Costs of Decommissioning A Reference Pressurized Water Reactor Power Station, NUREG/CR-0130 (1979), p. 2-2.

55. NRC, *Thoughts on Regulatory Changes on Decommissioning*, NUREG-0590 (1979), and "The President's Program on Radioactive Waste Management," The White House, 12 February 1980.

56. Peter N. Skinner, New York State Attorney General's Office, Testimony, NY Public Service Commission, Case 26974, 2 December 1977.

57. General Public Utilities Vice-President R.C. Arnold, Testimony before the New Jersey Public Utility Commission, 23 October 1978.

58. Reference 54.

59. W.J. Manion and T.S. LaGuardia, *An Engineering Evaluation of Nuclear Power Reactor Decommissioning Alternatives*, AIF/NESP-009 (1976); updated by LaGuardia in *Nuclear Safety*, *20* (No. 1), 15-23 (1979).

60. Reference 53, p. 172. General Public Utilities' cost of $101.1 million in 1978 dollars becomes $125/kW when multiplied by 1.125 (1978-79 industrial commodities inflation) and divided by the reference plant's 906-MW capacity.

61. Reference 59, T.S. LaGuardia.

62. See Henry Bermanis, "How Will The Cost of Decommissioning be Funded?," *Nuclear Engineering International*, *24* (No. 11), 21-23 (1979).

63. See Reference 62 or NRC, *Assuring the Availability of Funds for Decommissioning Nuclear Facilities*, NUREG-0584 (1979).

12

Nuclear And Coal Generating Costs

This chapter integrates the cost components into an overall comparison of nuclear and coal generating costs. Capital cost projections are taken from Chapter 10 and the other cost factors—fuel costs, capacity factors, etc.—from Chapter 11. Sensitivity analyses show the effects of deviating from the assumed values of different cost factors.

Section 12.1: Analysis Ground Rules

All costs are expressed in mid-1979 price levels. Where costs in 1978 dollars have been taken from the literature, they have been converted to 1979 values by adding 12.5%, the rate of inflation in industrial commodities from 1978 to 1979. (Readers may convert 1979 prices here to 1980 levels by adding 1979-80 inflation when that figure is published later in 1981.) Anticipated future *real* escalation, that is, price increases adjusted for inflation, has been projected relative to industrial commodities inflation.[a]

Where variable costs—fuel and operating and maintenance (O&M) costs—are assumed to escalate in real terms, they have been inflated from 1979 to 1988 price levels at their assumed real escalation rate and then "levelized" at this rate over 30 years of operation. Levelized averages give greater weight to costs incurred in the early years of plant operation to reflect the fact that ratepayers value money in hand more than money in the future, even in the absence of inflation. A weighting factor, or "discount rate," of 3.8% per year—the assumed inflation-adjusted cost of money for utilities making nuclear investments—is employed in the levelizing calculations. (Applying the 3.6% discount rate assumed for coal investments instead would have a negligible effect on calculated costs.)

Calculation of the levelized average cost of coal fuel during 1988-2017 is shown here for illustration. As discussed in Section 11.3, the delivered cost of

---

a. The "producer price index" of industrial commodities inflation used here has exceeded another commonly used inflation index, the GNP implicit price deflator, by about ½% per year in recent years.

Levelized Averages

The formula for levelized averages is:

$$\frac{\sum a_i /(1+r)^{i-1}}{\sum 1/(1+r)^{i-1}}$$

where a_i is the value of the cost factor in question in the "i'th" year of operation (here, i ranges from one to 30), r is the discount rate (equal to .038), and Σ indicates the summation of the terms for each year over the assumed period. Where constant escalation rates are assumed, a_i equals a_1 times $(1+e)^{i-1}$, that is, the value of the cost factor in the i'th year equals its initial-year value, a_1, times an inflation factor $(1+e)^{1-i}$ compounded on e, the real escalation rate.

Use of a computer or programmable calculator is advised for calculating levelizing factors. For i = .038 (used throughout this study), sample levelizing factors are: 1.128, for e = .01; 1.367, for e = .025; 1.677, for e = .04.

coal is assumed to escalate from its 1979 base of 1.20¢/kilowatt-hour (kWh) at an average of 2.3% per year in real terms (relative to industrial commodity prices). The first-year (1988) coal fuel cost, in 1979 constant dollars, is then 1.20¢ × $(1.023)^9$, or 1.47¢/kWh. Based on continued 2.3%/year real escalation and a 3.8%/year discount rate, a *levelizing factor* of 1.33 is derived. This is the ratio of the average fuel cost during 1988-2017 to the first-year (1988) cost, in real terms, based on the stated escalation and discount rates (see box). Multiplying this factor by the 1988 fuel cost yields a levelized average fuel cost of 1.96¢/kWh (1.47¢ × 1.33). Note that this figure is 63% greater than the actual 1979 fuel cost. The increase is the result of the assumed 2.3% annual real escalation between now and 2017.

A different but consistent procedure is applied to capital costs to reflect anticipated real escalation in construction prices prior to 1988 plant completion. First, projected capital costs at 1979 price levels without interest during construction (IDC)—$708/kW for coal and $1191/kW for nuclear (Tables 10.1 and 10.10)—are inflated by 1%/year for three and one years, respectively, to reflect anticipated real escalation to the assumed 1982 and 1980 construction start dates.[b] The "Comtois formula" (Table 10.10) is then applied to incorporate real escalation during construction and IDC, assuming the projected coal/nuclear construction periods in Table 10.2 (approximately six/eight years), the 3.6%/3.8% real costs of capital from Section 11.6, and the assumed 1% real annual escalation rate. The resulting Comtois factors, 1.148 and 1.226, respectively, are multiplied by the escalated costs to give capital costs in 1979 dollars of $838/kW for coal and $1460/kW for nuclear.

Section 12.2: Base Case Results

Table 12.1 presents cost components and total generating costs for the "base case" employed throughout this study. It includes no increase in nuclear capital costs specifically resulting from the Three Mile Island (TMI) accident, except perhaps for a slight increase to offset attrition in nuclear construction and the corresponding reduction in the application of new safety requirements. The table also incorporates the advanced pollution controls required to meet the extremely stringent emission standards projected for 1988 coal plants in Figure 7.2. (That is, both nuclear and coal capital costs are projected with the statistical regressions on 1972-78 capital costs.) It also reflects 2.3% annual escalation in the cost of coal fuel. Note that numbers with several significant digits are employed in Table 12.1 and throughout this chapter for computational accuracy but do not necessarily signify certainty in estimates.

The average nuclear generating cost in Table 12.1 is greater than the coal cost by .86¢/kWh, or 22%. Two geographical variants are also considered here: Northeast and West.

Northeast: This region, comprising New England and the Middle Atlantic states, has higher costs for construction, O&M, coal fuel, and decommissioning. The regional capital cost premiums found in the 1972-78 regressions—approximately 28% for nuclear and 14% for coal, relative to a Midwest base—appear to reflect higher wages and material costs, more difficult working conditions, and more stringent environmental and safety standards. They are applied here for future plants.[c] Northeast decommissioning costs rise in the same proportion as nuclear capital costs. O&M costs are increased by an arbitrary but probably conservative 15% for both plant types.[1] The coal fuel cost was boosted from that in Table 12.1 by 23%—the 1978 differential for

---

b. The Handy-Whitman index of construction prices increased during 1965-79 at an average rate of .64%/year for nuclear plants and .98%/year for fossil plants, relative to industrial producer prices.

c. Northeast and other regional capital costs are calculated directly from the nuclear and coal capital cost and duration regressions in Chapters 8 and 9, assuming construction in that region.

Table 12.1
Projected Costs, 1988 Plants
(in 1979 constant dollars)

	Nuclear	Coal
Unit Size	1150 MW	300 MW
Capital Cost	$1460/kW	$838/kW
Decommissioning	$138/kW	
Real Fixed Charge Rate	10.3%	9.8%
Capacity Factor	60%	70%
Capital Cost Fixed Charges	2.86¢/kWh	1.34¢/kWh
Decommissioning Fixed Charges	.21¢/kWh	
Fuel	1.09¢/kWh	1.96¢/kWh
Operating and Maintenance	.62¢/kWh	.62¢/kWh
TOTAL	4.78¢/kWh	3.92¢/kWh
Nuclear/Coal Cost Ratio	1.22	

Table 12.2
Northeast Nuclear And Coal Cost Factors
(where different from Table 12.1)

	Nuclear	Coal
Capital Cost	$1820/kW	$936/kW
Decommissioning	$172/kW	
Capital Cost Fixed Charges	3.57¢/kWh	1.50¢/kWh
Decommissioning Fixed Charges	.26¢/kWh	
Fuel	1.09¢/kWh	2.41¢/kWh
Operating and Maintenance	.71¢/kWh	.71¢/kWh
TOTAL	5.63¢/kWh	4.62¢/kWh
Nuclear/Coal Cost Ratio	1.22	

Table 12.3
Western Nuclear And Coal Cost Factors
(where different from Table 12.1)

	Nuclear	Coal
Capital Cost	$1401/kW	$1033/kW
Decommissioning	$132/kW	
Capital Cost Fixed Charges	2.75¢/kWh	1.65¢/kWh
Decommissioning Fixed Charges	.20¢/kWh	
Fuel	1.09¢/kWh	.96¢/kWh
TOTAL	4.66¢/kWh	3.23¢/kWh
Nuclear/Coal Cost Ratio	1.44	

New York, New Jersey, and New England.[2]

Table 12.2 shows projected Northeast costs where they differ from the national averages in Table 12.1. Total costs in Table 12.2 are 18% higher than in Table 12.1 for both nuclear and coal, preserving the 22% differential between the two plant types. The higher Northeast construction premium for nuclear plants offsets the higher Northeast coal cost.

West: Western coal plants have higher capital costs but lower fuel costs. The higher capital cost—26% relative to a Midwest base, based on past costs—probably reflects the region's more stringent environmental standards and the higher cost of electrostatic precipitators designed for western low-sulfur coal. Accordingly, western coal plants are further along the "pollution control curve" and should experience smaller cost increases than coal plants elsewhere during 1978-88. Nevertheless, the 26% western capital cost premium is applied here to provide a cost margin for any extraordinary measures to minimize emissions and other environmental impacts in clean air regions. Conversely, western nuclear plants are projected to cost slightly less to construct than the U.S. average shown in Table 12.1, which includes a high-cost northeastern component.

Coal delivered to western plants cost 51% less than the U.S. average in 1978,[3] the result of both low mining costs and the large proportion of mine-mouth plants. This differential is applied in Table 12.3, although future plants in some parts of the West, such as California, will incur substantial coal shipping costs. Nevertheless, the average western nuclear-coal cost differential of 44% shown in Table 12.3 appears sufficiently robust (twice the average projected for the U.S. as a whole and the Northeast) to accommodate such costs, especially considering the conservative coal capital cost projection.

Other Regions: Average costs for most other regions should not vary substantially from the national average in Table 12.1, based on past trends. A possible exception is the West South Central region (Texas and several neighboring states), where 1978 coal fuel and 1972-77 coal capital costs were 35% and 24% less, respectively, than the national averages.[4] (The former difference may narrow as Wyoming coal supplements local lignite.) In the South (South Atlantic and East South Central regions), coal plants had 14% higher fuel costs than the national average in 1978, but this was offset by their 20% lower capital costs.[5] Similarly, Midwest coal fuel and capital costs both equalled the national average. Accordingly, there appears to be no region where the projected differential between nuclear and coal generating costs is less than the 22% national average in Table 12.1.[d]

Section 12.3: Sensitivity Analyses

This section varies the cost components to measure the sensitivity of projected nuclear and coal generating costs to the assumed cost values. The base case is the national cost average in Table 12.1: coal, 3.92¢/kWh; nuclear, 4.78¢/kWh, or 22% higher. Calculated nuclear-coal cost ratios are shown at the end of each case.

Case 1. Plausible Three Mile Island Effect on Nuclear Capital Costs: The 1988 nuclear capital cost projection was developed from a statistical regression on 1972-78 reactor costs and makes little or no allowance for the additional regulatory impact of the Three Mile Island accident. That impact will be felt over a long period of time as the many items in the TMI Action Plan such as the degraded core rulemaking (Chapter 6) are researched, debated, and resolved. Although any cost estimates are speculative at this time, a 50% boost in the *increase* in direct capital costs (exclusive of interest during construction) projected for 1978-88 appears conservative in view of the extensive scope of the Action Plan and the accident's severe challenge to basic regulatory premises. This assumption would appear to allow for a modest slackening in the rate

---

d. There is, of course, an anomaly when the projected nuclear-coal cost differential is higher than the national average in some regions, but is nowhere lower. The explanation lies in the difference between the regional weights used to project the average 1988 coal *capital* cost and the regional shares of coal burned in 1978 which make up the national average used to project coal *fuel* costs. The difference results in a high-side projection of the national average coal generating cost. For example, the heavy (26%) capital cost premium for western coal plants was applied to 22% of 1988 plants—the West's anticipated share—but future coal costs were projected here from 1978 cost data of which the cheap coal burned in the West accounted for only 15%. The 1978 average coal fuel cost calculated with the same weights used to calculate future capital costs is $1.02 per million Btu—9% less than the true national average in that year.

of imposition of new standards if reactor cancellations reduce future nuclear capacity below the mid-1980 target of 150 gigawatts (GW) used to calculate costs with the regression equation. Its result would be to raise the direct cost of a typical 1988 reactor by one-sixth, from $1191/kW to $1388/kW (1979 steam-plant dollars, without IDC).

Under the same assumptions, TMI could cause construction durations to lengthen from the presently assumed 8.1 years to nine years, requiring addi-tional interest payments during construction that alone would add 2% to real capital costs. The average nuclear capital cost would be $1738/kW (1979 constant dollars, including IDC and 1%/year real escalation to 1990), adding .55¢/kWh to nuclear fixed charges and raising nuclear generating costs by 12% from the base case.

The author regards this case as extremely plausible, but it is not pre-sented as a ''base case'' due to the lack of a methodology for estimating the effect of TMI on capital costs. The associated nuclear lifetime generating cost of 5.33¢/kWh would be 36% higher than the coal base case cost.

Case 2. Plausible Three Mile Island Effect on All Nuclear Costs: In addition to the effects on capital costs postulated in Case 1, the TMI accident could add as much as .5 percentage points to the real rate of return required for nuclear investments and also reduce the average capacity factor of new reactors from the 60% projected in the base case to 55%. The former would add .64 percentage points to the annual real fixed charge rate and also increase capital costs by 2% through its effect on real interest costs. Combined with the addition to direct capital costs stipulated in Case 1, the result would be a nuclear capital cost of $1778/kW (1979 constant dollars) and fixed charges of 4.04¢/kWh, up from 2.86¢/kWh in the base case. Per-kWh decommissioning costs would rise proportionally with the increase in direct capital cost and the decrease in capacity factor, from .21¢/kWh to approximately .27¢, and O&M would increase by .06¢ per kWh due to the reduced capacity factor. The combined effect would be a 27% increase from the base case to 6.08¢/kWh (Nuclear/Coal = 1.55).

A more severe assumption of capital cost escalation due to TMI, involv-ing a doubling of the extrapolated 1978-88 increases in construction cost and duration and resulting in a $2086/kW cost (1979 constant dollars), is treated in Section 12.4.

Case 3. Nuclear Capacity Factor: Varying only the nuclear capacity factor from 60% to 55% increases base case nuclear generating costs by .34¢/kWh: .26¢/kWh for capital fixed charges, .02¢ for decommissioning and .06¢ for O&M (N/C = 1.31). Other possible nuclear capacity factors, 50%, 65%, and 70%, yield these respective nuclear-coal cost ratios: 1.41, 1.15, 1.08. A 78% nuclear capacity factor is required for nuclear-coal breakeven under base case assumptions. A 90% nuclear capacity factor is required assuming the TMI capital cost impact estimated in Case 1.

Case 4. Nuclear Fixed Charge Rate: As described in Case 2, a .5 percentage point increase in the required nuclear real rate of return, from 3.8 to 4.3 percentage points, would raise the annual real fixed charge rate from 10.3% to 10.94%. Separately, reducing plant life by five years from the assumed 30 would add .6 percentage points to the fixed charge rate, and interim replacements could easily require annual investments equal to .7% or .8% of capital costs instead of the assumed .5%. A one percentage point increase in the fixed charge rate, whatever its origin, would add .28¢/kWh to fixed charges on the capital cost and .02¢ to decommissioning charges, if carried over to that sector (N/C = 1.30).

Case 5. Decommissioning and Waste Disposal: In view of the virtually unbroken record of real cost overruns in nuclear work involving irradiated materials, decommissioning costs could conceivably be twice the $138/kW estimate (1979 constant dollars) even without any TMI-caused increase in the complexity of the reference reactor on which the estimate was based. Moreover, the effective fixed charge rate might be half again as great as the 8% assumed here if conservative financing methods are required to ensure that funds set aside for decommissioning are available at end of plant life. The combined effect would be a tripling of decommissioning charges, from .21¢/kWh to .63¢ (N/C = 1.33).

Waste-disposal costs—spent fuel storage, shipping, and burial—could also double if the cost assumptions employed in Section 11.4 prove insufficiently conservative. This would add relatively little, .16¢/kWh, to nuclear costs, however (N/C = 1.26).

Case 6. Nuclear Fuel: Continuing declines in projected nuclear capacity could soften the uranium market and reduce uranium ore costs to only $20/lb, with no escalation, shaving fuel cycle costs by .34¢/kWh. Concomitant increases in the economical enrichment tails assay could also reduce enrichment costs by as much as .1¢/kWh. The combined reductions would cut total nuclear costs by 9% (N/C = 1.11).

Case 7. Coal Fuel Costs: Coal generating costs are most sensitive to the future cost of coal. Real escalation at only 1% per year rather than the assumed 2.3%/year rate from the 1979 U.S. average cost of utility-burned coal would reduce coal costs by .48¢/kWh (N/C = 1.39). Conversely, real coal price escalation at 4%/year would add .90¢/kWh to coal costs, resulting in roughly equal nuclear and coal generating costs (N/C = .99).

These comparisons exclude the effects of the TMI accident on nuclear costs. Incorporating the nuclear capital cost increase assumed in Case 1 yields a 1.11 nuclear/coal cost ratio even with 4% annual coal price escalation. Adding the postulated effects of TMI on the nuclear rate of return and capacity factor in Case 2 gives a cost ratio of 1.26 with 4%/year coal cost increases.

Case 8. Coal Capacity Factors: Coal capacity factors might average 60% instead of the assumed 70% if advanced emission controls interfere markedly with performance, if control equipment failures lead to regulatory restrictions, or if much larger plants than 300 MW are employed and do not improve on past performance. Fixed charges on capital and O&M costs would rise by .28¢/kWh (N/C = 1.14), assuming fixed expenses account for half of O&M. Capacity factors of 65% and 75% yield ratios of 1.18 and 1.25, respectively. A 46% coal capacity factor is required for nuclear-coal break-even with the other base case assumptions in Table 12.1.

Case 9. Coal Capital Costs: Coal capital costs could exceed the $838/kW average projected here (1979 constant dollars), although Chapter 7 indicates that this figure would allow new plants to reduce emissions by two-thirds from the Environmental Protection Agency's (EPA) new 1979 standards. For plants not surpassing the EPA standards, the assumed cost includes liberal margins for both cost overruns on new control equipment and for presently unanticipated environmental requirements. Yet a 10% addition to capital costs would add only .13¢ to total coal costs in any event (N/C = 1.18). Similarly, a 10% decline would remove .13¢ (N/C = 1.26).

Section 12.4:   Plausible Extreme Cases

Extreme but plausible cases are shown in Table 12.4 and Figure 12.1, along with the base cases presented earlier and nuclear Case 1. The latter reflects the author's judgment of the plausible impact of Three Mile Island on reactor costs. When like cases (base, best, and worst) are matched against each other, the projected nuclear/coal cost ratios are, respectively, 1.22 (base), 1.27 (best), and 1.37 (worst). Thus, varying the cost assumptions within the ranges posited here tends to widen coal's projected advantage over nuclear.

Moreover, because the best nuclear case yields a higher projected cost than the base coal case, unanticipated escalation in the coal sector is required to create a nuclear advantage. Similarly, the worst coal case is only 10% more costly than the base nuclear case, whereas the worst nuclear case is 84% more expensive than the base coal case. Finally, average nuclear costs are subject to greater uncertainty, with a 51% spread between base and worst case costs, versus 34% for coal. (This is aside from the arguably greater uncertainty in costs among individual reactors due to wide variations in capacity factors.)

As stated above, these cases represent the author's judgment of *plausible* rather than possible bounds. Coal fuel costs could escalate faster than 4%/year

Figure 12.1

Nuclear And Coal Generating Costs, Plausible Cases

(1979 ¢/kWh)

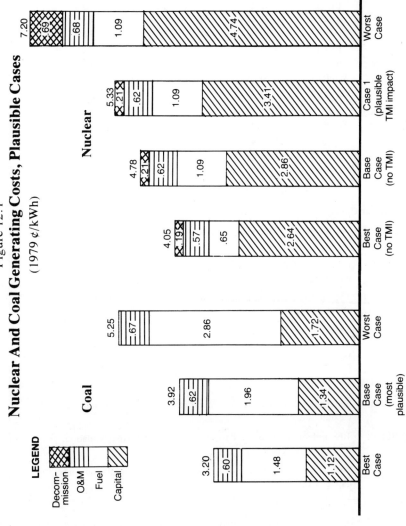

See Table 12.4 for assumptions employed in each case.

Table 12.4
Breakeven Cost Calculations

Nuclear Cases	Coal Cases	Projected Nuclear-Coal Cost Ratios	Paid-Off Share Of Nuclear Plant For Breakeven	Annual Real Coal Escalation For Breakeven
	BEST	1.49	55%	4.3%
BASE	BASE	1.22	30%	3.9%
	WORST	0.91	−16%	3.2%
	BEST	1.27	32%	3.1%
BEST	BASE	1.03	5%	2.6%
	WORST	0.77	−45%	1.5%
	BEST	1.67	62%	5.0%
CASE 1	BASE	1.36	41%	4.7%
	WORST	1.02	2%	4.1%
	BEST	2.25	84%	7.1%
WORST	BASE	1.84	69%	6.6%
	WORST	1.37	41%	5.3%

Nuclear Cases

BASE: See Table 12.1 (includes little or no TMI impact and 60% capacity factor)

BEST: Little or no TMI impact, 65% capacity factor, and reduced uranium and enrichment costs in Case 6

CASE 1: TMI adds 50% to extrapolated 1978-88 increases in construction cost and duration

WORST: TMI doubles extrapolated 1978-88 increases in construction cost and duration, adds 0.5% to real nuclear rate of return, and reduces capacity factor to 55%; decommissioning charges triple

Coal Cases

BASE: See Table 12.1 (includes 2.3%/year real fuel cost escalation and 70% capacity factor)

BEST: 10% reduction in capital cost, 1%/year real fuel cost escalation, and 75% capacity factor

WORST: 10% addition to capital cost, 4%/year real fuel cost escalation, and 60% capacity factor

---

Two right-most columns are alternative, not simultaneous, scenarios. Paid-off shares are calculated relative to nuclear capital costs for that case. Minus sign indicates that nuclear costs could exceed projections and still achieve breakeven with coal. Breakeven coal fuel escalation rate varies relatively little with different coal cases because coal costs are relatively insensitive to other coal-related variables. Nuclear Case 1 was included because of author's judgment that it is highly plausible. All figures assume adherence to utilities' completion of reactors with construction permits.

in real terms, but this was judged to be no more likely than the possibility that TMI would cause more than a doubling in the extrapolated 1978-88 increase in nuclear capital costs. Similarly, a rise in the interim replacement allowance or a shortening of reactor life could increase the nuclear fixed charge rate, but this appeared no more probable than a uranium price break below $20/lb (1979 dollars with no real escalation). Possible variation in annual O&M costs was also omitted because these costs account for only about 15% of total base costs. Most "second-order" cost impacts—e.g., effect of increased interest rates on nuclear fuel inventory costs, effect on real capital cost of longer construction durations for larger coal units—were ignored because they change total costs by only about 1%, far less than the myriad uncertainties in future costs.

Finally, it is of interest to calculate *breakeven costs* for various cases for two key variables: nuclear construction status, and the rate of real escalation in coal fuel prices from the 1979 U.S. average. The former variable is calculated as the percentage of the total capital cost of a new nuclear plant that a utility could pay for and then abandon, and still break even in lifetime generating costs by constructing a new coal-fired plant.[e]

The results, shown in Table 12.4, indicate that coal would be less expensive than nuclear power, even under the best nuclear case and the worst coal case, if real escalation in coal fuel prices dropped below 1.5% per year. Conversely, under the worst nuclear and best coal assumptions, coal's real price would need to rise by 7%/year (*i.e.*, increasing 16-fold over the next 40 years, adjusted for inflation) for the two plant types to have equal costs on a lifetime basis.

Similarly, for the cost cases judged especially plausible by the author (coal base case and nuclear Case 1), a utility could pay 40% of the capital cost of a new reactor and still break even by constructing a new coal facility instead of completing the nuclear plant. Even discounting any impact from TMI, as in the nuclear base case, a utility would need to have paid for 30% of a new reactor's capital cost for completion to be cost-justified, based on the assumptions employed here. This figure would need to be revised downward, however, in the event that cancellation of reactors under construction leads to a slackening in the rate of imposition of new regulatory requirements.

---

e. Equivalently, it is the percentage of new nuclear plant construction that the utility would have to receive for free in order to break even with a new coal-fired plant. Although the calculation excludes any prospective cancellation payments to the utility's architect-engineer, construction contractors, and suppliers, it also omits any credits from applying completed construction or purchased equipment to alternative coal plants at the site or nuclear plants elsewhere.

The calculation applies only to reactors undertaken today. Breakeven costs for a specific reactor under construction must be calculated from that reactor's projected capital cost based upon its individual parameters, including nuclear sector size, listed in Appendix 1.

References

1. DOE, Office of Nuclear Reactor Programs, *Update*, July/August 1980, Part III, Table 3, indicates that 1979 nuclear O&M costs—admittedly a thin sample—were approximately 40% higher for Northeast and 50% higher for mid-Atlantic plants than for other reactors.

2. EIA, *Cost and Quality of Fuels for Electric Utility Plants—1978*, DOE/EIA-0191 (78), indicates 1978 delivered coal costs of $1.371 per million Btu for New York, New Jersey, and New England; and $1.116 for all U.S. The Pennsylvania average of $1.163 for a far greater quantity of coal was excluded from this calculation, although the Northeast region as defined here includes Pennsylvania.

3. Reference 2 indicates $.551 per million Btu for Pacific and Mountain States.

4. Reference 2 indicates $.729 per million Btu for Texas, Oklahoma, and Arkansas. The capital cost differential is from Table 9.1. Both the fuel and construction cost differentials are based overwhelmingly on data for Texas.

5. Reference 2, with a $1.269 per million Btu average coal cost; and Table 9.1.

Appendix 1
Nuclear Data Base

Operating Reactors

The nuclear data base comprises 50 reactors. Forty-six are used in the cost analysis and 49 in the construction duration analysis. Criteria used to define and compile the data base are described in Section 8.1.

The data base is presented below. Fifty-two units are listed, including Diablo Canyon 1 and 2, incomplete units licensed during the same period as the other reactors. (Listed further below are the 82 other reactors with construction permits in order of the date of construction permit award.)

The following text describes information sources and important features of the data.

Reactor Name and Vendor: Reactors are listed in sequence of construction permit award from the Atomic Energy Commission, with one exception: Browns Ferry 3 is listed directly following units 1 and 2, despite receiving its construction permit 14 months later. The single letter following the plant name denotes the nuclear reactor vendor: B for Babcock & Wilcox, C for Combustion Engineering, G for General Electric, and W for Westinghouse.

Key Dates: Listed in succession are dates of reactor order (NSSS), construction permit application (CPAP), construction permit issuance (CPIS), and commercial operation (CO). All are drawn from *U.S. Central Station Nuclear Electric Generating Units: Significant Milestones* (DOE/NE-0030), except for the CO dates of St. Lucie 1 and Trojan, both taken from licensees' originally reported CO dates in NRC *Operating Units Status Reports* (Gray Books) prior to revision. The CO dates of four incomplete units that are included in the duration data base but are not in the cost data base are listed in parentheses as estimated by author absent startup delays caused by the Three Mile Island accident (see Section 8.2).

MW Capacity is unit's design electrical rating in megawatts (MW) as originally reported in NRC *Operating Units Status Reports.*

OPEX is cumulative number of operating reactor years for the U.S. commercial nuclear sector on date each unit received its operating license. It is employed only in an alternative regression model of capital cost discussed in Section 8.1

CumCap, or sector size, is the cumulative capacity of U.S. commercial reactors operating or under construction. It is determined by adding the reactor's own capacity to the cumulative capacity for the previously licensed reactor.

A-E is the number of reactors built by the architect-engineer of the plant in question, including that plant. Multiple units (see below) are assigned the same A-E number. The letter ''S'' denotes a case in which the utility acted as its own (self) A-E—a variable tested in an alternative regression model examined in Section 8.1. The A-E number used for all such cases in the primary regression (Table 8.1) was one, since each was the utility's first venture as its own reactor A-E.

CT (Cooling Tower) refers to the use of closed-cycle cooling towers, denoted by the letter ''T.''

Multiple Units, identical (or nearly so) plants built at a shared site within approximately two years of each other, are denoted by ''P'' for the first unit (prototype) and ''D'' for subsequent duplicate units. Brunswick 1 and 2 are, respectively, duplicate and prototype because of the utility's reversal of their construction sequence. Four first-of-a-pair plants with incomplete second units are denoted as ''I'' (initial units); they were assigned special status in the cost regression as ''dangling'' units because of unusually high costs probably associated with their incomplete partner units. Absent this special variable, the measured rate of cost increase with growth in nuclear sector size would have been larger than shown in Table 8.1.

Region is either Northeast (NE), Southeast (SE), Midwest (M) or West (W). There were no South Central reactors in the nuclear data base.

Costs, in succession, are plant cost in conventional, ''mixed current'' dollars (including interest during construction, or IDC), and in ''1979 steam-plant'' dollars, *without IDC*, both in thousands; and the same respective measures divided by plant capacity for expression as dollars per kilowatt. The procedure for converting mixed dollars to 1979 steam-plant dollars is described in Appendix 3. The data sources for the mixed-dollar costs are described below.

Note that, for multiple units, the 1979 steam-plant dollar costs actually used as the dependent variable in the regression analysis were the *average* of the costs shown for the members of that set of identical units. Thus, the cost of Oconee 1, 2, and 3, for example, was the average of their separate costs in 1979 steam-plant dollars—$324/kilowatt, $307/kW, and $299/kW—or $310/kW. The rationale for this averaging procedure is stated in Section 8.1.

Time is construction time—CO date minus CPIS date. It is the dependent variable in the duration analysis. Entries in parentheses are the construction times of four incomplete units estimated absent the Three Mile Island accident, as noted above. Not shown here, but tested and found to lack statistical significance as an explanatory variable in the cost analysis, was *licensing time:* CPIS minus CPAP

Deviations from Norm: These entries are percentage differences between a reactor's cost (in 1979 steam-plant dollars without IDC) or construction time and the cost or time predicted by the regression equations (Tables 8.1 and 8.4) on the basis of the reactor's MW capacity, A-E experience, sector size, and other statistically significant variables. The +12.9 and −2.0 entries for Peach Bottom 3 (unit 12 in the data base), for example, indicate that the reactor cost 12.9% more (in adjusted, constant dollars without IDC) but took 2.0% less time to build than expected for a 1065-MW, duplicate unit with a cooling tower in the Northeast which was the eleventh reactor built by an A-E and which brought cumulative nuclear capacity to 20,707 MW when it was licensed. The cost deviations correspond to the distances in Figure 8.1 between each reactor's adjusted cost and the 46-plant trend line.

Some deviation figures may appear anomalous until their characteristics (values of their explanatory variables) are considered. For example, although Vermont Yankee's cost of $693/kW (1979 steam-plant dollars without IDC) was far greater than that of other reactors licensed up to that time, its cost was 5.2% *less* than the predicted cost. The explanation is that Vermont Yankee's characteristics—small capacity, low A-E number, not a multiple unit, with a cooling tower, in the Northeast—are all correlated with higher costs, and therefore the plant would have been expected to have a very high cost in any event.

The deviation figures do not necessarily measure cost or time overruns, since the norm from which the deviations are measured—commercial reactor construction experience—itself is characterized by chronic overruns. In addition, use of the figures as a barometer of construction performance implies that characteristics which were correlated to costs, such as A-E experience or plant size, were outside the utility's control, when such was not the case. Moreover, the cost deviations do not reflect real interest during construction, which is proportional to construction time. Granting these limitations, however, the deviation figures are a good indicator of the skill with which utilities coped with the industry-wide problems involved in constructing nuclear plants in the late 1960s and the 1970s.

The construction time deviations for the four incomplete reactors are shown in parentheses, since they are based on hypothetical durations estimated without the Three Mile Island accident. No deviation is given for Farley 1, since it was excluded from the duration regression analysis for the reason given in Section 8.1.

Source gives the data source for each reactor's cost in mixed current dollars. The year refers to the issue number of *Steam-Electric Plant Construction Cost and Annual Production Expenses* (DOE/EIA-0033), a compilation of data from utility "Form 1" reports published for many years by the Federal Power Commission and now issued by the Energy Information Administration. For reasons explained in Section 8.1, the edition for the year following initial commercial service was used as the data source, with several exceptions. Where multiple units were completed a year apart, the first unit's cost was obtained from the report for its year of completion in order to measure the unit's cost before the second unit's cost was added. Some 1977 units' costs were obtained from the 1977 edition because the 1978 report was not published until December 1980. The ''1978'' entry for two 1978 units indicates that costs were compiled by the author from utilities' 1978 Form 1 reports.

Costs for five units were obtained by direct communication with the utility, as follows:

Browns Ferry 3: Larry Knott, Tennessee Valley Authority (TVA) power generation section, reported Browns Ferry station cost of $885,990,900 as of 30 September 1978 (end of TVA's 1978 fiscal year) (telephone communi-

Nuclear Plant

| | Reactor | Vendor | Dates | | | | MW | OPEX | CumCap |
			NSSS	CPAP	CPIS	CO			
1	Palisades	C	66.00	66.42	67.17	71.92	821	13	10621
2	Turkey Point 3	W	65.83	66.17	67.25	72.92	745	29	11366
3	Turkey Point 4	W	67.25	66.17	67.25	73.67	745	43	12111
4	Browns Ferry 1	G	66.42	66.50	67.33	74.58	1098	47	13209
5	Browns Ferry 2	G	66.42	66.50	67.33	75.17	1098	75	14307
6	Browns Ferry 3	G	67.42	67.50	68.50	77.17	1098	168	15405
7	Oconee 1	B	66.50	66.83	67.83	73.50	886	39	16291
8	Oconee 2	B	66.50	66.83	67.83	74.67	886	55	17177
9	Oconee 3	B	67.33	67.25	67.83	74.92	886	78	18063
10	Vermont Yankee	G	66.58	66.83	67.92	72.83	514	24	18577
11	Peach Bottom 2	G	66.58	67.08	68.00	74.50	1065	51	19642
12	Peach Bottom 3	G	66.58	67.08	68.00	74.92	1065	78	20707
13	Diablo Canyon 1	W	66.83	67.00	68.25	—	1084	—	21791
14	Three Mile Is. 1	B	66.83	67.33	68.33	74.67	819	70	22610
15	Surry 1	W	66.75	67.17	68.42	72.92	823	26	23433
16	Surry 2	W	66.75	67.17	68.42	73.33	823	38	24256
17	Fort Calhoun	C	66.75	67.25	68.42	73.67	457	45	24713
18	Prairie Island 1	W	67.08	67.17	68.42	73.92	530	51	25243
19	Prairie Island 2	W	67.42	67.17	68.42	74.92	530	87	25773
20	Cooper	G	67.25	67.50	68.42	74.50	778	62	26551
21	Pilgrim 1	G	65.58	67.42	68.58	72.92	670	28	27221
22	Kewaunee	W	67.08	67.58	68.58	74.42	560	60	27781

cation, 19 March 1979). Unit 1 and 2 costs were subtracted from total to obtain Unit 3 cost.

Crystal River 3: Letter from R.R. Hayes, vice president and controller, Florida Power Corp., 4 April 1979.

Indian Point 3 and Fitzpatrick: Discussions with staff of Power Authority of State of New York, March 1978.

Davis-Besse 1: Letter from J.R. Dyer, public relations, Toledo Edison, 14 June 1979.

Data Base

| A-E | CT | Mult. | Reg. | Costs (000) | | Costs ($/kW) | | Time | Deviations | | |
				Mixed	1979	Mixed	1979	(yrs.)	$/kW	Time	Source
5	T		M	146690	301132	179	367	4.75	+2.5	+2.7	1972
6		P	SE	108710	209590	146	281	5.67	+4.8	+24.7	1972
6		D	SE	126786	229320	170	308	6.42	+1.0	+16.0	1974
S	T	P	SE	276179	471816	252	430	7.25	+2.7	+0	1976
S	T	D	SE	276179	451853	252	412	7.84	−1.9	−11.4	1976
S	T	D	SE	333633	461835	304	421	8.67	−6.0	−3.3	see text
8		P	SE	155610	286929	176	324	5.67	−4.4	−3.7	1973
8		D	SE	160420	272075	181	307	6.84	−7.3	−4.3	1974
8		D	SE	160420	265016	181	299	7.09	−9.9	−1.7	1974
3	T		NE	184480	356022	359	693	4.91	−5.2	−4.1	1973
11	T	P	NE	376990	646767	354	607	6.50	+16.4	+11.8	1975
11	T	D	NE	376990	627579	354	589	6.92	+12.9	−2.0	1975
S		P	W	—	—	—	—	—	—	—	—
2	T		NE	400928	677737	490	828	6.34	+6.3	−17.8	1975
3		P	SE	146710	281054	178	341	4.50	+2.1	−22.6	1972
3		D	SE	255386	475497	310	578	4.91	+0.1	−30.3	1974
1			M	175800	319605	385	699	5.25	+7.7	+10.8	1974
1	T	P	M	233230	415304	440	784	5.50	−3.0	−0.9	1973
1	T	D	M	176977	292094	334	551	6.50	−4.2	−3.0	1975
3			M	269287	458797	346	590	6.08	+8.8	+18.2	1975
13			NE	239330	462038	357	690	4.34	+11.3	−6.6	1973
1			M	203389	349178	363	624	5.84	−6.5	+12.2	1975

| | Reactor | Vendor | Dates | | | | MW | OPEX | CumCap |
			NSSS	CPAP	CPIS	CO			
23	Salem 1	W	66.58	66.92	68.67	77.42	1090	172	28871
24	Salem 2	W	67.33	67.75	68.67	(79.92)	1115	—	29986
25	Crystal River 3	B	67.08	67.58	68.67	77.17	825	190	30811
26	Maine Yankee	C	67.08	67.67	68.75	72.92	790	32	31601
27	Rancho Seco	B	67.58	67.83	68.75	75.25	913	81	32514
28	Zion 1	W	67.08	67.50	68.92	73.92	1050	43	33564
29	Zion 2	W	67.50	67.58	68.92	74.67	1050	57	34614
30	Arkansas 1	B	67.25	67.83	68.92	74.92	850	72	35464
31	Cook 1	W	67.50	67.92	69.17	75.58	1090	87	36554
32	Cook 2	W	67.50	67.92	69.17	78.50	1100	246	37654
33	Calvert Cliffs 1	C	67.33	68.00	69.50	75.33	845	78	38499
34	Calvert Cliffs 2	C	67.33	68.00	69.50	77.25	845	172	39344
35	Indian Point 3	W	67.25	67.25	69.58	76.58	965	140	40309
36	Hatch 1	G	67.92	68.33	69.67	75.92	786	81	41095
37	Three Mile Is. 2	B	67.08	68.25	69.83	78.92	906	256	42001
38	Brunswick 2	G	68.00	68.50	70.08	75.83	821	94	42822
39	Brunswick 1	G	68.00	68.50	70.08	77.17	821	177	43643
40	Fitzpatrick	G	68.92	68.92	70.33	75.50	821	87	44464
41	Sequoyah 1	W	68.25	68.75	70.33	(80.08)	1148	—	45612
42	Sequoyah 2	W	68.25	68.75	70.33	(80.75)	1148	—	46760
43	Duane Arnold	G	68.08	68.83	70.42	75.08	538	65	47298
44	Beaver Valley 1	W	67.67	69.00	70.42	76.75	852	144	48150
45	St. Lucie 1	C	67.92	69.00	70.50	76.42	810	152	48960
46	Millstone 2	C	67.92	69.08	70.92	75.92	828	128	49788
47	Diablo Canyon 2	W	68.50	68.42	70.92	—	1106	—	50894
48	Trojan	W	68.83	69.42	71.08	75.92	1130	136	52024
49	North Anna 1	W	67.75	69.17	71.08	78.42	907	265	52931
50	North Anna 2	W	67.75	69.17	71.08	(79.92)	907	—	53838
51	Davis-Besse 1	B	68.75	69.50	71.17	77.83	906	208	54744
52	Farley 1	W	69.33	69.75	72.58	77.92	829	218	55573

A-E	CT	Mult.	Reg.	Costs (000) Mixed	1979	Costs ($/kW) Mixed	1979	Time (yrs.)	Deviations $/kW	Time	Source
S		I	NE	850318	1162439	780	1066	8.75	+4.2	+17.9	1977
S		D	NE	—	—	—	—	(11.25)	—	(+24.1)	—
3			SE	419080	582097	508	706	8.50	+20.9	+18.1	see text
5			NE	219230	418770	278	530	4.17	−26.6	−25.9	1973
14	T		W	343620	552327	376	605	6.50	+0.8	+4.9	1975
5		P	M	275990	491444	263	468	5.00	−6.6	−11.5	1973
5		D	M	291997	490132	278	467	5.75	−8.2	−15.9	1975
15			SE	238751	394748	281	464	6.00	−12.6	−3.9	1975
S		P	M	544650	854421	500	784	6.41	+4.4	−7.8	1976
S		D	M	451527	556374	410	506	9.33	+2.8	+10.6	1978
16		P	NE	430674	689726	510	816	5.83	+6.7	+10.9	1976
16		D	NE	335321	461459	397	546	7.75	+5.4	+22.0	1977
2			NE	570000	829042	591	859	7.00	−2.3	+0	see text
18	T	I	SE	390390	594667	497	757	6.25	−18.3	+8.5	1976
4	T		NE	715466	853680	790	942	9.09	−7.1	+9.5	1978
3		P	SE	389118	596315	474	726	5.75	−1.2	−11.5	1976
3		D	SE	318442	438237	388	534	7.09	−2.3	−9.6	1977
6			NE	419000	653173	510	796	5.17	−7.1	−13.2	see text
4	T	P	SE	—	—	—	—	(9.75)	—	(+22.8)	—
4	T	D	SE	—	—	—	—	(10.42)	—	(+8.6)	—
19	T		M	279928	453629	520	843	4.66	+5.2	+3.1	1976
7	T		NE	598716	855653	703	1004	6.33	−4.1	−5.0	1977
4			SE	486230	711866	600	879	5.92	+18.3	−7.8	1977
20			NE	426271	649323	515	784	5.00	−2.5	−6.4	1976
S		D	W	—	—	—	—	—	—	—	—
21	T		W	451980	683257	400	605	4.84	−16.3	−18.5	1976
8		I	SE	781739	966977	862	1066	7.34	+12.3	+16.8	1978
8		D	SE	—	—	—	—	(8.84)	—	(+16.6)	—
22	T		M	672452	866772	742	957	6.66	+23.6	+2.9	see text
23	T	I	SE	727426	922905	877	1113	5.34	+4.7	—	1977

Reactors Under Construction

Below is a list of the 82 reactors holding construction permits as of February 1981, arranged in sequence of construction permit award. Their unit size, architect-engineer experience, and cumulative nuclear sector capacity are listed to enable readers to calculate projected plant costs and construction durations using the nuclear cost and duration regressions (Tables 8.1 and 8.4). (Other parameters that correlated significantly with costs and construction times—Northeast location, multiple units, and cooling towers—are not listed.) Readers are cautioned that costs calculated with the regression are in 1979 steam-plant dollars without interest during construction, rather than the mixed-current dollars in which capital costs are usually expressed (Appendix 3 and Table 10.10 will aid in converting costs). The regressions also do not reflect the effect of the Three Mile Island accident or of deliberate stretchouts of construction on reactor costs and schedules.

The reactor numbers—68 through 153—reflect the 52 reactors listed in the complete data base and the 15 commercial reactors that precede the data base. The latter are, in construction permit sequence: San Onofre 1, Connecticut Yankee, Oyster Creek, Nine Mile Point 1, Dresden 2, Ginna, Millstone 1, Dresden 3, Indian Point 2, Quad Cities 1 and 2, Robinson 2, Monticello, and Point Beach 1 and 2.

The list of plants under construction shows four licensed units which utilities cancelled in 1980: Forked River, North Anna 4, Sterling, and Tyrone. Their capacities had been included in calculating the sector size of 149,648 MW used to project the cost of a standard 1988 nuclear plant in Chapter 10. Projected costs would be 1.6% less if calculated from the currently projected sector size of 145,421 MW which is the cumulative capacity for Yellow Creek 2, the last unit listed below.

Reactors Under Construction

	Reactor	CP Issue Date	MW	CumCap	A-E
68	Farley 2	08/72	829	56,402	23
69	Fermi 2	09/72	1093	57,495	7
70	Zimmer 1	10/72	810	58,305	8
71	Arkansas 2	12/72	912	59,217	25
72	Hatch 2	12/72	795	60,012	26
73†	Midland 1	12/72	811	60,823	27
74	Midland 2	12/72	811	61,634	27

	Reactor	CP Issue Date	MW	CumCap	A-E
75	Watts Bar 1	01/73	1177	62,811	6
76	Watts Bar 2	01/73	1177	63,988	6
77	McGuire 1	02/73	1180	65,168	1
78	McGuire 2	02/73	1180	66,348	1
79	Summer 1	03/73	900	67,248	4
80	WPPSS 2	03/73	1100	68,348	5
81	Shoreham	04/73	819	69,167	10
82*	Forked River	07/73	1070	—	—
83	LaSalle County 1	09/73	1078	70,245	9
84	LaSalle County 2	09/73	1078	71,323	9
85	San Onofre 2	10/73	1100	72,423	29
86	San Onofre 3	10/73	1100	73,523	29
87	Susquehanna 1	11/73	1050	74,573	31
88	Susquehanna 2	11/73	1050	75,623	31
89	Bailly	05/74	643	76,266	11
90	Beaver Valley 2	05/74	833	77,099	11
91	Limerick 1	06/74	1065	78,164	33
92	Limerick 2	06/74	1065	79,229	33
93	Nine Mile Point 2	06/74	1100	80,329	12
94	Vogtle 1	06/74	1110	81,439	35
95	Vogtle 2	06/74	1110	82,549	35
96	North Anna 3	07/74	907	83,456	13
97*	North Anna 4	07/74	907	—	—
98	Millstone 3	08/74	1158	84,614	14
99	Grand Gulf 1	09/74	1250	85,864	37
100	Grand Gulf 2	09/74	1250	87,114	37
101	Hope Creek 1	11/74	1067	88,181	39
102	Hope Creek 2	11/74	1067	89,248	39
103	Waterford 3	11/74	1113	90,361	5
104	Bellefonte 1	12/74	1213	91,574	8
105	Bellefonte 2	12/74	1213	92,787	8
106	Comanche Peak 1	12/74	1111	93,898	2
107	Comanche Peak 2	12/74	1111	95,009	2

	Reactor	CP Issue Date	MW	CumCap	A-E
108	Catawba 1	08/75	1145	96,154	3
109	Catawba 2	08/75	1145	97,299	3
110	Braidwood 1	12/75	1120	98,419	12
111	Braidwood 2	12/75	1120	99,539	12
112	Byron 1	12/75	1120	100,659	14
113	Byron 2	12/75	1120	101,779	14
114	South Texas 1	12/75	1250	103,029	1
115	South Texas 2	12/75	1250	104,279	1
116	WPPSS 1	12/75	1218	105,497	5
117	Clinton 1	02/76	933	106,430	16
118	Clinton 2	02/76	933	107,363	16
119	Callaway 1	04/76	1120	108,483	41
120	Callaway 2	04/76	1120	109,603	41
121	Palo Verde 1	05/76	1270	110,873	43
122	Palo Verde 2	05/76	1270	112,143	43
123	Palo Verde 3	05/76	1270	113,413	43
124	Seabrook 1	07/76	1200	114,613	7
125	Seabrook 2	07/76	1200	115,813	7
126	River Bend 1	03/77	934	116,747	15
127	River Bend 2	03/77	934	117,681	15
128	Hartsville A 1	05/77	1233	118,914	10
129	Hartsville A 2	05/77	1233	120,147	10
130	Hartsville B 1	05/77	1233	121,380	10
131	Hartsville B 2	05/77	1233	122,613	10
132	Perry 1	05/77	1205	123,818	5
133	Perry 2	05/77	1205	125,023	5
134	St. Lucie 2	05/77	810	125,833	6
135	Wolf Creek	05/77	1150	126,983	46
136*	Sterling	09/77	1150	—	—
137	Cherokee 1	12/77	1280	128,263	5
138	Cherokee 2	12/77	1280	129,543	5
139	Cherokee 3	12/77	1280	130,823	5

Appendix 1

Reactor	CP Issue Date	MW	CumCap	A-E
140* Tyrone	12/77	1100	—	—
141 Phipps Bend 1	01/78	1233	132,056	14
142 Phipps Bend 2	01/78	1233	133,289	14
143 Harris 1	01/78	900	134,189	7
144 Harris 2	01/78	900	135,089	7
145 Harris 3	01/78	900	135,989	7
146 Harris 4	01/78	900	136,889	7
147 WPPSS 4	01/78	1218	138,107	5
148 Marble Hill 1	02/78	1130	139,237	18
149 Marble Hill 2	02/78	1130	140,367	18
150 WPPSS 3	04/78	1242	141,609	11
151 WPPSS 5	04/78	1242	142,851	11
152 Yellow Creek 1	11/78	1285	144,136	16
153 Yellow Creek 2	11/78	1285	145,421	16

*Denotes cancelled unit.

†Midland 1 is a steam- and power-producing unit with a 460-MW electrical capacity. Its electrical-equivalent capacity is shown here.

Appendix 2

Coal Data Base

The coal data base comprises 116 coal-fired generating units. All are employed in the cost analysis, and 92 are used in the construction duration analysis. Criteria used to define and compile the data base are described in Section 9.1.

The data base is presented below. The following text describes information sources and important features of the data.

Plant Name, Utility, and Key Dates: Coal plants are listed in sequence of boiler order as compiled in the Kidder Peabody data base (see Section 9.1). Utility owners (or lead owner in the case of joint ventures) are also listed (utility names were excluded from the list of reactors in Appendix 1 because they are available from numerous sources). The CO date is the utility's declared date of initial commercial service as reported in *Steam-Electric Plant Construction Cost and Annual Production Expenses* (DOE/EIA-0033, annual).

MW Capacity is the unit's nameplate generator rating reported in the *Steam-Electric Plant* reports just cited.

Multiple Units, identical (or nearly so) plants built at a shared site within approximately two years of each other, are assigned one of three designations: "P" (prototype) for first units with a subsequent duplicate also in the data base, "D" (duplicate) for duplicate units, and "I" (initial) for first units whose follow-on partners were *not* completed in time to enter the data base.

Both prototype and duplicate units are referred to in this study as *multiple* units. As explained in Chapter 9, the costs in constant dollars of the members of a set of multiple units were averaged to smooth fluctuations in the allocation of costs within each set. (In a slight variant of this procedure for the Winyah station, the cost in constant dollars of unit 2 was made 26.5% greater than unit 1's cost to adjust for the fact that only unit 2 has a scrubber.) Multiple units averaged approximately 10% less cost than non-multiple units (see Section 9.1). Costs of so-called initial units did not differ from those of other one-of-a-kind units in the sample. Note that "add-on" units sharing a site with earlier units were not classified as multiple units unless they matched the capacity and general design features of their immediate predecessor.

SO₂ control devices, or scrubbers, are denoted by "S." Table 9.4 lists the 15 data base units with scrubbers along with SO_2 design removal efficiency and coal sulfur content.

CumCap, or sector size, is the cumulative capacity of utility coal units operating or under construction. It is determined by adding 95% of the unit's own capacity (allowing for retirements equalling 5% of new construction) to the cumulative capacity of the previous plant in the boiler-order sequence. Beginning with the 91st unit, the cumcap figure also incorporates the capacities of coal plants ordered prior to each new unit but completed in 1978, the first year following the sample period. Sixteen such units totalling 8804 MW (without the 5% reduction described above) are reflected in the sector-size figure for the last sample plant, although ten of the 16 enter the capacity total only for the last four plants in the data base.

Regions are Northeast (NE), Southeast (SE), Midwest (M), South Central (SC), and West (W).

Company variables that were correlated with capital cost were ownership by American Electric Power (AEP) or the Southern Company. The former, shown in the utility listings, are units numbered 12, 38, 46, 54, and 102. The latter, listed under the names of company subsidiaries Alabama Power, Georgia Power, Gulf Power, and Mississippi Power, are numbers 16, 24, 36, 37, 56, 57, 67, 77, 78, and 91.

Costs, in succession, are plant cost in conventional, as-reported "mixed current" dollars (including interest during construction, or IDC), and in "1979 steam-plant" dollars, *without IDC*, both in thousands; and the same respective measures divided by plant capacity for expression as dollars per kilowatt. The procedure for converting mixed dollars to 1979 steam-plant dollars is described in Appendix 3. Costs for four units, numbered 33, 73, 89, and 116, were reported by utilities *without* IDC, and therefore only inflation was subtracted in converting their costs to 1979 steam-plant dollars.

For multiple units, the costs employed in the regression analysis were the *averages* of the costs, in 1979 steam-plant dollars, of the members of each multiple-unit set. This procedure was also applied to duplicate (second or later) units whose unit 1 forerunners preceded the data base. Following are the boiler order date, commercial operation date, reported capital costs in mixed current dollars, and adjusted cost in 1979 steam-plant dollars of the eight non-data base units whose constant-dollar costs were averaged with those of their duplicate successors in the data base.

Unit	Boiler Order	Commercial Operation	Reported Costs ($000)	Adjusted Costs (1979 $ 000)
Amos 1	67.42	71.67	128,331	281,626
Bowen 1	67.67	71.50	92,525	204,949
Cayuga 1	66.67	70.75	80,028	185,115
Big Bend 1	66.75	70.75	71,286	164,903
Big Brown 1	67.67	71.92	86,153	185,883

Monroe 1	67.00	71.42	176,260	392,996
Stuart 1	65.92	70.75	92,864	214,462
Stuart 2	66.25	71.33	92,864	208,150

Time is project time from boiler order to commercial operation. It is the dependent variable in the coal duration analysis.

Deviations from Norm: These entries are percentage differences between the unit's cost (in 1979 steam-plant dollars without IDC) or project duration and the cost or duration predicted by the respective regression equations (Tables 9.1 and 9.7). A plus sign indicates that a plant cost more (or took longer) to construct than would have been expected on the basis of its sector-size value, location, use of a scrubber, etc. Several *caveats* expressed in Appendix 1 (nuclear plants) apply here: deviations for costs (but not for durations) of multiple units were calculated on the basis of the average constant-dollar cost for that multiple-unit set, not from the costs shown in the data base; cost deviations do not reflect higher real IDC for plants with longer construction times; in some instances, the deviations in construction times reflect a particular utility's need, or lack thereof, to complete plants rapidly to meet load growth; and the capital cost deviations are adjusted for the lower and higher costs, respectively, of plants built by the Southern Company and American Electric Power. Bearing these limitations in mind, the deviation figures are a useful measure of utilities' skill in constructing plants economically and expeditiously.

Not shown in the data base listings are several variables that were tested for, but failed to demonstrate, statistical significance in regressions on capital cost or construction duration. They are: supercritical boiler (design steam pressure), boiler manufacturer, use of a cooling tower, heat content of fuel used, and ownership by a publicly held utility. Listings for these variables are available by special arrangement with the author.

Also not shown is the source of each plant's cost in mixed current dollars. Costs were generally obtained from the edition of the federal *Steam-Electric Plant* report for the year in which a unit was declared commercially operational. Because the 1976 and 1977 editions of the report were not available when the author began to compile these data in mid-1978, costs for 1976 and 1977 units were obtained from utility "Form 1" reports on file at the Federal Energy Regulatory Commission. Exceptions to this procedure, in which the author relied upon direct communication with the utility, are as follows, tabulated by units' data base sequence numbers:

No. 4 (Montour 1): Capital cost, exclusive of gas turbine units, was reported in letter to author, 27 July 1978.

Nos. 5, 60 and 61 (Big Brown 2, Monticello 1 and 2): Separate capital costs for each individual unit and for Big Brown 1 (completed prior to data base period but needed to average constant-dollar costs with Big Brown 2) were

Appendix 2

reported in letter to author, 24 July 1978.

No. 31 (Stout 7): Capital cost, exclusive of gas turbine units, was reported in letter to author, 19 July 1978.

No. 63 (Smith 2): Capital cost not in FPC/EIA reports, was reported in letter to author, 17 July 1978.

No. 73 (Drake 7): Capital cost not in FPC/EIA reports, was reported in letter to author, 5 July 1979.

Nos. 89 and 116 (Winyah 1 and 2): Separate capital costs for individual units were reported in letter to author, 27 July 1978.

No. 94 (Duck Creek 1): 1976-reported cost was supplemented by $13,909,000 spent in 1977-78 to complete scrubber, as reported in letter to author, 5 March 1979.

No. 95 (Hayden 2): Capital cost not in FPC/EIA reports, was reported in letter to author, 16 August 1978.

No. 101 (Deely 1): Joint capital cost for Deely 1 and 2 (latter was completed after data base period) was reported as $246,998,389 in letter to author, 19 July 1978; author subtracted reported cost of $30,175,633 for railroad cars and railroad car maintenance facility, and halved remainder to obtain cost for Deely 1.

No. 102 (Cardinal 3): Capital cost not in FPC/EIA reports, was reported in letter to author, 16 August 1978.

No. 108 (Southwest 1): Capital cost not in FPC/EIA reports, was reported in letter to author, 19 July 1978.

No. 113 (Young 2): Capital cost not in FPC/EIA reports, was reported in letter to author, 31 August 1978,

Coal Plant Data Base

Plant and Utility	Dates Order	Finish	MW	Mult.	SO₂	CumCap	Reg.	Costs (000) Mixed	1979	Costs ($/kW) Mixed	1979	Time (yrs.)	Deviations $/kW	Time
1 Johnston 4 WY Pacific P&L	66.75	72.50	360			142342	W	64621	135051	180	375	5.75	−13.4	—
2 Monroe 2 MI Detroit Ed.	67.00	73.17	823	D		143124	M	122846	245689	149	299	6.17	+5.8	—
3 Cumberland 1 TN TVA	67.25	73.17	1300	P		144359	SE	219275	437839	169	337	5.92	+9.1	—
4 Montour 1 PA Penn P&L	67.58	72.17	806	P		145125	NE	141729	301367	176	374	4.59	−12.2	—
5 Big Brown 2 TX Texas Utils.	67.67	72.92	593	D		145688	SC	57834	116993	98	197	5.25	+5.9	—
6 Stuart 3 OH Dayton P&L	67.67	72.33	610	D		146267	M	81490	171176	134	281	4.50	+1.2	—
7 Labadie 3 MO Union Elec.	67.83	72.58	621	D		146857	M	125290	258869	202	417	4.75	+15.9	—
8 Centralia 1 WA Pacific P&L	67.83	72.92	730	P		147551	W	153060	309180	210	424	5.09	+3.2	—
9 Cayuga 2 IN Indiana P.S.	67.92	72.42	531	D		148055	M	69415	144843	131	273	4.50	−2.5	—
10 Cumberland 2 TN TVA	67.92	73.83	1300	D		149290	SE	171905	326314	132	251	5.91	+6.8	—
11 Coffeen 2 IL Central IL P.S.	68.00	72.67	617			149876	M	99155	203341	161	330	4.67	−7.3	+6.9
12 Amos 2 WV. AEP.	68.08	72.42	816	D		150652	M	102311	213194	125	261	4.34	−20.2	−12.9
13 Montour 2 PA Penn P&L	68.08	73.25	819	D		151430	NE	107348	211615	131	258	5.17	−14.5	+3.4
14 Mill Creek 1 KY Louisville G&E	68.25	72.58	356	P		151768	M	62166	127953	175	359	4.33	−5.6	+8.6
15 Eastlake 5 OH Cleveland Elec.	68.25	72.67	680			152414	M	109887	224803	162	331	4.42	−8.0	−1.4
16 Bowen 2 GA Georgia Power	68.25	72.67	806	D		153180	SE	92525	189285	115	235	4.42	+2.9	+3.0
17 Labadie 4 MO Union Elec.	68.25	73.58	621	D		153770	M	102700	197551	165	318	5.33	+12.6	+11.1

Plant and Utility	Order	Finish	MW	Mult.	SO₂	CumCap	Reg.	Mixed	1979	Mixed	1979	Time (yrs.)	$/kW	Time
								Costs (000)		Costs ($/kW)			Deviations	
18 Henderson 1 KY Big Rivers Coop	68.42	73.25	180	P		153941	M	36307	71301	202	396	4.83	+16.9	+35.7
19 Neal 2 IO Iowa P.S.	68.50	72.33	349			154272	M	49150	102625	141	294	3.83	-18.8	-4.5
20 Edwards 3 IL Central IL Light	68.50	72.42	364			154618	M	70132	145543	193	400	3.92	+10.3	-3.1
21 Baldwin 2 IL Illinois Power	68.50	73.33	635	P		155221	M	102718	200410	162	316	4.83	-3.1	+8.0
22 Monroe 3 MI Detroit Ed.	68.58	73.33	823	D		156003	M	122846	239444	149	291	4.75	+0.3	-6.6
23 New Madrid 1 MO Associated Elec.	68.67	72.75	650	P		156621	M	143907	291383	221	448	4.08	+21.5	-9.6
24 Gorgas 10 AL Alabama Power	68.75	72.42	789			157370	SE	98739	204328	125	259	3.67	-3.1	-8.4
25 Powerton 5 IL Commonwealth Ed.	68.75	72.67	893	P		158218	M	184042	374338	206	419	3.92	+4.5	-18.3
26 Conesville 4 OH Columb. & So. OH	68.83	73.42	842			159018	M	130356	251689	155	299	4.59	-19.0	-3.6
27 La Cygne 1 KS Kansas City P&L	68.83	73.42	873		S	159848	M	185146	357477	212	409	4.59	-12.1	-4.5
28 Centralia 2 WA Pacific P&L	68.83	73.50	730	D		160541	W	153060	293903	210	403	4.67	-2.0	-7.6
29 Harrison 1 WV Potomac Elec.	68.92	72.92	684	P		161191	M	166559	332149	244	486	4.00	+2.8	-13.5
30 Mt. Storm 3 WV Vepco	68.92	73.33	522			161687	M	160104	310680	307	595	4.41	+59.7	-0.1
31 Stout 3 IN Indianapolis P&L	68.92	73.50	471			162134	M	85275	163545	181	347	4.58	-7.0	+5.5
32 Harrison 2 WV Potomac Elec.	68.92	73.92	684	D		162784	M	111470	207748	163	304	5.00	+2.2	-0.6
33 Erickson 1 MI City of Lansing	69.17	73.67	160			162936	M	33747	63745	211	398	4.50	+21.6	+25.3
34 Monroe 4 MI Detroit Ed.	69.17	74.33	817	D		163712	M	113662	205282	139	251	5.16	-2.6	-0.9
35 Big Bend 2 FL Tampa Elec.	69.25	73.25	446	D		164136	SE	62995	122332	141	274	4.00	+25.2	-2.7
36 Watson 5 MS Mississippi Pwr.	69.25	73.42	578			164685	SE	66247	127143	115	220	4.17	-20.0	+7.4
37 Crist 3 FL Gulf Power	69.25	73.58	578			165234	SE	74514	141452	129	245	4.33	-11.2	+11.3

Plant and Utility	Dates Order	Finish	MW	Mult.	SO₂	CumCap	Reg.	Costs (000) Mixed	1979	Costs ($/kW) Mixed	1979	Time (yrs.)	Deviations $/kW	Time
38 Amos 3 WV AEP	69.25	73.75	1300			166469	M	363934	628930	280	525	4.50	+17.5	−14.6
39 Powerton 6 IL Commonwealth Ed.	69.25	75.83	893	D		167318	M	152372	245758	171	275	6.58	+1.0	+23.0
40 Cliffside 5 NC Duke Power	69.33	72.42	571			167860	SE	92966	190837	163	334	3.09	+2.1	−23.0
41 Culley 3 IN So. Indiana G&E	69.33	73.42	265			168112	M	54691	104841	206	396	4.09	+3.7	+2.4
42 Mich. City 12 IN No. Indiana P.S.	69.33	74.25	521			168607	M	128437	232779	247	447	4.92	+16.8	+9.1
43 Belews Creek 1 NC Duke Power	69.33	74.58	1080	P		169633	SE	168703	298823	156	277	5.25	−8.6	+16.1
44 Belews Creek 2 NC Duke Power	69.33	75.92	1080	D		170659	SE	180761	289125	167	268	6.59	−8.9	+34.2
45 Mill Creek 2 KY Louisville G&E	69.42	74.50	356	D		170997	M	50506	89859	142	252	5.08	−12.2	+10.5
46 Gavin 1 OH AEP	69.42	74.75	1300	P		172232	M	324604	567540	250	437	5.33	−9.8	−0.7
47 Roxboro 3 NC Carolina P&L	69.50	73.17	745			172940	SE	114118	222014	153	298	3.67	−10.6	−14.2
48 Sutton 3 NC Carolinia P&L	69.58	72.42	447			173364	SE	71922	147075	161	329	2.84	−1.4	−27.4
49 Ghent 1 KY Kentucky Utils.	69.58	74.08	557	P		173893	M	111693	204189	201	367	4.50	−4.8	−3.0
50 San Juan 2 NM New Mexico P.S.	69.83	73.83	329	P		174206	W	85753	158725	261	482	4.00	+2.4	−5.4
51 Navajo 1 AZ Salt River Proj.	69.83	74.33	803	P		174969	W	205845	368687	256	459	4.50	+1.2	−9.4
52 Stuart 4 OH Dayton P&L	69.83	74.42	610	D		175548	M	105292	187431	173	307	4.59	−9.6	−10.5
53 Harrison 3 WV Potomac Elec.	69.83	74.92	684	D		176198	M	98371	169081	144	247	5.09	−2.6	−2.9
54 Gavin 2 OH AEP	69.83	75.50	1300	D		177433	M	243083	399632	187	307	5.67	−11.4	−3.9
55 Comanche 1 CO Colorado P.S.	69.92	73.92	350	P		177766	W	94509	173650	270	496	4.00	−5.4	−7.4
56 Bowen 3 GA Georgia Power	69.92	74.33	952	P		178670	SE	129205	231170	136	243	4.41	−11.9	−0.4

Plant and Utility	Dates Order	Finish	MW	Mult.	SO₂	CumCap	Reg.	Costs (000) Mixed	1979	Costs ($/kW) Mixed	1979	Time (yrs.)	Deviations $/kW	Time
57 Gaston 5 AL Alabama Power	69.92	74.58	952			179574	SE	156542	275340	164	289	4.66	−0.3	+4.9
58 Navajo 2 AZ Salt River Proj.	69.92	75.25	803	D		180337	W	194548	325902	242	406	5.33	−0.7	−2.4
59 Navajo 3 AZ Salt River Proj.	69.92	76.25	803	D		181100	W	251406	387851	313	483	6.33	−1.0	+15.6
60 Monticello 1 TX Texas P&L	70.08	74.92	593	P		181663	SC	128772	220689	217	372	4.84	+3.6	+0.8
61 Monticello 2 TX Texas P&L	70.08	75.92	593	D		182227	SC	75131	118959	127	201	5.84	+3.4	+12.2
62 Henderson 2 KY Big Rivers Coop	70.42	73.92	180	D		182398	M	36307	66236	202	368	3.50	+5.3	−16.9
63 Smith 2 KY Owensboro Muni	70.42	74.17	265			182650	M	45129	81140	170	306	3.75	−23.8	−10.1
64 Bridger 1 WY Pacific P&L	70.42	74.67	561	P		183183	W	174823	303940	312	542	4.25	+10.2	−11.0
65 Huntington 2 UT Utah P&L	70.50	74.50	446	P		183606	W	136240	239456	305	537	4.00	+3.0	−12.8
66 Big Stone 1 SD Otter Tail Power	70.50	75.42	456			184039	M	163185	267897	358	587	4.92	+45.6	+6.7
67 Bowen 4 GA Georgia Power	70.50	75.83	952	D		184944	SE	130382	206846	137	217	5.33	−13.8	+9.2
68 Colstrip 1 MT Montana Power	70.50	75.83	358	P	S	185284	W	158543	251522	443	703	5.33	+3.8	+20.2
69 Bridger 2 WY Pacific P&L	70.50	75.67	561	D		185817	W	187444	301433	334	537	5.17	+9.3	−0.7
70 Rush Island 1 MO Union Elec.	70.50	76.17	621	P		186407	M	223335	344046	360	554	5.67	+17.5	+15.6
71 Mansfield 1 PA Ohio Edison	70.50	76.25	914	P	S	187275	NE	454152	694519	497	760	5.75	+18.4	+9.2
72 Mansfield 2 PA Ohio Edison	70.50	77.75	914	D	S	188143	NE	340437	452932	372	496	7.25	+18.0	+26.8
73 Drake 7 CO Colo. Springs	70.67	74.25	127			188264	W	22666	45185	178	356	3.58	−30.8	−3.8
74 Columbia 1 WI Wisconsin P&L	70.75	75.33	545	I		188782	M	144959	239116	266	439	4.58	+7.0	−5.1
75 New Madrid 2 MO Associated Elec.	70.75	77.42	650	D		189399	M	168675	230178	260	354	6.67	+8.1	+23.6
76 Bridger 3 WY Pacific P&L	70.83	76.67	561	D		189932	W	165716	242364	295	432	5.84	+7.8	+10.9

Plant and Utility	Dates Order	Dates Finish	MW	Mult.	SO₂	CumCap	Reg.	Costs (000) Mixed	Costs (000) 1979	Costs ($/kW) Mixed	Costs ($/kW) 1979	Time (yrs.)	Deviations $/kW	Deviations Time
77 Yates 6 GA Georgia Power	70.92	74.33	404	P		190316	SE	86293	152925	214	379	3.41	+39.4	−13.3
78 Yates 7 GA Georgia Power	70.92	74.33	404	D		190700	SE	86293	152925	214	379	3.41	+39.2	−20.0
79 Gibson 2 IN Indiana P.S.	70.92	75.25	668	P		191334	M	160276	265696	240	398	4.33	−11.8	−14.1
80 Baldwin 3 IL Illinois Power	70.92	75.42	635	D		191938	M	124493	203485	196	320	4.50	−15.0	−16.9
81 Olds 2 ND Basin Electric	70.92	75.92	460			192375	M	93852	146875	204	319	5.00	−23.0	+5.7
82 Gibson 1 IN Indiana P.S.	70.92	75.92	668	D		193009	M	115655	174371	173	261	5.41	−12.3	−1.3
83 Sherburne 1 MN No. States Power	70.92	76.33	720	P	S	193693	M	215222	324487	299	451	5.41	−22.4	+5.3
84 Miami Fort 7 OH Cincinnati G&E	71.08	75.33	557	I		194222	M	118901	195459	213	351	4.25	−15.9	−13.6
85 Neal 3 IO Iowa P.S.	71.08	75.92	550	I		194745	M	128421	200461	233	364	4.84	−12.8	−1.5
86 Rush Island 2 MO Union Elec.	71.08	77.17	621	D		195335	M	138473	192110	223	309	6.09	+14.2	+11.9
87 Schahfer 14 IN No. Indiana P.S.	71.17	76.92	521	I		195830	M	199557	282907	383	543	5.75	+29.5	+17.8
88 Big Bend 3 FL Tampa Elec.	71.25	76.33	446	D		196254	M	134308	201215	301	451	5.08	+12.2	+12.5
89 Winyah 1 SC So. Carol. P.S.	71.58	75.17	315	P		196552	SE	64851	120684	206	383	3.59	−3.9	−8.6
90 Spurlock 1 KY E. Ky. Power Co.	71.58	77.67	300			196838	M	110760	144476	369	482	6.09	+14.5	+37.3
91 Wansley 1 GA Georgia Power	71.67	76.92	952	I		198646	SE	201765	282411	212	297	5.25	−3.9	+12.1
92 Sherburne 2 MN No. States Power	71.83	77.00	720	D	S	199330	M	147997	204751	206	284	5.17	−23.8	−8.5
93 Daniel 1 MS Mississippi Pwr.	71.83	77.67	548	I		199851	SE	144009	186449	263	340	5.84	−6.6	+33.7
94 Duck Creek 1 IL Central IL Light	71.92	76.42	417		S	200247	M	218853	319940	525	767	4.50	+43.3	−5.2
95 Hayden 2 CO Colo.-Ute Elec.	72.17	76.67	275			200508	W	128565	181796	468	661	4.50	+23.6	+2.0
96 Comanche 2 CO Colorado P.S.	72.25	75.83	396	D		200885	W	90863	139706	229	353	3.58	−12.2	−29.8

Plant and Utility	Dates Order	Finish	MW	Mult.	SO₂	CumCap	Reg.	Costs (000) Mixed	1979	Costs ($/kW) Mixed	1979	Time (yrs.)	Deviations $/kW	Time
97 Gardner 3 NV Nevada Power	72.42	76.42	114		S	200993	W	69301	99713	608	875	4.00	+29.7	+5.9
98 San Juan 1 NM New Mexico P.S.	72.42	76.92	348	D		201324	W	108503	148177	312	426	4.50	-6.3	-9.8
99 Muskogee 4 OK Oklahoma G&E	72.42	77.58	572	I		201867	SC	140921	180404	246	315	5.16	-3.4	+2.3
100 Colstrip 2 MT Montana Power	72.50	76.58	358	D	S	202207	W	127220	179439	355	501	4.08	-1.6	-18.8
101 Deely 1 TX San Antonio P.S.	72.50	77.58	447	I		202632	SC	108411	138373	243	310	5.08	-5.4	+5.1
102 Cardinal 3 OH AEP	72.50	77.67	650			203249	M	346000	437863	532	674	5.17	+33.3	-0.1
103 Harrington 1 TX Southwest P.S.	72.58	77.58	360	I		203846	SC	98000	124707	272	346	5.00	+5.5	—
104 Conesville 5 OH Columb. & So. OH	72.58	76.83	444	I	S	204268	M	140821	192882	317	434	4.25	-19.8	—
105 Martin Lake 1 TX Texas Utils.	72.58	77.33	793	I	S	205021	SC	254643	331982	321	419	4.75	+0.9	—
106 Homer City 3 PA Penn. Electric	72.67	77.92	692			206975	NE	266147	326995	385	473	5.25	-4.7	—
107 Newton 1 IL Central IL P.S.	72.75	77.83	617	I		207561	M	232086	286652	376	465	5.08	+6.9	—
108 Southwest 1 MO Springfield Ut.	72.83	76.50	195	I	S	208168	M	72868	102232	374	524	3.67	-4.4	—
109 Welsh 1 TX Cent. & SW E.P.	72.83	77.17	558	I		208698	SC	128554	168562	230	302	4.34	-9.3	—
110 La Cygne 2 KS Kansas City P&L	72.92	77.33	685			209349	M	229715	295394	335	431	4.41	-1.3	—
111 Huntington 1 UT Utah P&L	72.92	77.42	446	D		209773	W	142107	181164	319	406	4.50	-5.1	—
112 Petersburg 3 IN Indianapolis P&L	73.08	77.83	574	I	S	210793	M	290614	354469	506	618	4.75	+11.8	—
113 Young 2 ND Minnkota Coop	73.42	77.33	477		S	212936	M	194477	244920	408	513	3.91	-7.6	—
114 Ghent 2 KY Kentucky Utils.	73.50	77.00	557	D		213779	M	130411	168878	234	303	3.50	-16.2	—
115 Lansing 4 IO Interstate Power	73.67	77.42	275			215622	M	99033	122528	360	446	3.75	+0.2	—
116 Winyah 2 SC So. Carol. P.S.	74.00	77.50	315	D	S	217348	SE	74279	102583	236	326	3.50	-9.3	—

Appendix 3

Conversion Of Capital Costs Into Constant Dollars

This appendix describes the procedure used in this study to convert capital costs from *mixed current dollars*, the basis in which cost data are tabulated by utilities, into *constant dollars*, a form suitable for analyzing cost trends.

The term "mixed current dollars" refers to the arithmetic sum of dollars spent in each different year of plant construction, unadjusted for inflation. The dollars are "current" because they were spent according to their then-current value. The sum is "mixed" because it combines dollars of varying values; because of inflation, a dollar spent in May 1976, for example, has a different (lower) value than a dollar spent in May 1971. Accordingly, the reported cost in mixed current dollars of a power plant constructed over a period of years is an amalgam of dollars of different vintages and, therefore, of different values. Although such a figure is indeed the plant's actual cost, and determines its fixed charges for amortization and taxes, it does not represent accurately the value of the resources that were required for construction.

Capital costs must be converted to constant dollars to obtain a true measure of value. A power plant's cost expressed in constant dollars represents what that plant would have cost if all expenditures had been made at one moment in time, when all dollars spent on the plant had equal value. The particular moment chosen to express costs is unimportant, since a plant's cost expressed in, say, 1979 dollars, will differ from the same plant's cost expressed in 1976 dollars, only by the rate of inflation from 1976 to 1979. Although constant-dollar cost figures are in a sense hypothetical, since power plants cannot be paid for and built instantaneously, such figures are not distorted by inflation. Thus, they can be employed to discern trends in costs that occur independently of price changes caused by inflation.

Unfortunately, utilities either do not calculate capital costs in constant dollars or do not make such information available. Accordingly, a calculational procedure must be developed and applied to the reported mixed-dollar costs to estimate constant-dollar costs. To understand the procedure, it is useful to think of current-dollar costs as the sum of three components:

initial-year cost: expressed in constant dollars pegged to the date construction began, without inflation or interest during construction;

inflation during construction: increases in costs attributable to inflation in the prices of construction materials, equipment, and labor;

interest during construction (IDC): dollars expended by the utility to pay off interest on the capital previously invested in the plant.

The procedure followed here estimates the costs of inflation and interest during construction and subtracts them from the total reported cost in mixed current dollars. The remainder, initial-year costs, is then converted easily to a mid-1979 basis, the point in time chosen here for expressing costs in constant dollars.

Ideally, the calculations would employ the particular rates of inflation and interest and the actual schedule of expenditures for each plant. Because these data too are unavailable, industry-wide estimates were developed using standard references and were applied to each plant.

Average inflation rates for construction factors were obtained from the Handy-Whitman Index of Public Utility Construction Costs, a semi-annual compilation of wages, material costs, and other factor costs in power plant construction. [1] Line 16 in the index, "total fossil production plant," represents all inputs to fossil plant construction and, therefore, was used as the measure of inflation in construction inputs to coal plants. Since a single index for nuclear plants was developed only recently, a composite index was calculated from five Handy-Whitman categories pertaining to nuclear plants, as shown in Table A3.1.

Because the Handy-Whitman index is compiled regionally, values of the index for nuclear and coal plants were derived as described above for each of the six U.S. geographical regions, and weighted by each region's share of the cost data bases. The shares used correspond to each region's percentage of plants in the sample and are shown in Table A3.2. Actually, the inflation rates yielded by the index vary little enough between regions that the degree of precision used here was probably unnecessary.

National average values of the Handy-Whitman index for nuclear and fossil plants were calculated in this fashion for each year from January 1, 1965 to January 1, 1979, and for the half-year to July 1, 1979. They are shown in Table A3.3. The ratio between successive indices gave the intervening inflation rate. 1970 inflation, for example, was calculated by dividing the January

Table A3.1
Weights Used To Calculate Composite Inflation Rate In Nuclear Construction

Handy-Whitman Line No. and Category	FERC* Account No. and Category	Weight**
12 Turbogenerator Units	314 Turbine Plant	.32
13 Electrical Equipment	315 Electrical Plant	.10
14 Miscellaneous Equipment	316 Misc. Plant	.03
18 Structures & Improvements	321 (Same)	.22
19 Reactor Plant Equipment	322 NSSS	.33

*Federal Energy Regulatory Commission account categories for reporting power plant costs.

**Weights were derived from an Energy Department estimate of the direct (uninflated) costs of a standard 1150-MW nuclear plant.[2]

Table A3.2
Geographical Weights Used To Calculate Composite Inflation

Region*	Nuclear	Fossil
North Atlantic	.30	.08
South Atlantic	.41	.19
North Central	.24	.53
South Central	--	.06
Plateau	--	.14
Pacific	.05	--

*Some Handy-Whitman geographical categories vary slightly from categories in study.

Table A3.3
Inflation And Interest Rates
Used To Adjust Costs

Year	Nuclear Inflation, %	Coal Inflation, %	Interest, %
1965	2.7	2.0	3.8
1966	2.2	2.6	3.9
1967	3.6	3.1	4.0
1968	4.4	3.8	4.3
1969	5.8	6.3	4.6
1970	8.1	7.2	5.1
1971	10.1	11.1	5.5
1972	4.4	3.7	5.7
1973	6.4	7.5	5.9
1974	17.8	25.1	6.3
1975	10.8	9.0	6.8
1976	7.9	6.9	7.0
1977	5.9	6.3	7.1
1978	9.3	10.8	7.5
1979*	4.7	4.4	**

*Inflation between January 1 and July 1 only.

**1979 interest rate is not needed since data bases end in 1978.

1971 index by the January 1970 value.

Interest rates for calculating interest amounts during construction are also shown in Table A3.3. They were obtained from data compiled for investor-owned utilities by the federal Energy Information Administration (EIA).[3] Again, averages were employed because of the difficulty of obtaining the actual interest rates applicable to each plant. Interest rates actually vary little among utilities, in part because the embedded average cost of capital used to calculate IDC changes more slowly than the marginal cost of capital. Furthermore, a check of publicly-owned utilities—municipals, electric co-operatives, state authorities, and the Tennessee Valley Authority—indicated that interest rates on their projects were virtually always within one percentage point of the average interest rate for investor-owned utilities. Thus, little accuracy is lost by using an average interest rate drawn from the investor-owned group.

Finally, a breakdown of the cash flow during construction is needed to calculate inflation and interest because inflation and interest rates vary from year to year. Actual cash flows were not obtainable for each plant, and therefore "typical" cash flow curves were applied to total plant costs to obtain annual cost breakdowns. Curves representing cash flow exclusive of interest and inflation during construction were taken from an Atomic Energy Commission publication.[4] They are of the form,

$$Y = Y_{max}(1 - \cos(\pi/2 \cdot X/X_{max})^a)^b,$$

where Y = cumulative cost expended after X years of construction

X_{max} = total construction time (measured from date of NSSS or boiler order to commercial operation)

Y_{max} = total cost

Values of a and b were obtained by William Mooz of the Rand Corporation by applying a curve-fitting algorithm to the AEC curves. They are: coal, $a=3.2386$, $b=1.2932$; nuclear, $a=4.0820$, $b=3.2495$. These values imply that half of expenditures exclusive of inflation and interest are made during the first 45% of the construction duration for coal plants and 53% for nuclear plants.

With the foregoing annual inflation and interest rates and cash flow curves, one can calculate *initial costs*: a plant's hypothetical cost if it could have been built instantaneously at the moment it was ordered, with no inflation or interest during construction. This is done with the following equation:[5]

$$TC = IC \cdot \sum_{i=1}^{T} a_i \cdot \prod_{t=1}^{i} INF_t \cdot \prod_{t=i}^{T} IDC_t$$

where TC = Total Completed Cost (in mixed current dollars)

IC = Initial Cost (in initial year's dollars, without inflation or interest during construction)

a_i = Fraction of Initial Cost spent in year i

INF_t = 1 + construction inflation rate in year t

IDC_t = 1 + interest-during-construction rate in year t

T = Year of Commercial Operation

The equation calculates the initial cost that, when disaggregated into year-by-year expenditures which are supplemented by inflation and interest during construction, will total to the final cost in mixed dollars. Inflation accrues from plant start to the moment each dollar is spent. Interest is incurred from the expenditure of each dollar to the date of commercial operation, at which time the plant enters the rate base and ratepayers begin to pay for it, relieving the utility of the need to borrow additional capital to finance the plant. The initial cost, expressed in constant dollars of the year in which the plant was ordered, was then converted to July 1, 1979 dollars by multiplying by each year's inflation factor from plant order through July 1, 1979.

The resulting costs are referred to throughout the text as "steam-plant dollars" because the Handy-Whitman deflator used to calculate them is a measure of inflation in construction factors for steam-electric (fossil and nuclear) plants. Prices in the steam-electric sector have increased slightly faster than the prices of other industrial commodities (by an annual average of .64% for nuclear construction and .98% for fossil construction, based on a comparison of the Handy-Whitman index and the industrial producer price index during 1965-79). Therefore, the measures of real 1971-78 cost increases derived in this study—142% for nuclear plants and 68% for coal plants—slightly understate the true rates of cost increase relative to industrial prices. This difference is eliminated from projections of future costs, however, by incorporating an assumed real future inflation rate of 1%/year in construction factor prices (relative to industrial prices) into the calculation of costs in 1979 constant dollars (see Section 12.1). In addition, the small difference between past inflation rates for nuclear and fossil construction is reflected, properly, in the calculated costs of 1971 and 1978 standard nuclear and coal plants in 1979 steam-plant dollars, since these costs were calculated by multiplying by the respective sector inflation rates through mid-1979.

References

1. Whitman, Requardt and Associates, *Handy-Whitman Index of Public Utility Construction Costs* (semi-annual, Baltimore, MD).

2. Telephone communication with H.I. Bowers, Oak Ridge National Laboratory, 31 July 1979.

3. EIA, *Statistics of Privately Owned Electric Utilities in the United States*, 1977 edition, from "Average Interest Rate" under Tables 12 and 13, "Interest on Long-Term Debt." 1978 figure was estimated by the author.

4. AEC, *Power Plant Capital Costs: Current Trends and Sensitivity to Economic Parameters*, WASH-1345 (1974), p. 24.

5. The equation was formulated by Lewis Perl in "A Comparison of the Projected Cost of Electricity from the Palo Verde Nuclear Plant and Selected Coal-Fired Alternatives," Part II, 11 July 1979 (National Economic Research Associates, New York, NY).

Appendix 4
Statistical
Methodology

This appendix discusses several issues pertaining to the statistical analyses used to discern capital cost trends. It contains: (1) a brief description of regression analysis; (2) an explanation of why a multiplicative form was used in the regression equations; and (3) a demonstration to show that the choice of regression models has only limited bearing on the calculated costs of standard 1971 and 1978 plants.

Section A4.1:   Regression Analysis

Regression analysis is the standard statistical technique for estimating the extent to which changes in the value of a ''dependent variable'' are related to changes in one or more ''causal'' or ''independent variables.'' In this study, the dependent variable is the capital cost or construction time of nuclear or coal plants. Independent variables that were found to have ''statistical significance'' in explaining changes in cost or duration included multiple units, Northeast location, unit size and architect-engineer experience (for nuclear plants), use of scrubbers (for coal plants), and the cumulative amount of nuclear or coal capacity operating or under construction in the U.S. (''sector size''), among others.

The statistical significance of a regression equation is best indicated by two measures in combination: t-ratios and r^2 values.

T-ratios: The t-ratio is the measure of the statistical significance of the relationship of the dependent variable (capital cost) to an independent variable. Technically, it is the ratio of the independent variable's ''correlation coefficient'' to its ''error coefficient.'' The higher the t-ratio, the smaller the probability that the true relationship between the dependent and independent variables differs from the relationship indicated by the correlation coefficient. The t-ratio therefore determines the statistical significance of a correlation coefficient.

Statistical significance levels depend on the size of the data sample as well as on the t-ratio. They may be obtained by consulting a table of "Student's t-test" in any statistics textbook. A standard 95% significance level has been used as the threshold of statistical significance throughout this study, with several minor exceptions. Significance levels are displayed for each variable in the regression equations in Chapters 8 and 9. The "95% confidence limits"— the values of a coefficient which bound its true value with 95% probability— are stated for each independent variable discussed in the two chapters.

r^2 *value:* The r^2 value, or "multiple correlation coefficient," of a regression equation is the correlation between the actual values of the dependent variable (*i.e.*, the capital costs of all nuclear or coal plants) and the values that are calculated from the equation for each data point, using the values of the independent variables for each point. The r^2 therefore measures the "goodness of fit" of the regression equation to the data from which the equation is drawn. It ranges from zero to one, with higher values indicating a better fit of the equation to the data. A graph or "scatterplot" of an equation with a high r^2 would show the actual data points close to the "trend line" of the equation, with relatively little "scatter" between the line and the points.

Section A4.2:   Form Of The Regression Equations

The regression equations in Chapters 8 and 9 have a multiplicative form, that is, capital costs are expressed as the product of the values and coefficients of the separate independent variables. The equations were deliberately expressed in multiplicative form rather than a linear, or additive, form. Multiplicative form makes it possible to express the effects of variables such as Northeast location or multiple status as *percentage* increments or decrements to costs, *e.g.*, multiple units are approximately 10% cheaper to build than other units. In a linear model, these variables would appear as *absolute* increments or decrements, *e.g.*, plus or minus $50/kW—an untenable formulation when costs of plants within the sample vary so widely, from $281/kW to $1,113/kW for nuclear plants (without interest during construction). With multiplicative form it is also possible to express the effects of increased architect-engineer experience or unit size as percentage reductions in costs—the conventional method of denoting scale or learning effects.

In addition, costs were expressed as a polynomial function of sector size (*i.e.*, cost was made proportional to sector size raised to a numerical power) because of the relationship between reductions in risk and increases in cost. First, it was assumed as a rough approximation that constant percentage increases in costs are necessary to effect constant percentage reductions in per-reactor accident risks or in per-plant emissions of pollutants. Such relationships are frequently observed in risk applications and pollution control technology. Then,

$$\frac{dC/C}{dR/R} = -k,$$

where C = cost, R = risk, k is a constant, and d is a mathematical symbol denoting small changes in a variable. Using integral calculus,

$$\int dC/C = \int -k \cdot dR/R, \quad \text{or,}$$
$$C = R^{-k}.$$

If, as argued in this study, society acts to maintain more or less constant total risk or emissions from the nuclear or coal sectors, we have

$$R = \frac{k}{MW},$$

where MW = the size of the respective sector and k denotes another constant. Combining the two equations, we have

$$C = \left(\frac{k}{MW}\right)^{-k}, \text{or,}$$
$$C = MW^{k}.$$

That is, plant cost equals (or is proportional to) sector size (MW) raised to a power. The particular values of k derived here, .58 for nuclear and .61 for coal (when the effect of scrubbers is measured separately), arise from the actual plant costs and values of sector size and other independent variables in the data samples.

Section A4.3:   Independence Of Costs Of 1971 And 1978 Standard Plants

This section presents an illustrative calculation demonstrating that the calculated costs of 1971 and 1978 hypothetical standard plants are only slightly sensitive to the particular regression model used to measure cost trends. The calculation shown is for nuclear plants; coal plants yield a similar result.

The equations in Tables 8.1 and 8.2 may be used to calculate costs for 1971 and 1978 nuclear plants. Table 8.1 presents the results of a regression equation in which capital costs were assumed to be related to the size of the nuclear sector (cumulative nuclear capacity) in addition to other independent variables such as unit size. In Table 8.2, capital costs were assumed to be related to the date each nuclear plant received its construction permit (CP date).

The two equations were applied to the characteristics of standard 1971 and 1978 plants shown in Table 10.3. The values chosen for CP date (a parameter not listed in the table) were 66.75 (*i.e.*, October 1966) for the 1971-completed plant and 71.25 (April 1971) for the 1978-completed plant. The former date is the date of CP award for the last several reactors licensed before the start of the data base; it corresponds to the value of 9800 MW used to represent sector size immediately preceding the sample. CP date for the end of the sample was taken as April 1971, although the last of the 46 sample plants to be licensed, Farley 1, actually received its CP in August 1972. However, this was 17 months after the immediately preceding plant, Davis-Besse, received its CP, and therefore April 1971, one month after Davis-Besse's CP date, was chosen as a more conservative value.

Costs calculated with these assumptions, exclusive of interest during construction, were as follows: for 1971 plants, $339/kW with the model using sector size and $389/kW using CP date; for 1978 plants, $798/kW using sector size and $848/kW using CP date. The costs with CP date are higher in both instances, by 15% for 1971 plants and by 6% for 1978 plants. When real IDC is added to costs, the measured average 1971-78 rates of increase in real nuclear costs are 125% using CP date and 142% using sector size. The cost differences are noticeable but not particularly significant. Costs obtained by using CP date are higher, moreover, than those obtained when costs are made a function of sector size. Accordingly, neither the very large 1971-78 real increase in nuclear costs nor the wide gap between the costs of 1978 nuclear and coal plants can be attributed to the use of the sector-size formulation to measure cost trends.

Appendix 5
Acronyms

ACRS	Advisory Committee on Reactor Safeguards
A-E	Architect-Engineer
AEC	Atomic Energy Commission
AEOD	Analysis and Evaluation of Operational Data (Office of)
AEP	American Electric Power Co.
AFR	Away From Reactor (Waste Storage Facility)
AIF	Atomic Industrial Forum
ANSI	American National Standards Institute
ASME	American Society of Mechanical Engineers
ATWS	Anticipated Transients Without Scram
B&W	Babcock & Wilcox
BWR	Boiling Water Reactor
CE	Combustion Engineering
CEP	Council on Economic Priorities
CFR	Code of Federal Regulations
DOE	Department of Energy
DOL	Department of Labor
ECCS	Emergency Core Cooling System
EEI	Edison Electric Institute
EIA	Energy Information Administration
EPA	Environmental Protection Agency
EPRI	Electric Power Research Institute
ERDA	Energy Research & Development Administration
GAO	General Accounting Office
GE	General Electric
GW	Gigawatt (one billion watts)
IAEA	International Atomic Energy Agency
IDC	Interest During Construction
IEEE	Institute of Electrical and Electronics Engineers
IREP	Interim Reliability Evaluation Program
KEA	Komanoff Energy Associates
kW	Kilowatt (one thousand watts)

LOCA	Loss of Coolant Accident
LWR	Light Water Reactor
MSIV	Main Steam Isolation Valve
MW	Megawatt (one million watts)
NNI	Non Nuclear Instrumentation
NRC	Nuclear Regulatory Commission
NSPS	New Source Performance Standards
NSSS	Nuclear Steam Supply System
O&M	Operating and Maintenance
OBE	Operating Basis Earthquake
PWR	Pressurized Water Reactor
QA	Quality Assurance
RCS	Reactor Coolant System
RG	Regulatory Guide
RHR	Residual Heat Removal (System)
SDV	Scram Discharge Volume
SEP	Systematic Evaluation Program
SSE	Safe Shutdown Earthquake
TMI	Three Mile Island
TVA	Tennessee Valley Authority
UCS	Union of Concerned Scientists
USI	Unresolved Safety Issue
W̲	Westinghouse